D1202210

Low Carbon Energy Transitions

Low Carbon Energy Transitions

Turning Points in National Policy and Innovation

KATHLEEN M. ARAÚJO

OXFORD
UNIVERSITY PRESS

Oxford University Press is a department of the University of Oxford. It furthers the University's objective of excellence in research, scholarship, and education by publishing worldwide. Oxford is a registered trade mark of Oxford University Press in the UK and certain other countries.

Published in the United States of America by Oxford University Press
198 Madison Avenue, New York, NY 10016, United States of America.

Library of Congress Cataloging-in-Publication Data
Names: Araújo, Kathleen M.
Title: Low carbon energy transitions : turning points in national policy
and innovation / Kathleen M. Araújo.
Description: New York, NY : Oxford University Press, [2017] |
Includes bibliographical references and index.
Identifiers: LCCN 2016044876 | ISBN 9780199362554
Subjects: LCSH: Renewable energy sources. | Energy policy. | Renewable energy sources—Social aspects. | Renewable energy sources—Economic aspects. | Biomass energy—Brazil. | Nuclear energy—France. | Geothermal resources—Iceland. | Wind power—Denmark.
Classification: LCC TJ808.A73 2017 | DDC 333.79/4—dc23
LC record available at https://lccn.loc.gov/2016044876

9 8 7 6 5 4 3 2 1

Printed by Sheridan Books, Inc., United States of America

CONTENTS

ACKNOWLEDGMENTS

The energy playing field continues to evolve. With it, new questions emerge about the interplay of technology, natural systems, and society. For those who encouraged my work in this area, I am deeply grateful.

It's impossible to adequately thank everyone who was involved in this project. For George, who made this journey with me, it's time for new adventures. For the many others who provided insights, reviews, or moral support, thank you. Special contributors included Sveinbjorn Bjornsson, Jon Ingimarsson, Bertrand Barre, Manoel Regis Lima Verde Leal, Jose Moreira, Jose Goldemberg, Roberto Schaeffer, Sergio Rezende, Bjarke Thomassen, Povl-Otto Nissen, Birger Madsen, Hans Christian Soerenson, Pierre Berest, Mycle Schneider, Frank Carre, Jean-Louise Nignon, Jacques Gollion, Gudni Johannessson, Arnulf Grubler, Richard Hirsh, John Holdren, Amanda Graham, Tijs van Maasakkers, Nancy Odeh, Bill Moomaw, Jeff Tester, Richard Lester, Arna Palsdottir, Madhu Dutta-Koehler, William Koehler, Paola Moreno, Larry Susskind, Ignacio Perez Arriaga, and Calestous Juma.

The idea for this book came to life while I was a student at MIT with the encouragement from family, friends, advisors, and peers. It benefitted from the generous support provided by the MIT Presidential Fellowship, MIT Industrial Performance Center, MIT Center for International Studies, MIT Brazil Program, the Martin Family Fellowship, and Fletcher's Energy, Climate, and Innovation Program. More recently, this work flourished with the steady championing of Jeremy Lewis. Any errors are solely my responsibility.

INTRODUCTION

If we don't change direction soon, we'll end up where we're heading....
—Chinese proverb cited by the International Energy Agency, *2011a*

The world at large is wrestling with important energy choices. There is a strong sense today that we need to manage energy differently. Priorities in resilience, security, jobs, and access emphasize a need to substitute low carbon energy for traditional fossil fuels. Nevertheless, no one is entirely clear about how to carry out such a shift at the national or international levels. A widely-held view is that national energy transitions of any significance take several decades, if not longer, and entail least-cost economics as a principal driver. Based on this line of thinking, only energy sources that are low-cost have a chance to take hold. Although this appears reasonable, it can miss opportunities for wider gains. Some say that change will require an acceleration of innovation. Yet what assures innovation is also not entirely certain.

A century ago, a person observing the energy playing field would have found substantially less fossil fuels being used and a limited niche industry in electricity. At that time, biomass and coal supplied the majority of the global energy mix, with technologies like automobiles, gas turbines, and airplanes still emerging.

Today, more than 85% of the world's energy is derived from fossil fuels (BP, 2017). The environment is also showing signs of stress as air quality reaches dangerous levels in some regions (particularly those with heavy coal use) and change in the climate redraws our maps and ecosystems. Security of the energy supply is also brought into question, particularly when geopolitics flare up or prices spike. As all of this occurs, the world's population continues along a path in which projected growth by mid-century may represent an increase from 7 to

Table I-1. Market Shares of Low Carbon Energy Studied

Country and Low Carbon Energy	1970	2015
• Icelandic geothermal energy in power	Negligible	29%[a]
• Brazilian ethanol in primary automotive fuels	~1%	34%[a]
• Danish wind power in electricity	Negligible	42%
• French nuclear power in electricity	4%	76%
• Icelandic geothermal energy in space heating	43%	~90%

[a] Reflects data for 2014.

SOURCE: Compiled with data from various sources: Brazil (Ministry of Mines and Energy/MME, 2016); Denmark (Energinet, n.d.); France (IEA, n.d.); Icelandic power and space heating (Orkustofnun, 2016; Ragnarsson, 2015).

9 billion inhabitants. Importantly, regions where growth is expected to be the highest are also ones where energy access is currently challenged.

Low carbon priorities now regularly feature in public discussions.[1] One need only look at calls for decarbonizing change made by the United Nations, World Bank, and World Economic Forum. In line with such priorities, this book considers low carbon energy transitions in prime mover countries. For the purposes here, *transitions* reflect the displacement of at least 15% of traditional energy sources with a low carbon alternative in a given energy mix (relative change), and increased utilization of the same, low carbon alternative in absolute terms by 100% or more. *Prime mover countries* are ones that accomplished this feat. Histories of Brazilian biofuels, Danish wind power, French nuclear power, and Icelandic geothermal energy are examined in depth, here, for the period principally since 1970 (Table I-1).

This book highlights the interplay of technology, natural systems, and society with underlying logic that is rooted in planning and management, policy and applied history, and broader, sociotechnical systems. The research recognizes history as a valuable and often missed tool for decision-making and planning (Neustadt and May, 1986; Schaeffer, 2007; Diamond and Robinson, 2010; Sinclair et al., 2016), and puts forward tools for theoretical and practical scoping of transitions. Models of national readiness will integrate material and human aspects of change in a way that can be applied to structural shifts in energy as well as other sectors, including information or biomedicine. Complementing the models of readiness is a framework based on sectoral intervention that

1. *Low carbon* is used widely to refer to activity that produces much less carbon. The concept is discussed in Chapters 1 and 2.

provides ways to consider induced and emergent change. In doing so, this book emphasizes turning points, while linking theory and practice.

What lies behind national shifts may surprise some. Quick explanations of costs serve a purpose, but the influences behind such costs are much less understood. There is also a tendency to dismiss energy choices as being driven by an abundance of the "right" resources. Such ideas and others are considered, highlighting how government can play a role but not always lead the change. Broadly, this book is designed to challenge how we think about national energy objectives, and how strategies can evolve in energy transitions.

Given the aims and coverage, this book will be of interest to policymakers and practitioners, as well as to students and citizens who think about energy options. For policymakers and practitioners, the book provides ways to consider energy system change and course corrections with perspective from contemporary examples. For members of industry and funding agencies, as well as for think tanks and inter- or nongovernmental organizations, this book provides in depth insight into pivotal junctures that can emerge with energy path realignments. For students and interested citizens who want to better understand energy paths, these histories shed light on theory and better practices in technology diffusion and learning.

Overall, my aim is to show how challenges and opportunities arise in connection with energy, as well as how choices in this regard are made. Chapter 1 provides an overview of the current, energy playing field, outlining the rationale for low carbon change. Chapter 2 examines ideas in theory and practice relating to systems change, innovation, and policy. Chapter 3 outlines new, conceptual tools and the research design. It then turns to relevant developments in the global context and provides a preview of the four countries' transitions. Chapters 4 through 7 provide in depth histories of national energy system change, with a special emphasis on policy and innovations. Chapter 8 comparatively evaluates findings in the context of overarching themes and explores limits for the research. Chapters 8 and 9 then draw inferences for policymakers and scholars. Chapter 9 concludes with promising directions for future research. Those wishing to learn more about specifics of the energy technologies will find a technology primer on geothermal energy, nuclear energy, biofuels, and wind power in the Appendix. Timelines of each country's sociotechnical history are also available there.

1

Rethinking Energy at the Crossroads

> We can't solve problems by using the same kind of thinking we used when we created them.
>
> —Albert Einstein

The discovery of oil in Pennsylvania in 1859 was a relatively inconspicuous precursor to what would become an epic shift into the modern age of energy.[1] At the time, the search for "rock oil" was driven by a perception that lighting fuel was running out. Advances in petrochemical refining and internal combustion engines had yet to occur, and oil was more expensive than coal. In less than 100 years, oil gained worldwide prominence as an energy source and traded commodity.[2]

Along similar lines, electricity in the early 1900s powered less than 10% of the homes in the United States. Yet, in under a half a century, billions of homes around the world were equipped to utilize the refined form of energy. Estimates

1. The term "discovery" of oil or petroleum is used loosely here. Prior to 3000 BC, recorded history indicates that oil was used as asphaltic bitumen in Mesopotamia (Giebelhaus, 2004). Later adaptations included its use in waterproofing of ships and in construction, in addition to applications in medicine, illumination, and incendiary devices. At the time of the Titusville discovery, other developments relating to petroleum were already under way in Azerbaijan and France (Smil, 2010).

2. As of 2015, global primary energy (i.e., the raw supply of energy) totaled 13,276 million tonnes of oil equivalent (Mtoe), with oil representing 33% (BP, 2017). Additional primary energy sources included: coal (28%), natural gas (24%), hydropower (7%), nuclear power (5%), and other renewables (3%) (BP, 2017). Sources of energy are discussed more fully later in the chapter.

indicate that roughly 85% of the world's population had access to electricity in 2014 (World Bank, n.d.*b*). For both petroleum and electricity, significant changes in energy use and associated technologies were closely linked to evolutions in infrastructure, institutions, investment, and practices.

Today, countless decision-makers are focusing on transforming energy systems from fossil fuels to low carbon energy which is widely deemed to be a cleaner, more sustainable form of energy.[3] As of 2016, 176 countries have renewable energy targets in place, compared to 43 in 2005 (Renewable Energy Policy Network for the 21st Century [REN21], 2017). Many jurisdictions are also setting increasingly ambitious targets for 100% renewable energy or electricity (Bloomberg New Energy Finance [BNEF], 2016). In 2015, the G7 and G20 committed to accelerate the provision of access to renewables and efficiency (REN21, 2016). In conjunction with all of the above priorities, clean energy investment surged in 2015 to a new record of $329 billion, despite low, fossil fuel prices. A significant "decoupling" of economic and carbon dioxide (CO_2) growth was also evident, due in part to China's increased use of renewable energy and efforts by member countries of the Organization for Economic Cooperation and Development (OECD) to foster greater use of renewables and efficiency (REN21, 2016).[4] In April 2016, 175 countries signed the Paris Agreement, which aims to slow the growth of greenhouse gases (GHGs), including CO_2, in the atmosphere in order to limit global warming to "well below" a 2°C or 3.6°F increase, relative to pre-industrial levels.[5] Despite an announcement in 2017 that the US would withdraw, the general global focus appears to be in tact.

Importantly, it is not just governments that are currently focused on low carbon change. Traditional energy companies also have carbon-based priorities in energy. Oil and gas companies and power utilities are being asked to include stress tests in their portfolio assessments to reflect carbon or climate impacts (Hulac, 2016). In June 2015, heads of some of the largest European

3. *Energy systems* provide services like heating, cooling, power, and transport. They consist of infrastructure, fuels, people, institutions (including markets), practices, and the ecosystems that enable the provision of such services. *Low carbon* refers to a path that utilizes notably less carbon. This differs from (but can overlap with) renewable energy, which is "any form of energy from solar, geophysical or biological sources that is replenished by natural processes at a rate that equals or exceeds its rate of use" (Moomaw, et al, 2011). For more detailed discussion, see Chapter 2 and Section 1.2.1 of the *Special Report on Renewable Energy Sources and Climate Change Mitigation* (SRREN) (Moomaw et al., 2011).

4. OECD country groupings are based on a classification system that was set up in 1961. Loosely defined, OECD states are industrialized countries and non-OECD states are developing countries. For a discussion of country classifications, see OECD (n.d.), UN (2008, 2014), Nielsen (2011), and Araújo (2014).

5. Greenhouse gases are discussed in the following section.

oil companies published an open letter calling for a carbon pricing system to be instituted worldwide (Geeman, 2015). Ten CEOs of major energy companies also pledged in 2015 to collectively strengthen actions and investments to contribute to the reduction of GHG intensity, including carbon, in the global energy mix (Rascouet and Chmaytelli, 2015; Reed, 2015; World Economic Forum, 2015).

Private-sector banks, such as Citigroup, and the insurance industry are also scrutinizing low carbon pathways. Citigroup analyzed the cost difference of global energy investment pathways by considering the status quo and a path that reduces carbon through less fossil fuel plus greater utilization of renewables and nuclear energy. In doing so, it found "we can afford to act." Specifically, there are marginal cost differences between the two paths through 2040, if fuller consequences are considered. A low carbon approach could be expected to equal $190.2 trillion, whereas a business-as-usual path would be $192 trillion (Channell et al., 2015). In addition, a path of inaction is associated by 2060 with an estimated $44 trillion in lost gross domestic product (GDP) on an undiscounted basis, not accounting for savings.[6]

RATIONALE FOR LOW CARBON CHANGE

Arguments for low carbon change are often based on rationales ranging from the need for safeguarding the environment and health to price flux and security. The following discusses aspects of these arguments.

Environment and Health

Fossil fuels are known for their links to the degradation of air, water, and land quality through emissions, spills, contamination, and extraction practices. At a global scale, energy (principally from fossil fuels) contributes a reported 68% of total GHGs, the accumulation of which is changing the composition of the atmosphere and the climate system (International Energy Agency [IEA], 2015*a*; see also Box 1-1). Among the GHGs emitted by the energy sector—namely CO_2, nitrous oxide (N_2O), and methane (CH_4)—CO_2 accounts for roughly two-thirds of all GHG emissions (IEA, 2015*b*).[7]

6. For a discussion of costs in energy system change, see Araújo (2016).

7. Life cycle assessments of energy systems also point to CO_2 emissions from cement usage in the construction of power plants.

Box 1-1

Greenhouse Gas Emissions

Greenhouse gas (GHG) emissions are increasingly recognized as primary determinants behind the radiative forcing of the atmosphere that is producing climate change (Intergovernmental Panel on Climate Change [IPCC], 2013, 2014). Atmospheric concentrations of CO_2, a principal indicator of GHGs, have risen from roughly 280 parts per million (ppm) in 1800 to 409 ppm in June 2017 (International Energy Agency [IEA], 2015*a* and 2015*b*; NOAA, 2017.).

Should trends continue, the CO_2 level is expected to rise substantially, producing more extreme weather events and an increase of 2–4°C in the average global surface temperature. A reference point of 450 ppm of CO_2 in the atmosphere is a working guideline for avoiding the more uncertain, and assumed to be the most disruptive, aspects of global warming. Table 1-1 indicates GHG emission intensities for various fuels on a per kilowatt hour (kWh) basis. Here, renewables and nuclear power have the lowest intensities, and fossil fuels have the highest, differing in many cases by multiple orders of magnitude.

Table 1-1. Life Cycle Analysis of Greenhouse Gas Emissions for Select Fuels (gCO_2 equiv per kWh)

Coal	Oil	Gas	Nuclear	Wind	Hydropower	Geothermal	Solar	Biomass
675–1,689	510–1,170	290–930	1–220	2–81	0–43	6–79	5–217	(633)–75

SOURCE: Based on a literature review of life cycle analyses for GHG emissions of power generation technologies (Moomaw et al., 2011).

Regionally, fossil fuel emissions can introduce precursors of acid deposition that disperse over thousands of kilometers to damage harvests, natural systems, and anthropogenic structures (Goldemberg, 2006*b*). Oil spills and gas flaring can also lead to the collapse of local fishing and farming, as well as the loss of habitat and biodiversity (Baumuller, Donnelly, Vines, and Weimar, 2011). Moreover, leakage and runoff of pollutants from coal mining or hydraulic fracturing of natural gas or oil can compromise soil and water aquifers (Environmental Protection Agency [EPA], n.d. and 2011*a*; Osbourne, Vengosh, Warner, and Jackson, 2011).[8]

8. The extent to which hydraulic fracturing pollutes water aquifers remains under debate and study (Bambrick, 2012; Macalister, 2011; Massachusetts Institute of Technology, 2011; Stevens, 2010; Urbina, 2011*a*, 2011*b*; Vaidyanathan, 2016; Yost, Stanek, DeWoskin, and Burgoon, 2016).

Specific to public health, an estimated 6.5 million deaths occur each year in connection to air pollution, with the total expected to rise absent change in the energy sector (IEA, 2016*b*). Fossil fuel emissions are singled out for their ties to respiratory disease, rheumatic disorders, cancers, and premature fatalities (Argo, 2001; Baumuller et al., 2011; Goldemberg and Lucon, 2010; United Nations Development Program [UNDP], UN Department of Economic and Social Affairs, World Energy Council [WEC], 2004). Such emissions include particulate matter, sulfur dioxides, nitrogen oxides, volatile organic compounds, carbon monoxide, and carbon dioxide, among pollutants.

In recent years, the ties between fossil fuel use and public health have assumed new importance. In China, for instance, where coal represents about two-thirds of the power mix (Energy Information Administration [EIA], 2015), air pollution episodes now occur regularly. In 2015, during the most severe episode, concentrations of primary pollutants in Beijing reached levels that were nearly 40 times greater than what the World Health Organization considers safe for 24-hour exposure (Finamore, 2016). This issue is leading to a restructuring in the country's energy sector toward less carbon-intensive gas and non-fossil power (Jianxiang, 2016).

A study by the U.S. National Academy of Science estimates that premature deaths linked to air pollution from fossil fuel in the United States equal $120 billion per year in health costs (National Academy of Science, 2010). Of the roughly 20,000 deaths per year cited in the study, the majority was attributed to fossil fuel emissions from power plants and vehicles. If the direct environmental costs of gasoline and diesel fuel were factored at the pump, gasoline and diesel fuel would be priced $0.23–0.38 per gallon higher.[9]

When comparing fossil fuel and renewable energy on a life cycle basis for power generation, the human health effects of renewable energy were found in a study for the United Nations Environment Program to be 10–30% of those from state-of-the-art fossil fuel–based power (UNEP, 2015). Environmental damage by pollutants such as particulate matter and toxic metals was also found to be 3 to 10 times less from renewables compared to fossil fuel systems (UNEP, 2015). In short, environmental and health effects of energy utilization are real and uneven.

9. This does not cover all effects, including those associated with climate change, pollution control devices, or oil combustion specific to travel by rail, sea, and air (National Academy of Science, 2010).

Table 1-2. Mineral Fuel Imports for Select Countries/Regions (2014)

	% National Imports (based on $)	Amount
China	16	$372 B
EU-28	27	$617 B
India	39	$177 B
Japan	32	$262 B
United States	15	$358 B

SOURCE: United Nations Comtrade [UN Comtrade], n.d.

Import Dependence, Wealth Transfer, and Local Economic Development

As a top-traded commodity, fossil fuel introduces dependence vulnerabilities. In 2014, oil, coal, and gas as well as their distillation products equaled roughly $3 trillion or 17% of total imports worldwide (UN Comtrade, n.d.).[10] For countries with considerable shares of these fuels in their overall imports, such as India at 39% or Japan at 32% in 2014 (Table 1-2), this reliance represents transferred wealth from domestic industries and provides points of leverage for exporter nations (Levi, 2010). If fossil fuel importer countries were to switch to locally-sourced renewable energy, the domestic community could benefit from favorable economic payback and innovation around a cleaner economy, in addition to the reduction of uncertainties associated with international trade.

Price Flux

Price fluctuations in fossil fuels present yet another area to watch for energy decision-makers, as links between economic activity and energy price volatility are fairly well recognized. A $10 per barrel increase in the price of oil, for instance, is estimated to slow the global economy by 0.5% per year (UNDP et al., 2004). In the past decade, price uncertainties for fossil fuels have reflected fairly substantial swings, with coal and oil representing a spread of more than a factor of three (Figure 1-1). With this kind of volatility, switching to locally-sourced energy, like renewables, can serve as a hedge against price flux, particularly for people who may not otherwise be able to secure energy at the higher prices.

10. This is based on reporting for mineral fuels in the UN Comtrade HS category, which also contains derivative elements like petroleum jelly and bituminous mixtures based on asphalt and the like (UN Comtrade, n.d.).

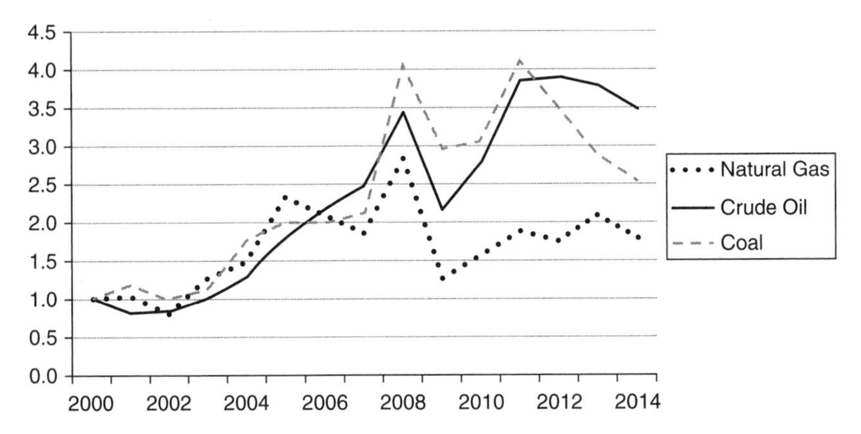

Figure 1-1 Change in fossil fuel prices (Base Year = 2000).
SOURCE: Adapted with data from BP (2015).

Looking strictly at oil, marker crude oil prices for Dubai, West Texas Intermediate (WTI), and North Sea more than doubled from a lower range at $24.04 to $28.60 to a higher range of $60 to $65.10 in the period between April 2015 and April 2016.[11] The currently low, yet dynamic prices are leading to a slowing of investment, with unprecedented cuts in upstream capital expenditure in addition to postponed or cancelled projects (Birol, 2016). For export nations like those in the Middle East, where oil revenues equaled roughly 30% of regional GDP in 2014, heavy reliance on such fuel revenues for public funds produces boom-and-bust cycles. This requires deep cuts in domestic expenditures in times of low prices, especially if special funds are not set aside. This can also trigger political instability, such as a strike in Kuwait to protest government cutbacks (Holodny, 2016). In such circumstances, the strategic use of locally-sourced, low carbon energy could serve as a hedge to minimize economic swings of uncertainty.

Subsidies

For many years, fossil fuels have been in a highly-favored position in terms of subsidies—a form of aid that is used to attain an economic or social goal (Carrington, 2015). In 2015, global subsidies for fossil fuel consumption were estimated at $325 billion, compared to that for renewables at $150 billion (IEA, 2016*f*). Through subsidization, prices are distorted, thus limiting consumers' capacity to judge scarcity and other considerations. Some may argue that the energy output per unit of subsidy makes fossil

11. Price highs occurred the week of May 4, 2015, and lows occurred the week of January 16, 2016 (IEA, 2016*e*).

fuels more attractive to support. Fossil fuels, however, are not new entrants to the energy landscape. This type of developmental aid could be used more effectively to enhance resilience or provide newer technologies a more even playing field.

Security

The security of an energy supply (i.e., the availability of sufficient and affordable energy) may be challenged by human error or attacks, political or cartel activity, natural disasters, or constraints of the delivery infrastructure. Although these issues could affect any energy system, a persistent form of energy insecurity that is associated with fossil fuels is instability of supplier regions. In recent years, for instance, political unrest, resource nationalism, and deliberate forms of supply disruption have been evident in some of the world's top oil and gas export nations.[12] Costs of safeguarding the international fuel supply, including routes and supplier stability, are often not factored into the calculus of energy options.[13]

Another aspect of energy security is the ownership of energy resources or reserves. Currently, national oil companies control roughly 90% of global oil reserves and 75% of production (Tordo, Tracy, and Arfaa, 2008, see also Figure 1-2), with a similar profile existing for natural gas. These state-owned energy companies (SOEs) have industrial aims that align more closely with the preferences of their respective national governments than do the aims of their private-sector counterparts (Marcel, 2005; McPherson, 2003; Stevens, 2003; United Nations Centre for Natural Resources, Energy and Transport

12. Resource nationalism or expropriation of oil and/or gas fields by the state has been evident in Venezuela, Bolivia, Ecuador, and Russia (British Broadcasting Company [BBC], 2006; Ingham, 2007; Johnson, 2007; Macalister, 2007). Some instances, particularly in Russia and Ecuador, may have occurred on the basis of contractual differences.

During the Arab Spring, revolutions and other major political uprisings occurred in countries of North and West Africa as well as in the Middle East, where fossil fuel exports are considerable (BP, 2017). Russia also has history of natural gas disputes with neighboring countries leading to delivery disruptions. A dispute in January 2009 with Ukraine, for instance, led to supply disruption in 18 other countries (R. Jones, 2009; Reuters, 2009).

13. When considering energy security and supply challenges, low carbon energy can also be affected, but in typically different ways. Bottlenecks as well as trade disputes associated with equipment have affected the adoption of wind and solar energy in recent years. The intermittency of supply also characterizes the availability of renewable energy resources like wind and solar power. However, this natural condition is increasingly being addressed with meteorological forecasting and balancing across fuels, geography, and time.

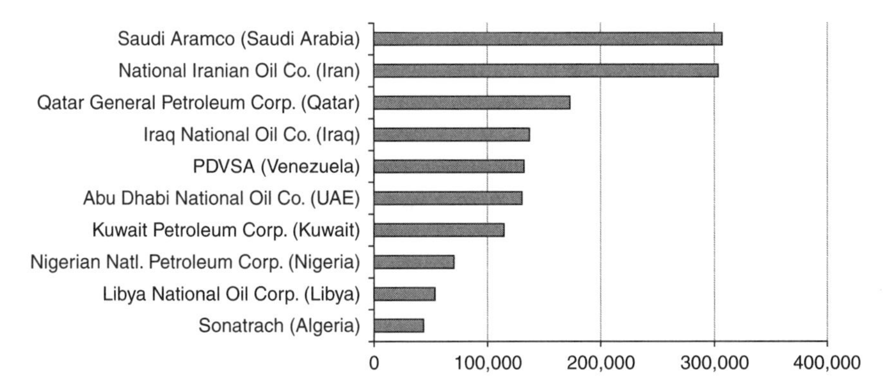

Figure 1-2 World's largest oil and gas companies based on oil equivalent reserves of liquids and natural gas (Million Barrels of Oil Equivalent/MBOE).
SOURCE: Adapted from Petrostrategies, as of July 18, 2012. All companies are at least partly, if not wholly, government-owned.

[UNCRET], 1980). This form of relationship between an energy supplier and a national government can produce societal benefits, such as support for affordable energy services for citizens. These ties can also be exploited, influencing the business plans of the SOEs, if the political agenda of a government or political unrest overshadows a company's aims. Although low carbon energy resources can also be controlled by the state, renewable energy forms are less prone to such security issues.

ENERGY SYSTEM CHANGE

When focusing on energy system change, it's important to expand on what is meant by an *energy system*. These systems are interconnected networks of people and institutions engaged in processes of energy exploration, production, transformation, delivery, and use within an enabling environment or ecosystem. Energy systems include inputs (i.e., fuel resources) and outputs or energy services that are linked by infrastructure and management systems, typically within a market (for a discussion of energy types, see Box 1-2).

Changes to such systems (i.e., energy transitions) can occur in the type, quality, or quantity of energy that is sourced, delivered, or utilized. These conversions can occur at any level, and typically entail co-evolutions in sociotechnical aspects such as user practices and market mechanisms. In recent decades, scholarly works on the subject of energy transitions have grown considerably. As new studies bridge disciplines and regions, annual publications have increased by more than a factor of 27 since 1970 (Figure 1-3).

Box 1-2

Sources of Energy

Primary energy is contained within natural resources including fossil fuels, uranium, and renewable energy. Unlike *final energy,* primary energy is mostly unrefined when it enters the energy system (Grubler et al., 2012). Final energy, such as electricity or gasoline, is available after processing, transformation, and distribution at the point of end use.

Primary energy includes a range of inputs. Feedstock for fossil fuels encompasses coal, natural gas, and oil—each of which is converted for use primarily through combustion. Elements such as uranium, plutonium, and thorium (and various isotopes, in some cases) generate nuclear energy through fission or fusion processes (Rogner et al., 2012). Renewable energy consists of sources such as hydro power, ocean and wave power, biomass, geothermal energy, wind power, and solar energy. Among the renewable sources (used interchangeably, here, with renewable energy technologies [RETs]), energy is derived essentially from solar radiation or the Earth's heat.

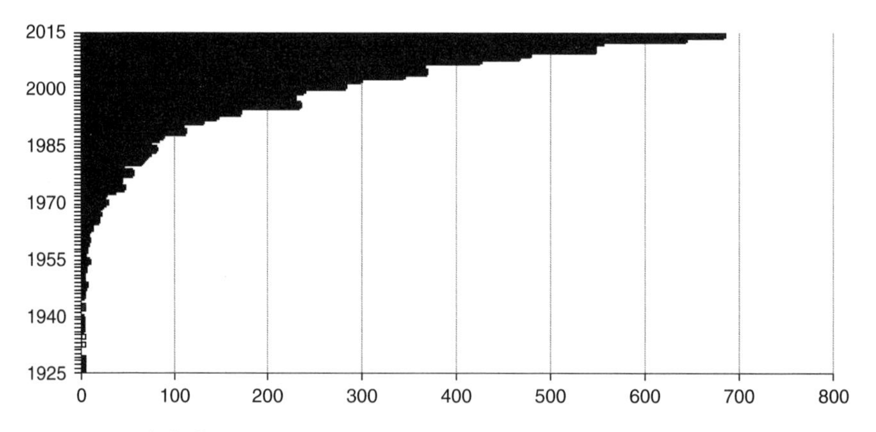

Figure 1-3 Scholarly writing on energy transitions, 1925–2015.
SOURCE: Compiled from Scopus, May 14, 2016, with "energy transition" and "energy system change" in the title, abstract, or keywords.

Contemporary thinking about energy transitions is deeply rooted in ideas from the 1970s and 1980s. During that period, writing began to increasingly address integrated conditions, emphasizing constraints in conventional resources that could be managed more strategically through alternative pathways like renewable energy, efficiency, and conservation (Anderer, McDonald, and Nakicenovic, 1981; Hafele, 1981; Lovins, 1976; Meadows, Meadows, Randers, and Behrens, 1972; Schumacher, 1973). At the time, there was a sense that deliberate change in energy systems was possible, but it would require that energy policies, corporate measures, and personal energy behavior be conceived in the context of broader energy challenges and not in terms of isolated aspects (Anderer et al, 1981).

A 1980 study by the Institute for Applied Ecology in Germany brought together these intellectual traditions with the concept of *Energiewende* or "energy transition" (Energy Transition, 2012; Krause et al., 1982). The study argued for a new energy path that would foster economic growth, yet be accomplished with efficiency. Looking at current efforts in Germany, the same change-oriented thinking of *Energiewende* continues today as the country produces more GDP with less energy (Morris and Pehnt, 2015). Further, Germany aims to reduce energy consumption to 50% of its 2008 usage by 2050, with 60% of the total comprising renewables. The country is doing so with new challenges and opportunities, as it leads in areas like highly-efficient building technologies that may become European Union (EU) standards.[14]

The above aims and ideas currently resonate with energy agendas in many regions of the world, including at the sub-national, national, and supra-national levels (REN21, 2017).

ALTERING THE ENERGY PLAYING FIELD

Adopting a plan to transform an energy system is no trivial undertaking. After all, energy infrastructure, practices, and industry are slow to change. Energy systems are traditionally characterized by limited competition and lengthy periods of research and development (R&D) investment (Flavin, 2008; Holdren, 2006*a*; Lund, 2006). Nonetheless, energy system change or transition is not new.

14. Germany's current efforts are linked to policy from September 2010 and fuller legislative measures from 2011 that represent a multi-pronged strategy to counter climate change, move away from nuclear power, reduce energy imports, strengthen energy security, stimulate a green economy and innovation, foster social justice, and support local economic development (Buchsbaum, 2016). The German power mix in 2015 consisted of 30% renewables (13% wind, 8% biomass, 6% solar photovoltaic, 3% hydropower), 14% nuclear, 52% fossil fuels, and 4% other (Appunn, 2016). Known for its initiatives in wind, solar, and biomass, among other areas, recent developments in Germany present research opportunities for the study of new transitions. For a discussion of aspects of the Germany transition, see Quitzow et al. (2016), Buchsbaum (2016), Pescia, Graichen, and Jacobs (2015).

At the country level, one can look at the transition from wood to coal that began in England in the 1500s and 1600s, when the growth of cities and deforestation practices produced conditions in which firewood had to be shipped greater distances (Brimblecombe, 1987; Fouquet, 2010; Landes, 1969; Rhodes, 2007). As the price of firewood increased, less wealthy citizens switched to coal while the nobility, including Queen Elizabeth I, maintained the practice of using wood because coal was considered to be less clean. When King James VI of Scotland assumed the throne of England and Ireland, he drew on knowledge about less sulfurous coal from Scottish practices to convert royal fuel from firewood to coal (Brimblecombe, 1987; Rhodes, 2007). This, in turn, influenced the nobility's view of coal, bringing their energy practices in line with other British citizens. Further developments in steam engines and canal systems extended the adoption of coal, with efficiency improvements in coal mining and steam-powered rail transport that enabled the creation of new markets (Brimblecombe, 1987; Rhodes, 2007; Fouquet, 2010).

Looking to more recent times, the British navy's shift from coal to petroleum highlights the significance of decision-makers co-evolving with changing conditions. In 1911, then British Home Secretary Winston Churchill opposed fuel switching for the British navy, seeing merit instead in a continued use of domestically-sourced coal. Yet as international tensions heightened with Germany, now First Admiral of the Navy Churchill changed his thinking, prioritizing naval tactical performance on the basis of power, efficiency, speed, and flexibility. This shift in strategic focus from domestic fuel sourcing to fleet performance meant that the British naval fleet would rely more heavily on oil imported from Persia (Churchill, 1928, 1968; Churchill and Heath, 1965; Yergin, 1991, 2011).

Considered at a global level, the primary energy mix has undergone fairly substantial inflections since 1850 (Figures 1-4 and 1-5). As energy use increased by roughly 20 times worldwide, the energy mix shifted from a reliance on biomass-based energy toward a mix of fossil fuels. Within this transformation came many, related shifts including catalytic cracking for refining oil, as well as the introduction of cars, electricity, and suburbanization, among factors.

Moving from historical examples of energy system change to forward-looking prospects, resource availability plays a key role when weighing energy options. Simple estimates of fossil fuels, for example, indicate a supply availability of roughly 50–115 years, based on reserve-to-production ratios of 50.6, 52.5, and 153.0 years for oil, gas and coal, respectively (BP, 2017).[15] Measured in somewhat different terms, estimates of low carbon energy potential indicate that the existing global

15. It is worth bearing in mind that changes in science, technology, and practice may extend the supply. See World Energy Council (2010*c*) for a discussion of reserves and resources, including proved, probable (indicated), possible (inferred), and undiscovered resources.

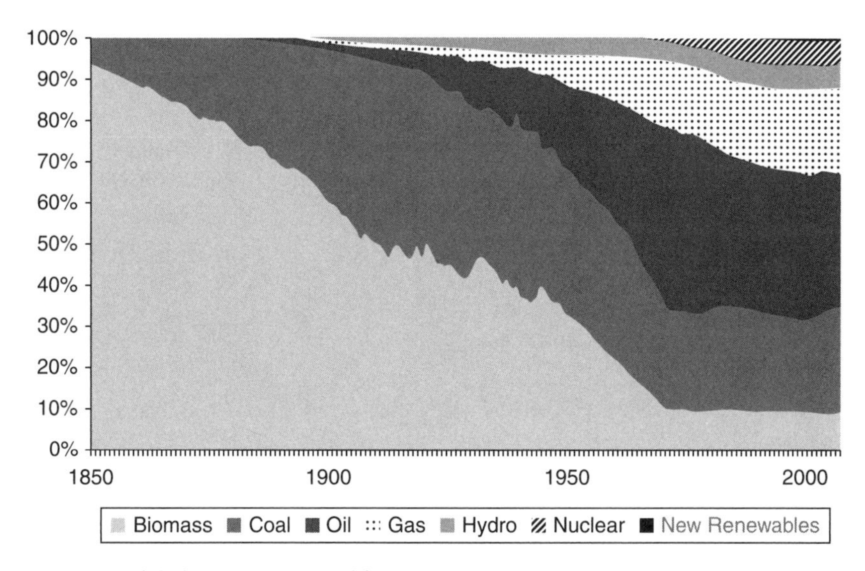

Figure 1-4 Global primary energy (share), 1850–2008.
From Araújo (2014). Adapted from Grubler, et al., (2012). NOTE: "New" renewables include technologies such as solar photovoltaic energy, geothermal power, and wind power. They do not include energy derived from traditional water or wind mills, wind-powered sea travel, solar water heating, and the like. The chart also does not reflect muscle power from animals and humans.

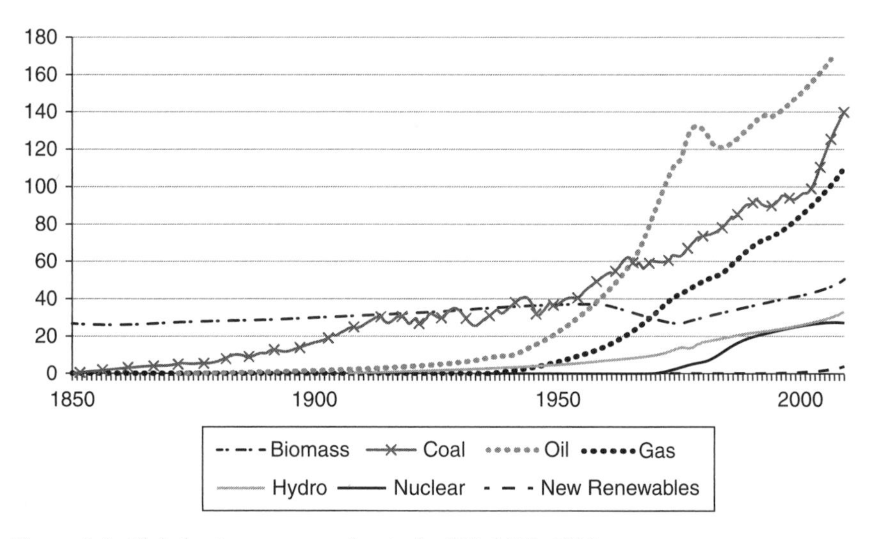

Figure 1-5 Global primary energy (exajoules/EJ), 1850–2008.
From Araújo (2014). Adapted from Grubler, et al., (2012). NOTE: "New" renewables include technologies such as solar photovoltaic energy, geothermal power, and wind power. They do not include energy derived from traditional water or wind mills, wind-powered sea travel, solar water heating, and the like. The chart also does not reflect muscle power from animals and humans.

Table 1-3. GLOBAL RESOURCE BASE FOR SELECT LOW CARBON ENERGY (EJ/YEAR)

Resource	2008 Use[a]	Technical Potential	Theoretical Potential
Hydropower	11.6	50	147
Biomass	50.3	276	2,900
Solar energy	0.5	1,575–49,837	3,900,000
Wind energy	0.8	640	6,000
Geothermal energy	0.4	5,000	140,000,000
Ocean energy	0.00		7,400
Nuclear energy	9.85	1,890[b]	7,100[b]
Total	73.45		

[a] 2008 use is taken from Moomaw et al. (2011) and based on direct equivalent accounting.
[b] Technical and theoretical potentials for nuclear energy are in EJ not EJ per year. These nuclear energy potentials reflect open cycle processes. If closed cycle processes were used with fast reactors, technical and theoretical potentials would equal 113,000 EJ and 426,000 EJ, respectively. The range of solar technical potentials reflects different assumptions pertaining to annual clear sky irradiance, annual average sky clearance, and available land area.

SOURCE: Adapted from UNDP et al., 2000, unless otherwise noted.

NOTE: Estimates, such as those by the International Energy Agency, the *Global Energy Assessment*, and the *Special Report on Renewable Energy Sources and Climate Mitigation,* provide related and, at times, different numbers based on assumptions and scoping. See, for example, discussions of the theoretical potential for geothermal energy in IEA, 2011*b*; Goldstein et al., 2012; Rogner et al., 2012.

---

utilization of energy is a small fraction of what could be effectively harnessed (Table 1-3).[16] The total amount of energy used worldwide in 2008, for instance, represented about 1% of the technical potential for geothermal energy and roughly 5% of the lower estimate for solar energy. Today's numbers remain similar.

What is often missed in energy discussions is the reality that renewable energy is widely available, essentially everywhere in varying combinations and potentials. By contrast, finite fuel sources like fossil fuels and uranium are not.

16. Precise definitions for technical and theoretical potential vary. *Technical potential* is frequently analogous to "resource," implying energy that can technically be extracted irrespective of economic feasibility. *Theoretical potential* refers to energy availability that is deduced as possible based on an understanding of the resource flows yet is not feasible to extract given prevailing technology and economic conditions. See United Nations Development Program [UNDP], UN Department of Economic and Social Affairs, World Energy Council [WEC] (2000) for a discussion of assumptions used for fuel potentials in Table 1-3 estimates. For a discussion of low carbon energy versus renewable energy and other forms, see Chapter 2.

Adopting a low carbon path that avoids an increase of 2°C would require on the order of $53 trillion in global, cumulative investment, according to one estimate (IEA, 2014). This includes costs for infrastructure and related expenditures associated with the energy supply and improved efficiencies by 2035. Although cost and benefit assessments have their limitations, it is reasonable to say that such a scale of investment will face challenges.[17] Financing can be seen as risky if the cost of capital is based on established returns of investment and newer technologies do not yet have a track record. Stranded assets can also complicate a shift if decision-makers limit their choice to legacy pathways due to unrecoverable, earlier investments. Going even further, players may choose to not engage because they see themselves as unable to capture all the benefits (Anex, 2000; Cowen, 2008; Dosi, Malerbra, Ramello, and Silva, 2006; Gravelle and Rees, 2004; Mitchell et al., 2011).

In such circumstances, *green financing* will be an important bridge.[18] Green bonds, for instance, offer ways to connect low-cost capital held by institutional investors with low carbon projects (World Economic Forum, 2013). Platforms for interaction between project developers and investors can also be developed as public financing agencies streamline risk mitigation (IRENA, 2016). Institutional investors, including pension funds, insurance companies, and sovereign wealth funds, also have opportunities to become critical players in the mobilization of such a wide-ranging initiative. In fact, efforts by groups like the United Nations Environmental Programme (UNEP), G20, International Monetary Fund/World Bank Group, and OECD are under way to align the global financial system with more sustainable aims (OECD, 2015*a* and 2015*b*, n.d.; UNEP, 2016; World Bank, 2015). Here, policy can play a critical role by focusing attention on prime areas where investments are needed, in addition to creating more stable and predictable investment environments.

Path dependence may be one of the strongest forces that impedes change. With this phenomenon, previous choices limit later options based on the inflexibility of sunk costs, the increased returns from continuing on the existing path, or the interrelatedness of technologies, among other factors (Arthur, 1989; David, 1985). Greg Unruh applied this idea in what he called *carbon lock-in*—conditions in which industrial economies have become entrenched in fossil fuel–intensive systems through the co-evolutionary development of technological and institutional processes driven by returns of scale that, in turn,

17. For a discussion of methods and scoping considerations in cost assessments of energy system change, see Araújo (2016). For more on barriers, see Organization for Economic Co-operation and Development (2015*a*); Brown, Chandler, Lapsa, and Sovacool (2008).

18. Definitions of green finance, like that of low carbon, are not fixed. See Lindenberg (2014).

create persistent market and policy failures (Unruh, 2000). This "vicious cycle" inhibits the diffusion of carbon-saving technologies despite cost-neutral or even cost-effective remedies that have apparent environmental and economic advantages (Unruh, 2000, citing Ksomo, 1987). Such self-reinforcing reliance on an incumbent carbon-dependent path is not necessarily permanent, but it can persist by creating systematic market and policy barriers to alternatives such as energy efficiency and renewable energy technologies (Unruh, 2000, 2002).

Today's newer entrants, like modern renewable energy, can encounter challenges in technological maturity that the fossil fuel industry overcame in previous eras.[19] In such cases, inherent advantages in the status quo may drive incumbent actors with vested interests to resist change. Resistance might then crystallize, with traditional players seeking to block structural change, technological progress, or the rise of industry challengers (Juma, 2016; Moe, 2015). This can be done by exerting pressure on governments to impose administrative procedures, taxes, trade barriers, or regulations in order to prevent new entrants from challenging the current power structures or undermining existing fee structures (Moe, 2015; Olson, 1982).

While resistance and organizational/institutional inertia will likely continue to impede some progress on low carbon adoption, rapid change is already evident in industries, such as banking, telephones, medicine, and computing, that have defeated similar odds. The underlying insight from such shifts is that agents of change are redefining the playing field rather than accepting lock-in as an ongoing status quo (Bardach, 1977; Garud and Karnoe, 2001, 2003; Garud, Kumaraswamy, and Kamoe, 2009; Kingdon, 1995; March and Olsen, 1989).

CENTRAL QUESTIONS

Given the convergence of low carbon priorities and the challenges facing wider utilization of low carbon energy, my aim in this book is to provide in depth perspective on how four, leading countries of advanced, low carbon energy technologies shifted their national energy systems in the period since 1970. I do so by examining the following, key questions:

19. The use of the term "modern renewables" recognizes that wind-, solar-, geothermal-, biomass-, and water-based energy sources have been used for centuries. Traditional uses include wind energy applications in sea power. This study focuses on more, contemporary applications.

- How can national energy transitions be explained in terms of inflection points; key interventions by government, industry and civil society; and structural change?
- To what extent do the patterns of change align and differ in the four, energy transitions that are examined?
- What role does policy play, particularly with innovations, in the cases that are considered?[20]

In answering these questions, I also explore elements of cost, societal acceptance and human development, industrial progress, carbon intensity, and natural resources.

Whatever the reason for switching to low carbon energy—to foster resilience, improve access, reduce import dependence, or address the business case for change—understanding underlying dynamics may provide timely insights.

20. *Policy* is considered here as explicit or implicit rules and (in)action of public entities. *Innovation* is seen as enhancements through novel or recombinatory ideas and applications in science, technology, and other areas, including societal practices.

2

Beyond Malthus

Evolution is a sequence of replacements.

—ELLIOTT MONTROLL, *Physicist (1978)*

This chapter explores the evolving understanding of carbon and sustainability since the 18th and 19th centuries. Relevant applications of influential ideas are then identified with respect to knowledge, innovation, policy, and meta-level change.

CARBON AND THE GREENHOUSE GAS EFFECT

More than 100 years ago, Swedish scientist Svante Arrhenius hypothesized about the onset of ice ages and interglacial periods by considering high latitude temperature shifts (NASA Earth Observatory, n.d.). Applying an energy budget model and ideas of other scientists, like John Tyndall, Arrhenius argued that changes in trace atmospheric constituents, particularly carbon dioxide, could significantly alter the Earth's heat budget (Arrhenius, 1896, 1897; NASA Earth Observatory, n.d.).

Today, science indicates that the global, average surface temperature has continued to rise alongside the increase in greenhouse gases. Among global GHGs, CO_2 emissions have increased by more than a factor of 1,000 in absolute terms since 1800 (Figure 2-1 and Box 2-1). During that time, global carbon emissions found in the primary energy supply increased by roughly 6% per year (Grubler, 2008*a*). This growth in carbon emissions from energy is significant because CO_2 from fuel combustion dominates global GHG emissions (IEA, 2015*a* and 2015*b*; IPCC, 2013). As noted earlier, 68%

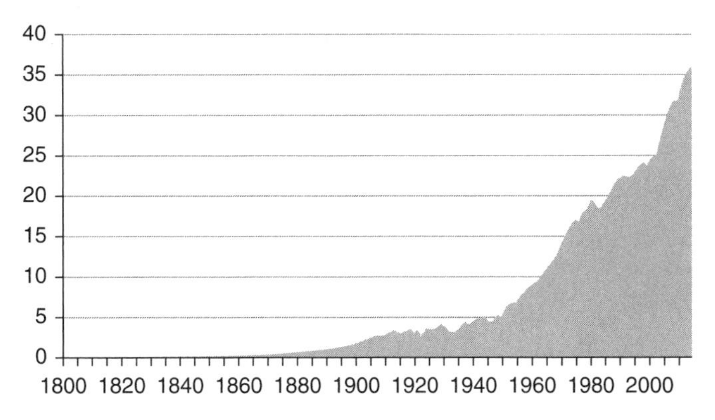

Figure 2-1 Global CO_2 emissions ($GtCO_2$), 1800–2013.
Adapted from Boden, T. A., Marland, G., and. Andres, R. J. (2016). *Global, Regional, and National Fossil-fuel CO_2 Emissions.* Carbon Dioxide Information Analysis Center, Oak Ridge National Laboratory. Oak Ridge, TN: US Department of Energy. doi 10.3334/CDIAC/00001_V2016.

Box 2-1

The Kaya Identity

The *Kaya identity* expresses carbon dioxide emissions as the product of the carbon intensity of energy (CO_2/E), energy intensity of economic activity (E/GDP), economic output per capita (GDP/P), and population (P) (Kaya, 1990):

$$CO_2 = (CO_2/E) * (E/GDP) * (GDP/P) * P.$$

For further discussion of CO_2 in the context of systems change, see Kaya, Nakicenovic, Nordhuas, and Toth (1992) and Kaya and Yokobuchi (1993).

---

of the global GHGs that are attributed to human activity are linked to the energy sector; namely, fuel combustion and fugitive emissions (IEA, 2015*a*). Within this share, 90% consisted of CO_2 (IEA, 2015*a*).

In contrast to the rise in absolute numbers, carbon emissions per unit of output in the global primary energy supply has decreased 36% overall or by slightly less than 0.2% per year over the past two centuries (Grubler, 2008*a*).[1]

1. Other carbon intensity metrics include carbon per unit of gross domestic product and per person. The differences in metrics must be factored in discussions about carbon reductions.

This subtle decarbonizing pattern in the energy mix is explained by the faster growth rate of energy use in relation to the rate of carbon emissions from that use. The delinking of energy utilization and carbon emissions occurred in part with the introduction of less carbon-intensive fossil fuel sources, like natural gas, in which a higher hydrogen-to-carbon ratio is evident (Gibbons and Gwin, 2009; Grubler, 2004, citing Marchetti, 1985).[2] Delinking is also fostered by the introduction of nuclear energy and the increased utilization of renewables in the latter half of the 20th century. Because global energy demand is projected to rise approximately 32% by 2040, choices influencing relative shares of renewable energy technologies (RETs) and nuclear energy will be important for these trends (IEA, 2015*d*).

SUSTAINABILITY

Well before Svante Arrhenius developed calculations on CO_2, British writer Thomas Malthus laid the conceptual ground work for ideas on sustainability in *An Essay on the Principle of Population*, first published in 1798 (1999). In his now classic writing, Malthus emphasized limitations of natural resources on societal growth. Building on Malthus' ideas, the concept of sustainable development emphasizes interdependent priorities valuing the environment, economics, and society across generations (World Commission on Environment and Development [WCED], 1987). Driven in part by complexities of competing priorities, the concept of sustainable development is widely used, yet also subject to criticism for its looseness of definition and form of gauging (Bartlett, 1997/1998; Daly, 1996; Taylor, 1992).

In today's discussions about energy, it is not uncommon to hear *sustainable* used interchangeably with *renewable, clean*, and *low carbon*. Although overlap exists between the meanings, differences matter. *Sustainable* refers broadly to durability and in more rigorous definitions is described as enduring intergenerationally in a way that does not unduly undermine society, the economy, or the environment. *Renewable* energy, as mentioned in Chapter 1, typically refers to energy forms that naturally regenerate. A more nuanced definition would stipulate that regeneration occurs in a manner that exceeds or matches the draw-down of the energy source. *Clean* energy is widely used in the context of fuel use that does not pollute. By contrast, *low carbon* energy signifies an approach that emits less carbon, yet the term opens questions about scope. Does the term refer to the life cycle of a fuel or to one particular stage,

2. The hydrogen–carbon ratios or H/C for various fuels are coal (0.5–1:1), oil (2:1), natural gas (4:1).

such as production or consumption? A number of examples will help clarify these distinctions.

- Natural gas is sometimes classed among low carbon fuels to differentiate it from more carbon-intensive coal and petroleum-based fuels. With this rationale, natural gas constitutes a low carbon fuel that is not renewable.
- Nuclear energy is characterized by some as a form of clean energy, if one discounts nuclear waste and focuses on the limited emissions of its power generation (excluding construction, mining, etc). Following this line of thinking, nuclear energy might be described as clean energy that is not renewable.
- Perhaps surprising for some, renewable energy can be managed in ways that are not sustainable, low carbon, or clean. The extraction of geothermal energy, for example, can radically draw down heat or steam from its source. In such a case, the energy source might be renewable, but not sustainable in practice. Somewhat differently, the production and use of biomass can result in varying levels of pollution, including CO_2. Depending on how biomass is managed, it can be renewable, but not low carbon or clean.[3]

What affects many of the above distinctions is the way a specific type of energy is managed from end to end (not just its production). Here, the succinctness of definitions is also challenged by continuous changes in technology and management practices. If such terms are used in international conventions, attention is required in how the concepts are defined. For this study, low carbon energy includes non-fossil fuels, namely renewable and nuclear energy. The terms "sustainability" and "durability" of energy transitions are used interchangeably.

KNOWLEDGE, INNOVATION, AND POLICY THEORY

Knowledge, innovation, and policy are powerful forces of transformation. Certain ideas on these forms of change can guide studies of energy transitions.

3. Definitions in the *Special Report on Renewable Energy* (SRREN), Section 1.2.1 distinguish renewable from sustainable energy, and exclude some forms of slow-growing bioenergy. Bioenergy is renewable, but may or may not be sustainable in practice. It also is high in carbon relative to other renewable resources like solar and wind (and nuclear). Further distinctions could focus on carbon intensity (Communications with B. Moomaw, 2016).

Knowledge

Knowledge has a role to play in systems change. Economist and historian of technology change Joel Mokyr wrote that it was not inventors or socioeconomic factors that drove the Industrial Revolution, but rather people exchanging knowledge (2002). Such knowledge can be thought of in terms of its source or level of authority. What some may call "established knowledge," for example, is typically tested through mainstream, disciplinary investigation and accepted by scientific peers. By contrast, local knowledge "does not owe its origin, testing, degree of verification, truth, status, or currency to distinctive professional techniques, but rather to common sense, causal empiricism, or thoughtful speculation and analysis" (Lindblom and Cohen, 1979). The latter form aligns with ideas on learning and innovation that emphasize the importance of user insights in extending the knowledge frontier (Johnson, Lorenz, and Lundvall, 2002; Lundvall, 1985, 1988; Von Hippel and Tyre, 1995). The pivotal nature of knowledge is also reflected in the context of adaptive capacity and the agility of a country to evolve (Smil, 2010), all of which factor in energy systems.

Innovation

Innovation, defined broadly as adaptations to improve performance and/or quality, can also be instrumental for energy transitions. Classic views of innovation highlight a linear progression in which technological development occurs in three stages: (1) invention, when an idea first emerges; (2) innovation, the first practical application of the invention; and (3) diffusion, when the innovation is dispersed widely for use (Schumpeter, 1942/1975). More recent views of the innovation life cycle map the inception of an idea and its incubation through testing and prototyping to niche market development, through to widespread diffusion with feedback loops, links, and overlap throughout the cycle (Grubler et al., 2012; Kline and Rosenberg, 1986; Nakicenovic, Grubler, and Macdonald, 1998/1999). Both approaches envision processes that principally occur in traditional institutional settings, yet there is a growing awareness that innovation and the agents of such change can extend well beyond industrial laboratories and academic settings (Lundvall, 1988, 2010; von Hippel, 2005, 2010) and do not need to be rigidly sequenced (Sovacool and Sawin, 2012). This book recognizes innovation in energy systems as an improvement that can occur with conventional and unconventional paths, and which enhances the quantitative or qualitative utilization of energy, including the sourcing, conversion, application and use, distribution, and final disposal. Such scoping allows for shifts that improve the system in less obvious ways, such as through its governance and financing practices.

Theory-building on national innovation systems (NIS) emphasizes the interactions among institutions and other elements within a country that produce systemic feedbacks and constructive adaptations in the advance of a technology (Edquist, 2005; Freeman, 1987; Lundvall, 1992; Nelson, 1993; Ridley, Yee-Cheong, and Juma, 2006). This body of theory recognizes that shared language, culture, and institutions at the country level can serve to frame important conditions for innovation systems. Richard Nelson and Bengt-Åke Lundvall developed two primary lines within NIS literature. Richard Nelson's applied analysis highlighted the role of national research and development systems (Edquist, 2005, citing Nelson, 1993), whereas Lundvall's more theoretical approach emphasized the role of learning in user–producer interactions and the home market for economic specialization (Edquist, 2005; Lundvall, 1992). Given the national scoping of this study, a wide view of innovation is used, acknowledging that developments can arise from a change in technology, products, processes, or practices tied to learning and experimentation, serendipity, and breakthroughs from any sector for a given country.

Technology[4]

Technology change provides another lens for understanding energy transitions. In the most basic terms, the introduction of a new technology can influence how energy is sourced, delivered and used. The concept of technology change in neoclassical economics has centered on (1) the relationship between supply and demand, (2) performance in production for which technology is an input, and (3) research and development, however these do not account for the unplanned and less precise elements of development (Mokyr, 1990; Mytelka and Smith, 2001; Nelson and Winter, 1982).[5] Evolutionary economics provided a critical point of departure for this thinking by emphasizing how natural selection and competition can be critical drivers, rather than the profit maximization and market equilibrium emphasized in neoclassical economic theory (Nelson and Winter, 1982).

4. Technology often includes hardware and software, such as equipment and computer applications, but it can also mean products, processes, devices, and practices (Grubler et al., 1999*a*, 1999*b*). For our purposes here, technology is defined as hardware, software, and material inputs (i.e., energy and raw materials), using the term "system" to encompass the broader conceptualizations of technology. Related dimensions, such as knowledge and practices, are considered under separate labels.

5. According to the neoclassical economics schools of thought, technology is viewed as an intermediary factor in relation to the basic factors of production: labor and capital (Hadjilambrinos, 2000). Technology change, then, derives from the need to improve resource utilization (Hadjilambrinos, 2000, citing Cohendet et al., 1991, Gilbert, 1985, and Moroney and Trapani, 1981).

Today, change in technology is often viewed in the context of incremental or radical/disruptive shifts (Abernathy and Utterback, 1978; Dosi, 1982; Grubler, 1998; Grubler, Nakicenovic, and Victor, 1999*a*, 1999*b*). *Incremental change* implies slight modifications to existing technology, like the addition of a catalytic converter to an automobile, whereas *radical* or *disruptive change* refers to substantial adaptations in one technology or when one technology supplants another. An example of a more radical form of disruptive change was evident in the emergence of automobiles that replaced carriages. This change opened new directions for practices, access and infrastructure. Joel Mokyr argued that incremental and disruptive changes do not need to be mutually exclusive, but can overlap since most macro-level inventions build on the accumulation of micro-level ones (1990).

Greg Unruh differentiates transition stages with a taxonomy that includes end-of-pipe (incremental), continuous (nondisruptive), and discontinuous change (disruptive or radical) (2002). The addition of an intermediary stage allows for an enhancement or upgrade to the existing architecture that repositions the prevailing technology trajectory along a more sustainable pathway (Unruh, 2002; Berkhout, 2002). While bridging legacy and novel technologies, this middle type of shift maintains inherent limitations since nothing is fundamentally changed about the technology or the institutions themselves (Berkhout, 2002).

Frank Geels and Johan Schot have theorized about the structure of technology change (Geels and Schot, 2007). Unlike many, related models that focus on the intensity of systemic disruption, Geels and Schot's approach differentiates processes to include transformation, reconfiguration, substitution, and realignment/dealignment. These concepts will be useful to bear in mind as additional theory-building on energy systems change is proposed later in this book.

Frames and Policy

There is a saying that great opportunities are often disguised as unsolvable problems. How we perceive conditions matters for the way in which we respond, with perspective being influenced by experience, philosophy, and power, among other factors (Allison, 1969; Allison and Zelikow, 1999; Schoen and Rein, 1995). Thinking in terms of frames, the challenges of an energy shortage, for instance, can be viewed negatively. Yet those same conditions also present windows of opportunity to modernize and improve the overall system. In such cases, focusing events may serve as inflection points for broader change (Birkland, 1997, 1998; Birkland and Warnement, 2013; Kingdon, 1995; Zahariadis, 1999). For policymakers wanting to develop more resilient energy strategies, infrastructure replacement following storm damage, for example, can serve as a point for integrated assessment and course correction aligned with longer-term aims (Baumgarten and Jones, 1993; Jones and Baumgartner, 2005; Howlett, Ramesh, and Perl, 2009).

Whether one sees shifts in energy paths as an opportunity or challenge, policy design and implementation eventually come into play. Hood's taxonomy of policy instruments suggests that one could approach governmental action by focusing on governing resources, like information, authority, finance, and organization (Hood, 1968). Bemelsmans-Videc, Rist, and Vedung, by contrast, offer a more simplified way of envisioning policy in terms of regulations, incentives, and information (2005). Irrespective of the approach, national policy styles differ, based at least partly on domestic idiosyncrasies in institutions and culture (Howlett, 2002; Linder and Peters, 1989). Writing on policy mixes and interaction effects speaks to the importance of good alignment of policy tools with conditions, aims, governance approach, and resources (Guerozoni and Raiteri, 2015; Howlett and Raynor, 2013; Kern et al., 2017). There is a sense, however, in the context of sustainability transitions, that policy mixes must more fully account for the dynamic settings in which energy systems reside, including real-world complexities, explicit incorporation of process, and strategic dimensions (Rogge and Reichardt, 2016). Effective implementation of a feed-in type of market premium policy, for example, should include upper and lower price limits, and clear guidance on triggers for policy review to minimize disruption. Policies, such as these, will be discussed more in Chapter 7 and 8.

THEORIES ON META-LEVEL CHANGE

Structure, Function, and Connection Points

Theories on meta-level change provide another set of important foundations for theory-building on energy transitions. Large technical systems, techno-economic paradigms, and multilevel perspectives are among the more well-known contributions.

History of science writing on large technical systems (LTS) conceives of complex and seamless webs that include not only the physical infrastructure, but also economic, legal, and social elements that can manifest in organizations, rules, and other elements (1983, 1998, and 2012). According to Thomas Hughes, systems builders, like Thomas Edison with electricity, focus on fostering the coherence of their technical systems within the social environment. As the systems and the environment mutually influence the other, the system grows and advances with a momentum that includes re-enforcing contributions from actors and inventions (Hughes, 1983, 1989).[6] When the system matures, it becomes resistant to change. *Reverse salients* may develop, in which components of the system create lag or are out of step. An example of this can be seen in the use of floppy disks for nuclear weapons systems. The mostly obsolete storage files require costly measures to maintain (BBC, 2016). When such a reverse salient is not rectified within an incumbent system, the condition can become radical, bringing about a new and competing system (Hughes, 2012).

Ideas on *techno-economic paradigms* (TEP) build on long-wave theories of business cycles to offer complementary ideas about change. As outlined by scholars, including Chris Freeman, Carlota Perez, and Francisco Louca, TEPs are seen as configurations of interlocking technologies, processes, economic structures, and beliefs that endure based on gains from key factors (Freeman and Louca, 2002; Freeman and Perez, 1988; Perez, 2009*a*), but which can transition with a technological revolution (Freeman and Perez, 1988; Twomey and Gaziulusoy, 2014). This approach emphasizes how *new logic*, including research rationale and norms, replaces earlier thinking over the course of five or six decades to shape the modernization of existing industries alongside newer entrants (Freeman and Louca, 2002; Perez, 1985, 2004*a* and 2009*b*). Flux is seen as being minimized through links between political, business, and cultural trajectories. In the context of sustainability, a new TEP could emerge in line with this aim, through novel information and communications (Perez, 2009*b*).

Described as a "middle range theory," the *multilevel perspective* (MLP) draws on sociology, evolutionary economics, and science, technology, and society studies (STS) to explain sociotechnical transitions toward sustainability (Geels, 2005; 2006). In conjunction with the work of Frank Geels, Johan Schot, and others, the MLP approach focuses on a sociotechnical system of nested niche, regime, and landscape levels in which each level provides different kinds of coordination and structuration to activities in local practices (Geels, 2002; Geels and Schot, 2007; Grin et al., 2010).[7] Niches are seen as the locus for radical novelties where innovations can accumulate. Co-evolving interactions are critical among technology, user practices, markets and industrial networks, policy, scientific understanding, cultural meaning, and infrastructure (Geels, 2005, 2011). According to this line of thinking, major change is produced by the *realignment of trajectories within and between the various levels.*

6. See Hirsh and Sovacool (2006) for additional discussion.

7. *Niches* are incubation spaces that are shielded from mainstream market selection (Geels, 2006, citing Schot, 1998). *Regimes* include cognitive routines, patterned development, regulatory structure, lifestyles related to technology systems, and sunk investment in equipment, infrastructure, and competency (Geels and Schot, 2007; see also Nelson and Winters, 1982; Unruh, 2000). *Landscapes* refer to the broader, external environment and include macro-economic conditions, culture, and macro-political developments (Geels and Schot, 2007).

Closely aligned with the MLP is the study of *strategic niche management* (SNM) and *transitions management*. This body of work considers ways to nurture socially desired aims and technological innovation (Schot et al, 1994; Kemp and Loorbach, 2006; Raven et al, 2010; Raven, 2012; Smith and Raven, 2012). Experimentation in niches is an area of particular focus for SNM, overlapping with features of the next model to be considered: *technology innovation systems* (TIS).

TIS theory, as put forward by Marko Hekkert and others, brings a function-based approach to understanding how systems perform as innovations are generated, diffused, and used (Hekkert et al., 2007; Jacobsson and Bergek, 2004; Raven, 2012). Seven subfunctions are outlined: entrepreneurial steps, knowledge development and exchange, guidance of a search, market formation, resource mobilization, and the counteracting of resistance/establishing legitimacy (Hekkert et al., 2007). The caliber of subfunction attainment and the relationships among the subfunctions influence whether a transition occurs. This body of work has been extended with related analytical tools, namely a mapping tool for subfunctions (Grin et al., 2012; Negro, 2007; Negro et al., 2007) and a typology of interactions between the functions that enable a transition (Suurs and Hekkert, 2008). TIS is seen as a powerful way to evaluate the internal strengths and weaknesses of a specific sociotechnical system, yet some transition scholars note that more could be elaborated on the timescales. Recently, TIS theory began to focus on the system's external environment (Grin et al., 2010). Important, emergent work has also identified ways to bridge MLP and TIS (Markard and Truffer, 2008).

A more institutional alternative for scoping energy system change is found in the analytical studies of the Transitions Pathways for the Low Carbon Economy Research Consortium. With it, Timothy Foxon and Ronan Bolton described a conceptual framework that envisions three ways that a decarbonized future can be attained: (1) centralized government, (2) market rules, and (3) a "one thousand flowers" to enable change (Bolton and Foxon, 2015; Foxon, 2013; Foxon et al., 2010). The centralized government path is one in which the national government "exerts strong influence over the energy system in order to deal with the trilemma of security, costs, and emission reduction targets," where technology push occurs with a focus on large centralized technologies (Bolton and Fox, 2015). By contrast, the market rules path is one in which a liberalized market framework prevails, with large energy utilities as the dominant investors. The third path or "one thousand flowers" sees a decentralized approach in which nontraditional investors in the energy system play a leading role with more distributed technologies

(Bolton and Foxon, 2015). Naturally, these are guiding conceptual constructs. Reality is often more nuanced (see Chapter 3).

Related theory-building also considers technology life cycles within a broader system. For this, Arnulf Grubler and co-authors (2012) outline an *energy technology innovation system* (ETIS) by locating a transformative energy ecosystem in the context of knowledge, technology characteristics, and actors/institutions. This framework highlights interacting stages of nonlinear development, namely research, development, demonstration, market formation, and diffusion. As the name suggests, this approach is predicated on innovation, providing some means for tracking metrics and considering elements like the loss of knowledge (Grubler and Wilson, 2014).

Another, more recent framework looks specifically at *complex established legacy sectors* (CELS). Focusing on the legacy infrastructure, such as that in transport, power, and the like, William Bonvillian and Charles Weiss explain that technology, economic, political, and social paradigms create barriers to desirable technology innovations (2015; Weiss and Bonvillian, 2013). Observing that dynamic shifts often are stymied by a mismatch of broader social goals and incentives that reinforce existing pathways, Bonvillian and Weiss outline a framework of obstacles to the market launch of innovation. Such barriers encompass perverse subsidies, pricing and cost structures; established infrastructure and institutional architecture that impose regulatory hurdles or other disadvantages to new entrants; politically-powerful, vested interest backed by public support; a financing system that is not suited for the development timeline of capital-intensive legacy-sector innovations; public habits and perceptions attuned to current technology; knowledge and human resource structure that are oriented to legacy sectors; aversion to innovation; and market imperfections that go beyond those faced by other innovations (Weiss and Bonvillian, 2013; Bonvillian and Weiss, 2015). As with the preceding ETIS framing, this approach focuses on innovation. Much of the path dependence features of this framework are relevant to the deployment of nonincumbent forms of energy in such systems.

Looking across all the preceding frameworks, one finds different levels as well as dimensions of analysis. The social embeddedness of the system is a common feature, underscoring how disruptive change involves a range of actors and actions well beyond the lab or the field. Some of the models have an economy or market-centered orientation, like that of the TEP, CELS, and Transition Project paradigms. Others focus more on structure, alignment, or functions, in which outcomes are influenced by a larger set of a social inputs. As systems thinking evolves in conjunction with energy, innovation, and sustainability, many new interdisciplinary connection points will emerge for problem-solving and theory-building.

Timescales, Pace, and Complexity

Theories on meta-level change may explain broader, conceptual aspects of energy transitions, yet academic and policy questions also emerge in connection with practical aspects of energy transitions. Here, the timescales, the pace, and the complexity of change are among such considerations.

Specific to timescales of transitions, one can ask what duration is needed for notable change to occur (Sovacool, 2016). Cesare Marchetti and Nebojsa Nakicenovic answered with two-parameter logistic function analysis indicating that the duration to shift market shares for a particular energy type from 1% to 50% for the global primary energy mix is 50–100 years (Marchetti and Nakicenovic, 1979).[8] Benjamin Sovoacool revisited this question roughly three and a half decades later, finding that definitional demarcation and differences are not always clear in academic writing and that the pace of a transition (national or global) may be less about what occurred and more about how the transition was characterized (Sovacool, 2016). In focusing on national-level energy transitions, Vaclav Smil emphasizes time frames ranging from just a few years to more than a century (Smil, 2016). Grubler and co-authors have reiterated Marchetti and Nakicenovic's earlier points on longer timeframes, yet also indicated that the point of focus today may be about what is required for an energy transition and less about how long it takes (Grubler, Wilson, and Nemet, 2016).

In connection with timescales, one can also ask what is known about the pace of change and the players. Arnulf Grubler and Charlie Wilson indicated that the rate of change for the global primary energy substitution has dramatically slowed since the mid-1970s, with early-adopter countries reaching a higher market saturation level as later adopters convert (i.e., scale up) more rapidly but not as extensively (Grubler, 1996; Wilson, 2009; Wilson and Grubler, 2011). Jose Goldemberg pointed to *leapfrogging*, arguing that later adopters, such as newly-industrializing countries, have an opportunity to sidestep many of the issues associated with early use by advancing to next-generation technologies (Goldemberg, 1998). In line with this, late adopters are not encumbered by the infrastructure and sunk investments of first movers, so may benefit from prior learning to advance directly to newer, more superior technologies. This form of Darwinian interplay between first and later adopters can define the dominance not only of market shares, but of the technology design (Grubler, 1998; Nakicenovic and Grubler, 1990; Suarez and Utterback, 1995; Utterback

8. In reaching this conclusion, Marchetti and Nakicenovic applied Fisher and Pry's model (1970); the predictability of this has been questioned for excluding hydropower (Smil, 2017).

and Abernathy, 1975; Utterback and Suarez, 1993), both of which matter for industrial leadership.

Daniel Kammen pointed out that leapfrogging is one of many ways to address locked-in technology. Hybridization is another, in which the new and old technologies are used concurrently, allowing continued learning and refinements to new technologies as older ones are sustained (2004). An example of this is seen with developments in steam and gas-powered generators during the 20th century. The early successes of steam technology in the power sector could have led to a full phase-out of gas turbine technology. However, gas turbine technology continued to advance independently in the aviation industry, with subsequent spillovers into the power sector. Eventually, gas turbine technology became the main component of the hybridized combined-cycle system in power generation.

Kammen goes on to argue that market transformation, which may entail public sector funding, is necessary to assist with the commercialization of clean energy technology (2004). This recognizes two rationales for a government to intervene with clean energy technology: namely, that traditional energy prices do not account for social costs or other externalities, and that private firms are unable to appropriate the benefits of precommercialization investment. Going a step further, Kammen indicates that the social rate of return on research and development is greater than private returns are, so private firms may be less inclined to invest to increase social welfare (2004). In line with this kind of thinking, public support of low carbon energy may be justified on the basis of multiple rationales.

Looking more squarely at diffusion rates, Everett Rogers (1995) theorized that the pace of change depends on:

- The relative advantage of a change versus the status quo it supersedes
- Compatibility or coherence of change with norms, needs, and understanding
- Complexity of a change
- Opportunities for testing a change
- Observability or clarity of positive results.

Perceived complexity of change appears, for example, to be negatively associated with the rate of adoption (1995). Grubler, Nakicenovic, and Victor adopted a different lens by distinguishing between factors that slow or accelerate transitions. Factors that slow diffusion can include market size (i.e., larger markets require more time), technology complexity, and infrastructure (Grubler et al., 1999*b*; Smil, 2010). By contrast, factors that can accelerate diffusion include the pre-existence of niche markets that allow early testing, and the comparative advantage of the new technology in areas such as performance, efficiency, and costs (Grubler et al., 1999*b*). While these ideas warrant further study with energy transitions, a reasonable consensus exists among writers on technology diffusion in terms of how determinants of transition rates can differ over time (Grubler, 1997; Rogers, 1995).

CONCLUSION

Accelerating change, particularly through innovation, is now regularly invoked as the solution for sustainability issues, yet diffusion research indicates that the large-scale adoption and substitution of new energy technologies can take 50–100 years or more. It is at this point that a closer inspection of actual transitions will shed more light.

3

Research Design, Scoping Tools, and Preview

This chapter outlines the design of the current study. It discusses my underlying logic for scoping energy system change with theory-building in the form of (1) a framework on intervention that operationalizes insights from the previous chapter and (2) conceptual models of structural readiness. A brief review then follows of related, global developments to provide broader context for the cases. The chapter concludes with a preview of the transitions that will be discussed in depth in subsequent chapters.

RESEARCH DESIGN

This book draws on my research of four national energy system transitions covering the period since 1970. I selected a timeframe that reflected a common context of international events which preceded as well as followed the oil shocks of 1973 and 1979. Such framing allowed me to trace policy and technology learning over multiple decades for different cases. I completed field work for this project primarily between 2010 and 2012, with updates continuing through to the time this book went to press.

Selection of Prime Mover Cases

I selected cases from more than 100 countries in the International Energy Agency (IEA) databases. The ones that I chose represented countries which demonstrated an increase of 100% or more in domestic production of a specific, low carbon energy and the displacement of at least 15 percentage points in the energy mix by this same, low carbon energy relative to traditional fuels for the country and sector of relevance. I utilized adoption and displacement metrics to consider both absolute and relative changes. Final cases reflect a diversity of energy types and, to some extent, differences in the socio-economic and geographic attributes of the countries. The technologies represent some of the more economically-competitive substitutes for fossil fuels. It's important to emphasize that the number of cases was neither exhaustive nor fully representative. Instead, the cases reflect an illustrative group of newer, low carbon energy technologies for in depth evaluation.[1]

Each of the cases shares certain, basic similarities. These include a national energy system comprised of actors, inputs, and outputs with systemic architecture connecting the constituent parts in a complex network of energy-centered flows over time—including extraction, production, sale, delivery, regulation, and consumption. Each country has a constitutionally-mandated set of institutions for effectuating policy. Moreover, decisions and actions related to energy are manifest in energy production, consumption, and import/export patterns embodied within a national energy trajectory.

The cases also differed in important ways. First, each focused on a discrete form of low carbon energy (namely, biofuel, nuclear energy, geothermal energy, and wind power). In addition, cases also reflect some differences in energy services and sectors. The countries vary in the scale of land mass, population, and economy. The approaches to governance in each country, as well as development paths and culture, have also differed over time. Recognizing these distinctions, I drew in this study on the "most different systems approach," which tests to see whether a relationship between variables is replicated across a wide variety of diverse settings. If such a relationship is observed, there are grounds for arguing that a causal link exists between the variables of different settings (Huitema and Meijerink, 2009, citing Hopkins, 2002).

Methods and Data

For the research, I used a number of primary methods, including: comparative case analysis, process mapping, historical record review, and semi-structured interviews. By design, I adopted a principally qualitative approach

1. At the time of case selection, there were no feasible cases of solar technology that met case criteria. Solar thermal water heating and solar photovoltaics now reflect areas for new cases. Hydropower is another important, low carbon area to consider that reflects a more traditional energy pathway.

to examine unexplored and more nuanced dimensions of energy transitions that quantitative modeling can miss. I drew on grounded theory to inductively evaluate national energy systems as a unit of analysis (Glaser and Strauss, 1967; Martin and Turner, 1986) and to develop explanations.

Recognizing that system transitions are multi-faceted in nature (Foxon, 2011; Freeman and Louca, 2002; Hughes, 1983; Laird, 2013; Nye, 1992; Sarrica, Brondi, Cottone, and Mazzara, 2016), I included science and technology, infrastructure, policy and politics, economics, and societal developments as preliminary foci. I also, however, maintained flexibility in the framing and questions, to allow for new categories of influences and more complete explanations that emerged.

In addition to data collection through semi-structured interviews (discussed next) and a review of historical records, I extracted energy, economic, environmental, and societal information from a number of databases. Primary database sources included those of the IEA, the International Atomic Energy Agency (IAEA), and United Nations along with World Development Indicators, International Monetary Fund, and Carbon Dioxide Information Analysis reporting. I also accessed historical documentation of energy policy, socioeconomic developments, and industry patterns to evaluate more qualitative, societal dimensions of the national energy trajectories. To test the reliability of this information, I cross-referenced across source types and sectoral inputs, where possible. Notable differences in underlying explanations are indicated, as needed.

I began this research with data trend analysis of the mixes, production, consumption and trade of energy for each country. I overlaid these trends with chronologies of related developments in (1) science and technology, (2) industry and infrastructure, (3) the economy and environment, and (4) governance and society. The integrated timelines provided in depth histories for evaluating the sequencing and causality of changes.[2] With this approach, I inspected shifts in energy trends more closely to identify underlying factors. Drawing upon preliminary findings from the above analysis, I explored the central research questions of this study more fully with interviews, historical records and additional case analysis.

2. Variance methods were used to explain change in terms of the relationship between independent and dependent variables, whereas process methods were used to explain outcomes on the basis of sequences that lead to such outcomes (Poole, 2004; also citing Mohr, 1982; Poole et al., 2000). To illustrate the distinction, a study of industrialization with the variance method might focus on the strength of causality between regional economic drivers and industrial advance, whereas the process approach may consider whether the sequence of economic drivers would produce the same level of industrialization absent technology change at various points (Poole, 2004; also citing Mohr, 1982; Poole et al., 2000).

Interviews

A key aspect of the study entailed more than 30 semi-structured interviews per case, totaling over 120 overall (see the "Interviews" section following the Appendix for affiliations). I selected interviewees through a systematic literature review and snowball sampling to represent different sectors, related expertise, and (in some cases) philosophical orientation in connection to energy. These individuals included members of government, academia, industry, scientific laboratories, and broader civil society. By design, their input reflected a range of sectoral and temporal perspectives to account for varying frames of reference on the influences of change (Allison, 1969; Schon and Rein, 1995). I completed interviews in person, by telephone, and via email—often in combination. Questions included two principal formats: targeted questions focused on specific detail and ranking factors of influence, and open-ended questions centered on underlying explanations.

Criteria for Evaluation: Weighting of Influence

To evaluate key determinants of the transitions, I assessed influence on system change in a number of ways. I noted, for instance, the extent to which a driver or barrier was indicated across different sectoral explanations and methods of review, as well as the level of its logical fit with observed changes in the data and conditions. If a robust policy change, for example, (1) was emphasized in historical records and interviews as having a positive effect on the adoption of low carbon energy, (2) logically resonated with observable conditions, and (3) appeared to be substantiated by data showing capacity increases, then I classified the policy as a key driver, absent other explanatory indications. I did not expect to find unanimous validation across all sources, and emphasized notable differences, when found.

Early Thoughts on Multi-technology, Multi-sector, and Multi-development Levels of the Study

Limits are discussed more fully in Chapter 8. However, one point is worth highlighting here. The study looks at energy system change across different technologies, energy sectors, and national contexts. The heterogeneity of the cases means that certain features (e.g., end-use technology) are less conducive to comparison across cases. Broader insights, however, are still useful in the comparison of discrete, national transitions occurring during the same period.

SCOPING FRAMEWORK: INTERVENTION

Public discussions about transformative change often revolve around questions that ask whether a transition should be driven by government or the market. If done by a government, the approach may include mandates. In contrast, a market-led approach is shaped by actions of the private sector, which may or may not be steered by policy. This traditional dichotomy misses changes, however, that emerge without government inducement or market-led action (Ostrom, 1990). New knowledge and innovations, as well as changes in societal rules, institutions, and access, are among other possible catalysts.

As described in the previous chapter, energy system change can be evaluated with a conceptual framework that recognizes broad influence through centralized government versus market rules versus private-sector innovation (the "one thousand flowers" approach) (Bolton and Foxon, 2015). This general framing is fairly versatile and aligns with policy scoping like that outlined by Vedung (2005). However, such conceptualizations can go further by more explicitly accounting for non–market drivers and sectoral contributions with a more robust and generalizable classification of sources and modes of intervention. Figure 3-1 provides a framework that features these.

As a scoping tool, this framework accounts for the above-noted levels of influence and interplay of government, industry, and civil society in an energy transition. Given the special emphasis on policy in this book, the framework also differentiates levels of government intervention to include "deployed" at one end of the spectrum (i.e., carried out by governmental actors) through to "emergent" at the other end of the spectrum, representing essentially no governmental action.

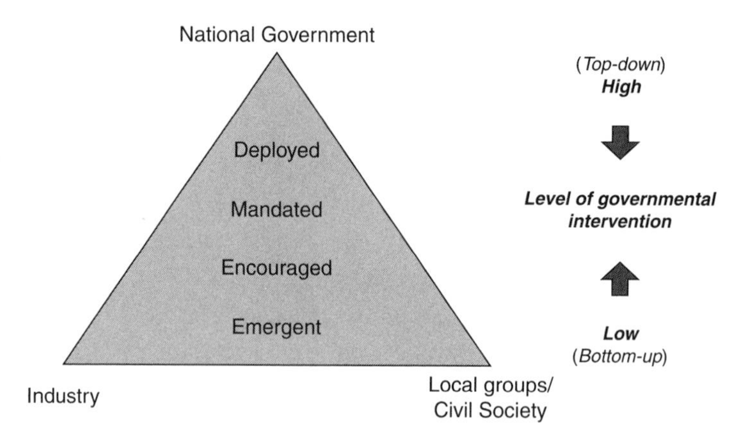

Figure 3-1 Intervention framework.

In the middle of this framework, one will find "mandated" change reflecting regulatory enablers. By contrast, "encouraged" change is achieved with measures like incentives, information, persuasion, or leading by example. In line with the integrated yet broader scoping of this framework, influences go beyond market characterization to capture nonmarket aspects. Table 3-1 differentiates the groupings within this basic conceptual framework. This framework recognizes that hybrid combinations of intervention often exist and can pivot fairly quickly. In connection with shifting mixes of intervention, the roles of sectoral actors naturally can change over time. Moreover, levels and influences of government can be quite distinct at the local, regional/state, national, or supranational levels. For the purposes here, non-national forms of government are grouped with civil society. This will matter, particularly in the Icelandic case, where municipalities were key agents in seeking geothermal-based district heating in the 1970s.

Table 3.1. Intervention Types

Types of Intervention	Description
Deployed	In deployed change, a government directly implements a shift to a new form of energy through, for instance, the construction of relevant infrastructure and the physical provision of energy.
Mandated	Mandated change entails government putting regulations in place or contracting with third parties to implement change. Vehicle emissions standards are an example of a policy tool that does this.
Encouraged	For change that is encouraged, government provides incentives, information, persuasion, or leadership by example. If a senior politician publicly demonstrates that his or her home is a zero-carbon design (i.e., carbon free generally on a net basis) or if people are provided with resource potential maps of renewable energy sources that highlight untapped energy, these exemplify ways to encourage change.
Emergent	Emergent change occurs for reasons other than active deployment or inducement by a government. Actors in industry or civil society/local groups may alter practices to meet individual interests, creating a shift. Such a shift can be found, for example, in an end-user's choice to diversify energy sources in response to service disruptions, safety concerns, or fuel price volatility.
Hybridized	This form of shift occurs via a combination of elements involving deployed, mandated, encouraged, and/or emergent change. A hybrid example would include change that emerges from industry, but that is also amplified by government policies designed to facilitate such developments.

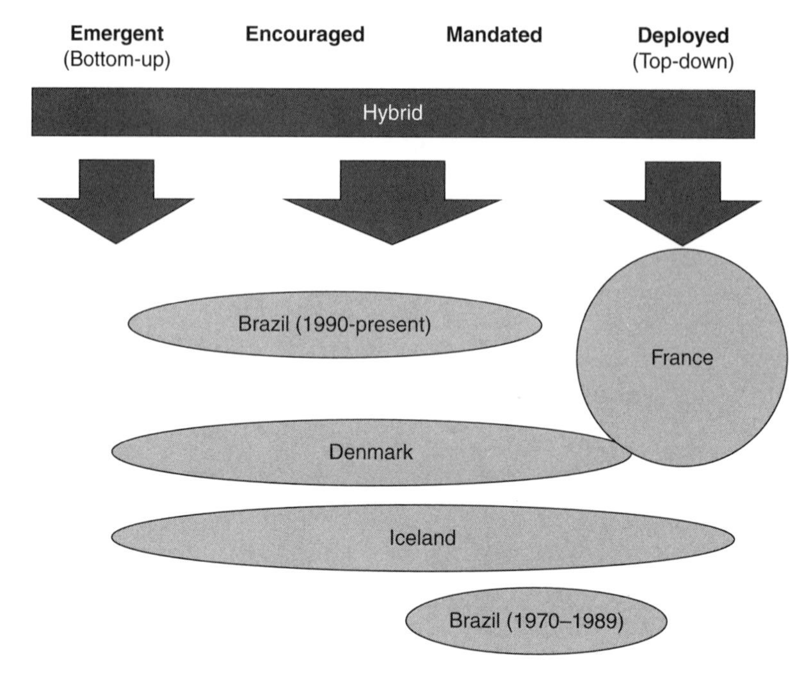

Figure 3-2 Mode of intervention for the four energy transitions.

Variations in sectoral combinations can have distinctly different types of political and financial implications. A "pure mix" of government, industry, and civil society inputs throughout a transition, for example, differs markedly from a hybrid approach in which government drives the early stages of an energy shift and industry carries progress the rest of the way, as will be seen in the case of Brazil.

Drawing on the preceding framework, the following chapters will explain transitions in terms of intervention by source and type as shown in Figure 3-2. The French nuclear case will reflect an example of deployment, whereas the Brazilian biofuels transition will have a distinctly different character in the early and later phases, exhibiting an initial, top-down approach that takes a bottom-up turn. Denmark and Iceland both will be shown to be hybridized shifts with a range of intervention levels. Between the latter two, Iceland's geothermal shift will have a stronger governmental deployment component relative to the Danish steering of its wind transition.

MODELS: STRUCTURAL READINESS

Another way to understand energy system change is in terms of structural readiness. This approach focuses on the material properties of the transition, as well as on institutions and the adequacy of capabilities. The approach aligns with ideas of mission readiness and strategic planning utilized in health (Cummings and Strikova, 2007; World Health Organization [WHO], 2006; Nuclear Energy Agency [NEA], n.d.), disasters (Brown, 1979; IAEA, 2015; NEA, 2009), and defense planning (Moton, 2000). Readiness resonates with thinking about the importance of institutional capacity and nimbleness (Bradshaw, 2014; Mytelka et al., 2012; Smil, 2010). Surprisingly, a readiness approach has not been explicitly used much to date in the context of energy transition writing.[3] Conceptual models are proposed here, based on: the underlying capacity of indigenous expertise; the presence of relevant infrastructure, markets, and industry; domestic fuel resources; and the contemporary existence of a smaller version of the energy pathway under consideration.

Indigenous expertise recognizes that a better quality of knowledge about a given technology can contribute to more advanced system functioning. This derives from exposure, experience, and/or study.[4] Decision-makers in industry and policy may gauge this with jobs that assume a level of competency or with accredited qualifications as a proxy. For the purposes here, expertise is not quantified by the number of formally-accredited people or number of jobs, but instead looks at whether the scientific and local knowledge detailed in Chapter 2 reflects an appropriate domestic basis for growth and other change.

Infrastructure, in the context of this study, equates to the physical components or material basis of a human-constructed system that provides services like energy, typically with a networked nature (Loorbach, Frantzeskaki, and Thissen, 2010). In energy systems, infrastructure can include components such as power plants and fuel-refining mills, as well as delivery means, like grids and pipelines, and end-use technologies, such as vehicles.

The presence of appropriate *markets and industry* also has obvious relevance for a national energy transition. If industries and markets already exist for a given fuel and technology, then transition time and costs may be less.

In examining feedstock/fuel, this study focuses on the concepts of availability of an indigenous resource—and clear access to it—to inform the analysis of a nation's energy self-sufficiency. Some may contend that the diversity of the supply from various sources (indigenous or otherwise) offers another way

3. A search of "readiness" and "energy transition" in the scholarly database Scopus through the end of 2015 produced 34 entries, of which fewer than a quarter referred to energy transitions in sociotechnical terms. A related Internet search in June 2016 produced limited results mostly pointing to readiness in terms of cooperation of energy executives and with respect to a new report on the topic (Cuff, 2016; Lang and Lang, 2014). Across both searches, readiness largely focused on technology, market, or managerial readiness, but did so separately. If integrated, these ideas overlap fairly well with the dimensions of the readiness models that are put forward here.

4. For a discussion of knowledge generation, codification, spillovers, learning, and depreciation, see Grubler et al. (2012).

of considering this aspect (Luft, 2012). The concept of self-sufficiency will be examined comparatively in Chapter 8.

The contemporary presence of an *existing energy pathway* toward a particular form of low carbon energy serves as a link between and proxy for many of the preceding dimensions since less start-up time and costs associated with learning can be expected. It is operationalized, here, by determining whether a specific fuel of interest is part of the country's contemporary energy mix, albeit in a smaller share, at the initial point of the period being studied.

Taking the preceding, constituent parts together, structural change can then be evaluated based on (1) expanding an existing pathway, (2) repurposing an existing pathway, or (3) developing a new path. Figure 3-3 illustrates these options.

Expansionary model: The expansionary model displaces traditional fuels by building out system capabilities and increasing utilization of a smaller but already existing energy pathway. Some adjustments may occur with infrastructure or technology, yet markets, industries, and expertise would already largely be in place. Fuel sourcing derives primarily, here, from domestic suppliers. An example would be the build-out of the existing natural gas infrastructure with fracking.

Repurposing or reconstitutive model: The reconstitutive model differs from the preceding model in that technology, industries, and/or markets must be repurposed before substantial scaling of a form of energy can occur. The "newer" energy might not yet be in use or is of limited utilization to date. The reconfiguring aspects typically will entail significant redevelopment of infrastructure, institutions, and practices, as well as widespread supplier

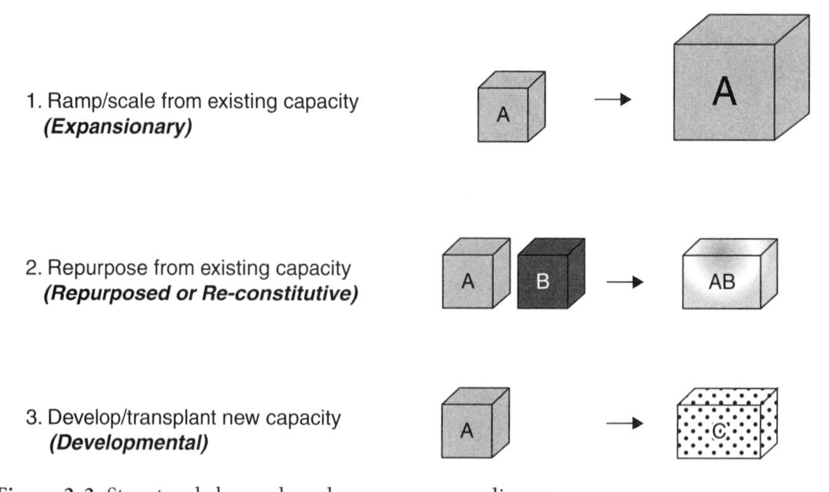

Figure 3-3 Structural change based on resource readiness.

and user learning, diversification of existing industries, and possibly the creation of new types of market players. An example, here, can be found with the system reconfiguration tied to increased distributed generation in Germany.

Developmental model: The developmental model introduces new technologies, industries, and/or markets for a given form of energy that is essentially not in domestic use at the initial point of the studied time period . This discontinuous approach can create or borrow technology and markets at some point in the studied time period. In doing so, an industry develops. In contrast to the other two models, this one is characterized more by the creation of new expertise and forms of infrastructure, substantial uncertainty, and the emergence of new kinds of market players. An example may be seen currently in the initial adoption of nuclear energy in the United Arab Emirates (IAEA, n.d).

As with the preceding source-intervention framework, the models of structural readiness will be used to explain the transitions in the case chapters that follow. Chapter 8 will revisit these conceptual tools, supplying additional comparison.

GLOBAL CONTEXT

National energy system change does not typically occur in a vacuum. Figure 3-4 highlights trends in the global energy supply, population, carbon dioxide emissions from fuel combustion, and gross domestic product that embody relevant developments. One can see, for instance that the earth's population nearly doubled

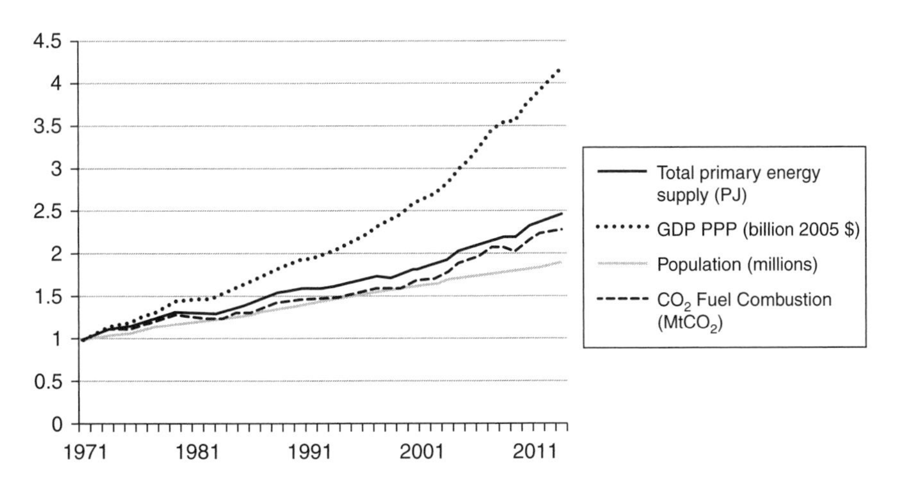

Figure 3-4 Global change indicators: 1971–2013 (Base year = 1971).
SOURCE: Data from IEA (n.d.).

Table 3-2. International Developments in Energy, the Environment, and Society

	1970	1980	1990	2000	2010+
Society and Economy	Rise of international nongovernmental organizations (NGOs), participatory development, privatization, and digital communications				
Environmental Summits and Advances	Stockholm Summit, UNEP established	Stockholm +10 Summit, Montreal Protocol, WCED Report, IPCC established	Rio Summit, Kyoto Protocol, UNFCCC, MDCs	Johannesburg Summit, Kyoto Protocol enters into force, European Emission Trading System	Kyoto I ends, Rio+20 Summit, Sustainable Energy for All, SDGs, Paris Agreement
Major energy events	Oil crises, Shimantan Dam Flood, TMI accident	Oil price collapse, Chernobyl accident, Exxon Valdez accident	Bieudron rupture, Nigerian National Petroleum Corp. pipeline rupture	Oil price flux, Emergence of fracking, Sayano-Shushenskaya accident, Itaipu outage, Northeast US/Canada blackout, Prestige oil spill, Apagao	Oil price flux, Deepwater oil spill, Fukushima accident

NOTE: For a discussion of methodologies in evaluating energy risk and accidents, see Sovacool (2008*a*), Felder (2009), Sovacool, Kryman, and Laine (2015), and International Association of Oil and Gas Producers (2010).

from 3.8 billion in 1971 to 7.1 billion in 2013, as total primary energy and CO_2 from fuel combustion increased by more than a factor of 2. Total primary energy rose from 231.2 (EJ) to 566.9 EJ, as CO_2 grew from 13,995 metric tons of CO_2 ($MtCO_2$) to 32,190 $MtCO_2$ (IEA, 2016*a*). During this period, the global economy also increased by more than a factor of 4, from roughly $20.7 billion to $86.3 billion (in 2005 US$). The figure highlights the decoupling of growth patterns in GDP, energy utilization, carbon emissions, and population to varying degrees.

In tandem with these quantitative changes, a number of historical developments are worth highlighting (Table 3-2). The rise of international nongovernmental organizations (NGOs), for instance, occurred as digital communications and participatory development became more mainstream (Turner, 2010, citing Union of International Associations, 2008; Cornwall,

Box 3-1
Liberalization and Restructuring

Liberalization (a form of deregulation) and restructuring are processes that often occurred in tandem in many regional power markets. Both liberalization and restructuring are usually intended to increase competition, new market entrants, and operational as well as technological innovation while decreasing consumer costs. These aims, in practice, are not always fully attained and broader public goods needs can be missed. Liberalization opens the market to forces of supply and demand, reduces the level of government intervention, and limits the barriers to market entry (Berg, 2005; Jamison et al., 2004). Restructuring entails the disaggregation of a vertically-integrated company's functions with the unbundling of services and centralized decision- making, in which owners and operators of newly configured companies become discrete entities (Berg, 2005; Jamison et al., 2004).

2002; Freeman and Louca, 2002; Osmani, 2008; Tufte and Thomas, 2009; Watkins and Tacchi, 2008). Environmental NGOs also grew in influence and abundance as international environmental summits and agreements occurred with more frequency. Trends in privatization (i.e., the transfer of ownership or management of public assets to the private sector) appeared largely in the 1980s and became prominent in the 1990s and 2000s often with liberalization and restructuring (see Box 3-1). Substantial privatization occurred in 1997 with infrastructure and energy across nearly all regions (Kikeri and Kolo, 2005). These "waves" impacted industries, including the oil, gas, and power sectors. Additional, energy-related events and developments, like accidents, price flux, and the rise of hydraulic fracturing, have varying levels of influence in the cases that follow.

The oil price shocks of the 1970s, in particular, will be seen as critical triggers in the transition histories of Chapters 4–7. With the first oil shock of 1973–74, prices quadrupled in a matter of months as members of the Organization of Petroleum Exporting Countries (OPEC) curtailed the oil supply in connection with political positions on the Yom Kippur War (Yergin, 1991) (Figure 3-5). The second oil shock of 1979 produced a doubling of prices over the course of a year as output decreased in conjunction with the Iranian Revolution and the Iran–Iraq War.

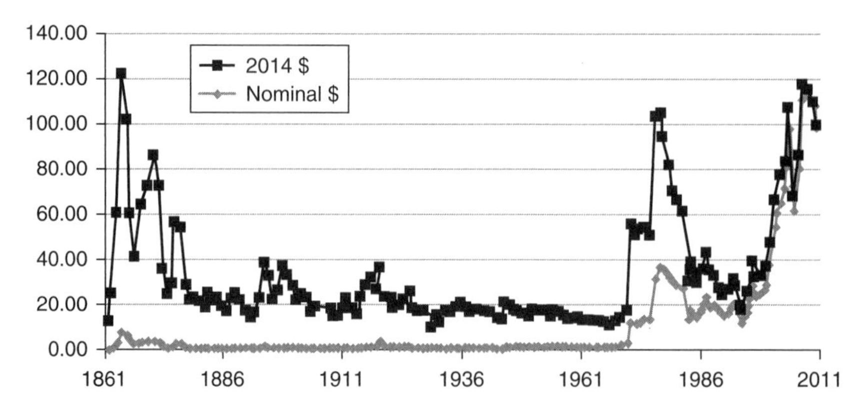

Figure 3-5 Historical oil prices ($).
SOURCE: BP data, Statistical Review 2015.
NOTE: 1861–1944 US Average; 1945–1983 Arabian Light posted at Ras Tanura; 1984–2014 Brent dated.

PREVIEW OF THE TRANSITIONS

Previews of the four countries in this study offer a preliminary sense of similarities, differences, and relative scales. Table 3-3 shows select information for energy, population, economic, CO_2, and governance indicators.

In terms of scale, Brazil and France reflect the larger countries with respect to GDP, energy consumption, population, and CO_2 from fuel combustion compared to Denmark and Iceland. Specific to final energy per capita, Iceland consumes roughly 7 times more energy than the global average,[5] whereas Brazil aligns fairly closely with the global average. For final energy consumption per dollar of GDP (2005 US$ purchasing power parity [ppp]), Iceland consumes roughly double the global average. Specific to governance, Denmark is represented by a monarchy, whereas the other countries of this study have various forms of republics. Brazil's federative republic came into effect in 1985, following a peaceful transfer from military rule. France established its 5th Republic (semi-presidential) in 1958 (CIA, n.d.). Iceland gained independence in 1944 and has a parliamentary republic (also semi-presidential) (CIA, n.d.).

Figure 3-6 depicts the transition curves for the low carbon shifts of this study. One can see at the outset that Iceland has what looks to be a gradual transition, whereas Brazil and France have front-loaded and robust shifts. Denmark has a later-staged and smaller scale energy transition.

5. The reason for this will be covered in Chapter 4.

Table 3-3. Select Indicators

Country	Iceland	Denmark	France	Brazil
Area (km2)	103,000	43,100	643,801	8,516,000
Population (Millions) (2016 estimate)	0.3	5.6	66.8	205.8
Pop per km2	3	130	120	24
TPES (TJ) (2013)	246,400	730,450	10,606,140	12,295,930
Electricity consumed (TWh) (2014 estimate)	17	33.3 (2015 est.)	415.4	518.0
Electricity per capita (MWh/cap) (2013)	55.4	6.04	7.38	2.58
GDP ppp (Billion US$) (2016 estimate)	16.2	265	2,699.0	3,081
TPES per capita (TJ/cap) (2013)	0.77	0.16	0.13	0.06
CO2 from energy consumption, (million Mt) (2013)	3.3	42	385.6	535.0
CO2 per capita (tCO2/cap) (2013)	6.26	6.91	4.79	2.26
Climate	Glacial	Temperate	Temperate	Tropical
Characteristics related to energy transition	Very small population, sparse, homogeneous and wealthy	Small, homogeneous population	Medium-sized, heterogeneous population	Large, heterogeneous population, developing economy, large wealth gap
	Major heating needs	Medium population density	Medium population density	Vast, sparsely populated areas
	Large geothermal and hydropower resources	Dwindling fossil resources, good wind resources	No fossil resources left, limited hydro, significant biomass potential	Huge natural resources (hydro, biomass, oil, etc.)
		Interconnected with Norwegian hydropower		

Source: Communications with Barre (2016), International Energy Agency (IEA, n.d.), and Central Intelligence Agency (CIA, n.d.).

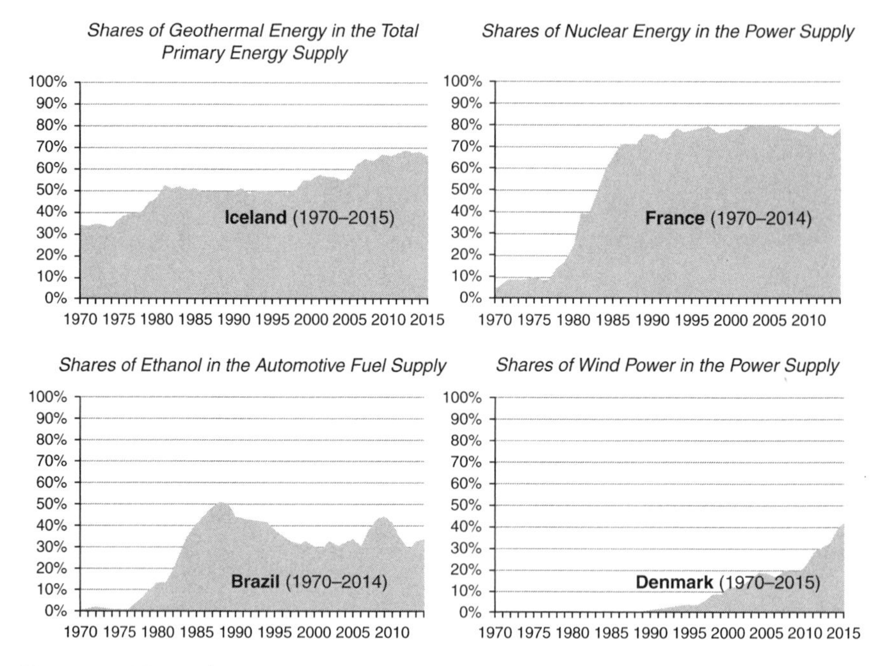

Figure 3-6 National transition curves.
SOURCE: Chapters 4–7.

COVERAGE

In the following case chapters, sociotechnical histories of the energy transitions are outlined with co-evolving policies highlighted. Each case chapter includes a section that examines the given, energy transition with readiness and intervention frameworks from Chapter 3. Case chapters also include a review of key innovations and adaptations enabling the specific transition. Innovations were defined earlier, however it is worth elaborating, here, on adaptations that improve a system's functioning. These are specific developments, such as entry into a new market, that may not rise to the level of an "innovation" per se, but that can contribute along related lines. Drivers and barriers are also covered in each case chapter. As with innovations and adaptations, drivers and barriers are key factors that were raised in interviews as having significance, and were cross-checked in the literature and data, when possible. Notable transition developments relating to costs, societal acceptance and human development, as well as industry are similarly explored.

4

Icelandic Geothermal Energy

Shifting Ground

We see Iceland as the world's lab for a decarbonized future.

—Ingibjorg Solrun Gisladottir, *former Icelandic foreign minister and mayor of Reykjavik (Montavalli, 2008)*

INTRODUCTION

Today's energy sectors hold different potentials for saving on energy, carbon, and other greenhouse gases (GHGs). Buildings, for instance, represent more than 40% of energy use worldwide and one-third of GHGs (United Nations Environment Programme [UNEP], n.d.*a*). Improvements in heating, cooling, and powering of buildings, as well as industrial processes, can deliver substantial and cost-effective savings. In line with this, geothermal energy represents a more unusual form of renewable energy in that it can directly contribute to heating, cooling, and electricity services. Unlike a number of its counterparts, geothermal energy can provide a more stable and renewable form of energy that is largely unaffected by weather.

The chapter focuses on geothermal energy adoption in Iceland, "a little country that roars," according to UNFCCC Executive Secretary Christina Figueres (Iceland Monitor, 2014), when discussing leadership in renewable energy use and related action.[1] In developing its renewable energy leadership, Iceland has wrestled, like many countries, with tradeoffs in energy, the environment, and economic development. The chapter highlights the interplay of these interests and explores

1. The UNFCCC is the United Nations Framework Convention on Climate Change.

the innovative engineering and industrial spillovers in Iceland's geothermal adoption.

PROFILE FOR ICELANDIC GEOTHERMAL ENERGY

Iceland is a country of roughly 333,000 people, and is a global leader in renewable energy use (Islandsbanki, 2010; Ministry of the Environment, 2010; Statistics Iceland, 2017). Two-thirds of the country's primary energy consists of geothermal energy, with roughly nine out of ten Icelandic homes heated by the fuel source and a quarter of the country's electricity powered by it (Orkustofnun, 2015; Ragnarsson, 2015).[2] The nation leads globally in terms of geothermal heat capacity per capita[3] and serves as a principal source of international training and consulting on geothermal energy, with a diverse industrial cluster that has developed around the technology (Gekon, n.d.; United Nations University Geothermal Training Programme [UN-GTP], n.d).

The country's low carbon development pathway reflects choices and debate about how to manage its natural resources and allow for foreign investment. Iceland began the 20th century as one of the poorest nations in Europe and is now a top-ranked country in the United Nations Development Program's Human Development Index (Hannibalsson, 2008; United Nations Development Program [UNDP], 2015). Iceland's GDP was estimated to be $16.2 billion in 2016, with GDP per capita at $48,100 (ppp) (CIA, n.d.). This represents the smallest economy among the countries of this study, yet the highest GDP per capita of the cases. Currently, nearly 77% of the national power supply is consumed in export-oriented manufacturing, principally by aluminum smelter subsidiaries of foreign-owned companies (Ministry of the Environment, 2010; Orkustofnun, n.d.). For anyone who may ask what the world can learn from a small nation and its energy choices, this chapter provides answers.

BACKGROUND

Iceland has a long relationship with geothermal energy. National sagas—oral histories of the 10th and 11th centuries that were later written down—indicate that early regional settlers built homes in the 9th century near hot springs (Einarsson, 1991; Icelandic Sagas, n.d.). Settlers used the area geysers for cleaning

2. Hydropower reflects the other major form of energy, at 19% of the primary energy supply (Orkustofnun, n.d.).

3. This excludes heat pumps (REN21, 2016).

and cooking in "the Bay of Steam," what is today's Reykjavik (Einarsson, 1991; Icelandic Sagas, n.d.). By the 13th century, sulfur was extracted from Icelandic geothermal steam for export and use in products like gunpowder (Einarsson, 1991). Centuries later, geothermal energy was used in salt production and to support gardening (Kristjansson, 1992; Loftsdottir, A., and Thorarinsdottir, 2006).

At the beginning of the 20th century, new activity related to Iceland's geothermal resources was evident (Einarsson, 1991). The utilization of geothermal energy for heating houses commenced in 1908 when Stefan Jonsson (i.e., farmer, carpenter, author, salesman, and entrepreneur) piped steam from hot springs to his farm (Thordarson, 2008; Figure 4-1). Exploration for suitable geothermal resources was spurred by increased demand for electricity as well as knowledge that geothermal utilization was occurring elsewhere (Jonasson and Thordarson, 2007; UNDP et al., 2000). By the 1920s, schools, hospitals, and institutions around the country were being sited near geothermal regions to leverage the natural resource (Einarsson, 1991). In the following decade, Iceland introduced what may have been the world's first large-scale, municipal district heating service fueled by hot springs that were accessed with a 3-kilometer pipeline (Einarsson, 1991; Jonasson and Thordarson, 2007; Loftsdottir and Thorarinsdottir, 2006).

Figure 4-1 Some of the first houses in Iceland to be heated by geothermal water with a geothermal borehole at Sudurreykir in 1908.
CREDIT: Reykjavik Energy.

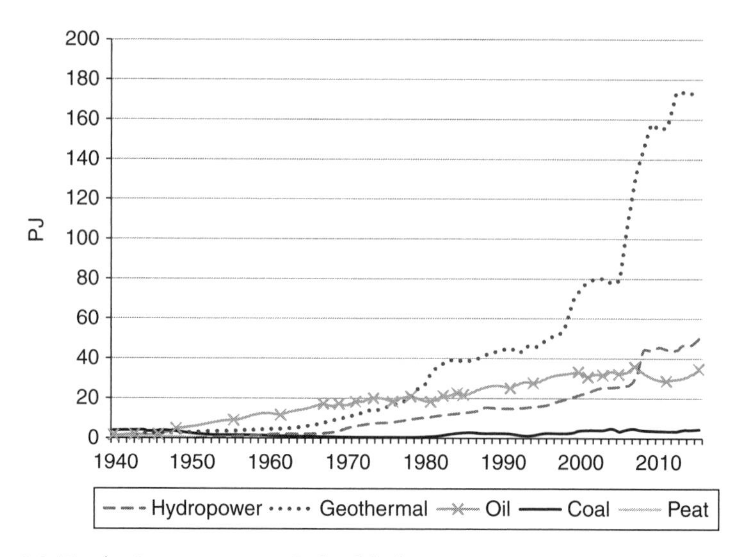

Figure 4-2 Total primary energy in Iceland (PJ).
SOURCE: Compiled with data from Orkustofnun (n.d.).

As the population of Reykjavik increased (more than doubling in the 1940s alone), energy use scaled, first with oil, followed by geothermal energy and hydropower (Figure 4-2; Jonasson and Thordarson, 2007; Loftsdottir and Thorarinsdottir, 2006).

In the years that followed, geothermal energy closely intersected with national priorities. The Icelandic government instituted the Electricity Act in 1946, creating the State Electricity Authority, an entity charged with advancing the knowledge of geothermal and hydrologic resources (Loftsdottir and Thorarinsdottir, 2006). The State Drilling Company was then formed in 1947, followed by the Act on the Geothermal Energy Fund in 1961 to facilitate geothermal resource studies and development. Combined with other energy funding, the Geothermal Fund was used to minimize risk in energy resource development by providing low-interest loans that typically covered 60% of total drilling costs (Ingimarsson, 1996). By mid-century, the Geothermal Department was established within the State Electricity Authority, and a rural electrification program was launched (Ingimarsson, 1996; Thordarson, 2008).

Considering Iceland's conditions more broadly, the economy was never far from Iceland's energy agenda. Battles with the British over fishing rights and fluctuations in the fish market negatively impacted the domestic economy. Since the country possessed significant, renewable energy resources, the Icelandic government looked to energy-intensive manufacturing as a way to overcome its economic challenges (Del Giudice, 2008; Hjalmarsson,

2009; Landsvirkjun, n.d.*a*; Loftsdottir and Thorarinsdottir, 2006). A government-sponsored task force identified new markets in the 1960s that could be powered by indigenous energy resources. Shortly thereafter, Swiss aluminum producer Alusuisse became one of the first foreign companies to agree to site operations on the island (Landsvirkjun, 2009; Logadottir, 2012).

In line with new energy planning, Landsvirkjun was formed as a public company in 1965, to build and operate power generation plants (Landsvirkjun, n.d.*a*). Prior to Landsvirkjun's formation, the national government and municipalities had loosely overseen Icelandic electrification, finding the financing of such projects to be a challenge. With the creation of Landsvirkjun, the power company was structured for 50:50 state and municipality ownership. The World Bank provided roughly one-third of the investment capital and was heavily involved in designing the energy company's structure (Davidsson, 1986; Landsvirkjun, 2009, n.d.*a*). The Bank emphasized that Landsvirkjun should set up operations so that projects would be undertaken through international tenders with external engineering consultants engaged in project management and subcontractor oversight (Landsvirkjun, 2009).

Amid Landsvirkjun's start-up, the government signed a 45-year contract to supply electricity at low rates for a soon-to-be-sited aluminum smelter with Alusuisse. Conditions for this and related contracts would affect energy development decisions in Iceland at domestic and international levels for many decades to come.

By the end of the 1960s, Iceland's energy path appeared to be pivoting in new directions. The first industrial application of Icelandic geothermal heat became operational with the Kísiliðjan Diatomite Plant.[4] This plant was one of the world's largest industrial users of geothermal energy, utilizing heat to dry diatomaceous earth sourced from Lake Myvatn. Construction of the country's first commercial geothermal power plant was also under way at Bjarnarflag in northeast Iceland. Meanwhile, the country's first foreign-owned aluminum smelter, Swiss-owned Icelandic Aluminum Company Ltd. (ISAL), came online with power from the new 210 MW Burfell hydropower plant built to service increased demand and a long-term, take-or-pay contract (i.e., guaranteed purchase) (Halfdanarson, 2008; Valfells et al., 2004). Recognizing the importance of research and development (R&D) for its energy strategy, the national government also established a program with the Orkustofnun (National Energy Authority, successor to the State Electricity Authority) to study high-temperature geothermal fields.

4. This plant produced 28,000 tons of diatomite filters for exports on an annual basis from 1967 until its closure in 2004 (Orkustofnun, n.d.).

MODERN TRANSITION: 1970–THE PRESENT

By 1970, Iceland had an early track record in geothermal energy, yet critical pivots in the 1970s and 1990s enabled the country to become an exceptional leader in the technology. During this period, Iceland's economy would restructure toward more services and manufacturing, and less marine and agricultural products (Statistics Iceland, 2017). The country would also reduce its energy imports as a share of net energy use from 54% in 1970 to 11% in 2014 (World Bank, n.d.*b*). Along the way, questions about environmentalism and trade would test the national and local communities' resolve. Innovations would emerge, not only in science and technology advances, but in industries that complement energy services.

Mobilization in the 1970s

By some accounts, Iceland was still very much a developing country in the early 1970s, as nearly all of the nation's export revenue was tied to fishing and other seafood exports (Statistics Iceland, 2017) (see "Driver and Barriers" section for more detail on the economic structure). The country was progressing with a 60-year track record of geothermal energy utilization in heating (40 years on a commercial basis), with geothermal power generation recently launched for commercial use. The fossil fuels that had dominated Iceland's primary energy supply during the first half of the 20th century now reflected a much smaller share

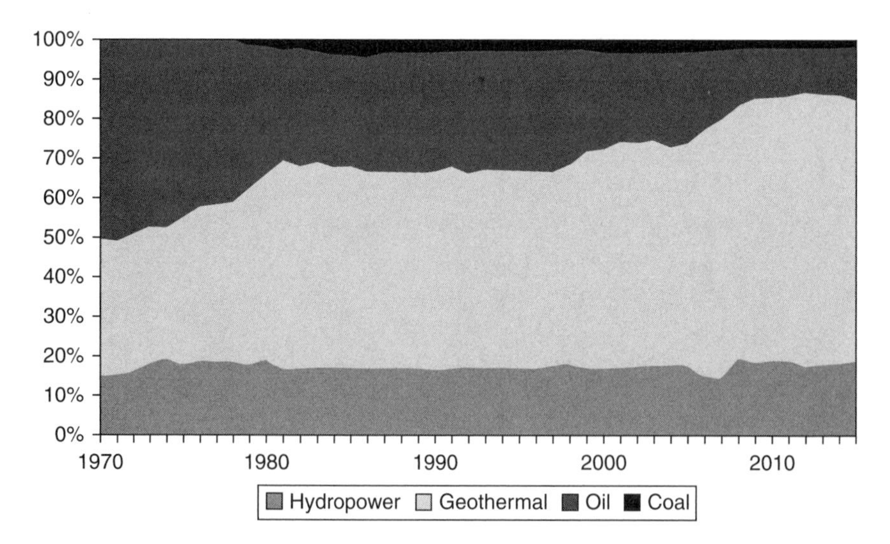

Figure 4-3 Total primary energy (%).
SOURCE: Data adapted from Orkustofnun (n.d.).

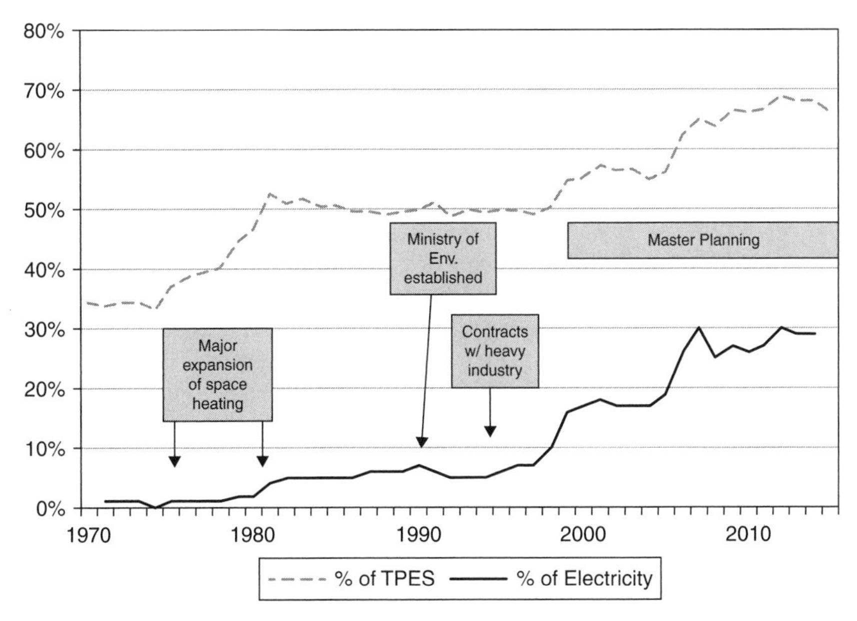

Figure 4-4 Geothermal energy development (% of geothermal energy in the total primary energy supply and electricity).
SOURCE: Compiled with data from Orkustofnun (n.d.) for TPES, and IEA (n.d,) for electricity.

of the overall energy mix (Figures 4-2, 4-3, and 4-4). At the time, geothermal energy use outpaced that of coal and would do the same with oil in just a matter of years.[5]

In 1970, energy, economics, and the environment were all visible in Iceland's public priorities. At that time, the Nordic country joined the European Free Trade Association (EFTA). Public discord also erupted over the way that Iceland's natural resources were being valued in the country's energy-intensive development. Installation of the Laxa-Myvatn hydropower project highlighted, for instance, the tradeoffs in local and national interests (Olafsson, 1981). Director General of the Orkustofnun Jakob Bjornsson weighed in on the debate by arguing for resource development in the context of economic progress:

> If the Icelandic nation has ambitions of living . . . with comparable material standards . . . to the best found elsewhere in the world, it needs to make maximum use of the country's resources . . . there may be some that take the view that economic progress on a par with other countries is not

5. Coal in the period 1980 to the present was not used as fuel but for electrodes in metal smelters (communication with S. Bjornsson, 2016).

> necessarily a goal worth aiming for. Against such a view there is nothing to say. But is it something the majority of the nation will go along with. (Magnason, 2008, citing Bjornsson, J., 1970)

Nobel Prize-winning novelist and Icelandic native, Halldor Laxness, challenged the basis for such thinking, saying:

> The problem is the unquestioning faith of the people at the National Energy Authority [Orkustofnun] in filling this country with endless metal smelters. For it presents a grave danger to the land and its community when a group of men in suits, at the say-so of their slide rules, sets out to obliterate as many of the places we hold sacred as they can in as short a time as possible, drowning familiar settlements in water (twelve kilometers of the Laxa valley are set to be submerged, according to their plans), and given the chance, declaring war upon everything that lives and draws breath in Iceland. (Magnason, 2008, citing Laxness, 1971)

Director General Bjornsson's view would be echoed in ideas on resource-based development (Barbier, 2005; Sachs and Warner, 1997), which, for Iceland, meant tapping energy resources for heavy (mostly foreign) industry. The approach was seen as a way to protect the economy from extreme fluctuations already experienced with the fishing industry. There was also a sense of some urgency to act, predating the oil crisis: the growth of nuclear technology could well have become inexpensive within the next 20 years, rendering Icelandic geothermal energy uncompetitive for future foreign investors (Bjornsson, S., 1970). Deep interest in protecting the environment would remain key, however, and become more widely evident in the years ahead.

Meanwhile, change within the government was visible. The Centre-Right Independence Party that had led Iceland during the 1960s was followed by a succession of shifts in office holders between 1971 and 1983. With ensuing pivots of administrations, the energy agenda was recast to prioritize the interconnection of power networks and to meet increased demand through smaller power plants.

At the point of the first oil shock, 52% of the country's total primary energy consisted of oil imports (IEA, 2016*a*). In line with this, roughly 43% of the Icelandic population used geothermal energy for heating (Bjornsson, S., 2006) and approximately 1% of electricity was derived from geothermal energy (IEA, 2016*a*). With the global oil shocks of 1973 and 1979, Iceland's national government, municipalities, and energy companies mobilized to eliminate the majority of oil being used in space heating by substituting with geothermal energy. Geothermal energy utilization to date, included the low-temperature resource found in non-volcanic hydrothermal areas (<130°C) (Arnorsson, 1975; Bjornsson, S., 1970; for a discussion of geothermal resource categories, see the "Technology Primer" in the Appendix). Higher temperature geothermal resources in the volcanic hydrothermal areas would be used later in electricity and industry (Arnorsson, 1975; Bjornsson, S., 1970).[6] In regions where geothermal water was not found, electricity would replace oil in heating.

By the mid-1970s, geothermal activity was found with deep drilling at 1,500–2,000 meters. Hydraulic pressure stimulated existing fractures and the injection of cold water caused thermal contraction of the rock plus thermal cracking. These measures would increase the permeability and yield of geothermal energy in support of heating needs for Reykjavik and surrounding communities (Arnorsson, 1975). Development in outlying areas was similarly occurring. In the Sudurnes of Southwest Iceland, for instance, Grindavik's town council and others knew of localities that were displacing oil in district heating with geothermal energy. In light of this, drilling for geothermal resources commenced in the Sudurnes. When a high-temperature geothermal reservoir (240°C) containing water with large amounts of minerals and salinity was struck, engineers applied heat-exchange methods to harness the resource. The Orkustofnun advised the community on an appropriate plant design, and area municipalities elected individuals to oversee the implementation of the plant. Shortly thereafter, the Icelandic Parliament (Althingi) passed Act No. 100/1974, establishing Hitaveita Sudurnesja to produce and oversee distribution of geothermal-based hot water and heating services for the region. Company ownership would be shared between the national government (40%) and the seven municipalities that were involved.

As geothermal studies and development continued, a project to install 60 MW_e of capacity for power generation at Krafla ran into difficulty in 1976 in the northern region of Iceland, when volcanic activity impeded progress. The volcano injected corrosive magmatic gases into the better areas of the reservoir, requiring roughly 15 years to recover (Communications with S. Bjornsson, 2016).[7] With the change in physical state of the region, the Krafla project was adjusted to install half of the planned capacity. The remaining 30 MW_e of capacity would later be installed in 1997.

Despite challenges in geothermal development, like that at Krafla, advances were evident. Geothermal heating plants at Svartsengi and Nesjavellir were adapted to function as combined heat and power (CHP) suppliers that would more efficiently harness the geothermal energy resource. Sales of CHP-based

6. See Appendix, "Country Timelines", and Bjornsson, S. (1969). Aaetlun um Rannsokn Hahitasvaeda [A Program for Exploration of High Temperature Geothermal Fields]. Skyrsla, Orkustofnun, Reykjavik.

7. The best wells are still delivering acidic steam (Communications with S. Bjornsson, 2016).

geothermal generation would initially be limited by geothermal company charters and, according to some, Landsvirkjun's monopoly.[8] The Parliament later revised the rules, enabling companies like Hitaveita Sudurnesja to expand more fully into power generation.

One of the more unique developments in Iceland's geothermal adoption began in 1976, when the Blue Lagoon was formed with discarded geothermal fluid from the Svartsengi geothermal plant being released into a lava field. Local swimmers claimed health benefits from the mineral rich waters, and word spread. A spa would eventually be formed in addition to a treatment center for psoriasis. This Lagoon would later become Iceland's most popular tourist attraction (see also "Innovations and Adaptations" section for more detail).

On the educational front, the Orkustofnun of Iceland launched the Geothermal Training Programme in partnership with the United Nations University in 1979. In the ensuing decades, this program would distinguish itself by providing specialized, postgraduate training in geothermal resource development to individuals from developing countries. Beginning with two fellows in 1979, the center educated 647 scientists and engineers from 60 countries as of 2016 (UNU-GTP, n.d).

By the early 1980s, Landsvirkjun became a full-fledged national power company, acquiring the Krafla geothermal plant and the transmission network. In line with the company's expansion, its ownership was modified to reflect a breakdown that included the state (50%), the municipality of Reykjavik (45%), and now the municipality of Akureyri (5%). By 1984, the national power grid was integrated under Landsvirkjun, with all major power stations connected to the grid (Loftsdottir and Thorarinsdottir, 2006).

Progress, however, was not without criticism. The fact that the Swiss smelting plant seemed to generate no profits raised concerns over the power pricing and taxation agreement with the firm.[9] After some investigation, Alusuisse's ISAL plant was found to have reported ore prices in Iceland that were 54% higher than those leaving Australian mines (Skulason and Hayter, 1998). This price differential was publicized, and, in 1983, Alusuisse agreed to pay $10.4 million in unpaid taxes (Skulason and Hayter, 1998, citing Davidsson, 1989). In 1984, price formulas were renegotiated linking power and aluminum prices

8. Differences in views exist over the extent of this limitation. Specific to the charters, the parliament approved the creation of municipal energy companies to focus on heat and hot water services, not power.

9. Iceland power pricing for aluminum smelters reflected some of the lowest numbers globally at 6.45 mills per KWh. Other low-cost power locations included Canada (8–16 mills per KWh), Brazil (13–17 mills per KWh), and Australia (15–19 mills per KWh) (Skulason and Hayter, 1998, citing Skulason, 1994, adapted from OECD 1983).

more closely (Skulason and Hayter, 1998, citing Kirchner, 1988). Concerns such as these over the domestic presence of foreign-owned plants led to political demands that Icelanders be able to retain majority ownership of such companies. This issue would resurface in later years, resonating with public views on independence and Iceland's long history of foreign rule.

By 1984, Iceland's modern, geothermal energy transition was already observable. With the expansion of the heating infrastructure, the share of the Icelandic population using geothermal energy in space heating had risen from 43% in 1970 to 83% (Loftsdottir and Thorarinsdottir, 2006). By contrast, use of geothermal resources in power generation would not become significant until the late 1990s.

Limited Inroads: Mid-1980s–Early 1990s

Following the expansion of the heating infrastructure and the integration of the national power grid in the 1970s and early 1980s, energy-related policy would shift for a period to focus more fully on attracting foreign investment that could utilize the domestic energy supply. Record high inflation was among the pressing conditions driving this initiative.[10] The collapse of global oil prices also factored, as imported oil became favorable for use relative to domestically-sourced geothermal energy. Unfavorable economics of the period eventually led the national government to assume the debt for a number of district heating companies that relied on geothermal energy.[11]

10. According to Anderson and Gudmundsson (1998), Iceland's inflation between 1945 and 1973 was never above 15% for more than 2 years. Yet between 1974 and 1984, inflation averaged 49%, peaking in 1982–84. Reasons for inflation were characterized as the domestic response to the international oil crisis and negative supply shocks reinforced by strong devaluation bias. Recovery was attributed to tighter monetary and exchange polices, income policy focused on inflation reduction, and sweeping structural reforms (Sedlabanki Islands, 2005).

11. A committee appointed in 1990 by the Ministry of Industry to report on the equalization of energy prices indicated that the Icelandic Treasury absorbed part of the debt of the geothermal district heating services in the period 1983–90 for:

- Egilstadir and Fell: ISK 6 million (1983) and 2.5 million (1984)
- Svalbardseyri ISK 0.5 million (1983)
- Hrisey ISK 1 million (1983)
- Sudureyri ISK 4.5 million (1983) and 2 million (1984).
- Akranes and Borgarfjordur ISK 220 million (1987)
- Akureyri ISK 100 million (1987)
- Vestmannaeyjar ISK 89 million (1987)
- Hella and Hvolsvollur ISK 48 million (1987) and 20 million (1990) (Interview, 2012, referencing the committee report).

Around the time of the Ukrainian nuclear accident at Chernobyl in 1986, renewed interest in alternative power presented Iceland with opportunities for discussions with individuals in the United Kingdom about the prospect of exporting Icelandic electricity through a subsea cable.[12] New feasibility studies were completed indicating that such a project was technically possible. Costs of Icelandic energy exports were found to be equivalent to that of coal and nuclear. For the United Kingdom, however, the pending liberalization of its power sector meant substantial uncertainty for long-term contracts, so no deal was made. Studies and discussions of this type have continued through to the time of this writing.

Despite promotional efforts by the Icelandic government to attract foreign investment, a period of investment stagnation took hold. It was around this time that Landsvirkjun was also criticized for implementing the Blanda hydropower project without significant new demand for electricity. As a result, no additional geothermal or hydropower projects would be added until the mid-late 1990s, when additional needs were demonstrated.

In an effort to rekindle foreign investment, the government of Iceland established a professional marketing office in 1988 to bridge the activities of the Ministry of Industry and Commerce, and Landsvirkjun.[13] Promotional materials were circulated throughout Europe, emphasizing the "lowest energy prices" being found in Iceland together with the lowest wages and taxes (Magnason, 2008; Palsdottir, 2005). Discussions to negotiate new contracts for aluminum and other heavy industry also took place, but contracts were slow to materialize. The market for global metals declined amid a global recession, with supply "dumping" by former Soviet states also having a dampening effect on prices (Mackay and Probert, 1996; Skulason and Hayter, 1998, citing Palsson, 1993).

In 1990, the government established the Ministry of the Environment, and published a white paper highlighting the need for a long-term plan in energy use (Interview, 2011; Steingrimsson, Bjornsson, and Adalsteinsson, 2007, citing Government of Iceland, 1997). These changes would set the stage for more extensive development in the ensuing period.

Acceleration, Planning, and Rethinking: The Mid-1990s Onward

By the mid-1990s, a multitude of new forces could be seen in Iceland's geothermal energy transition. In contrast to the previous period of inactivity,

12. Orkustofnun had reviewed similar proposals in 1975 and 1980 for a deal with Scotland (Palsdottir, 2005). The first proposal was introduced at least as far back as the 1960s (Landsvirkjun, n.d.*a*).

13. This formalized marketing activities that had already been under way (Interview, 2012).

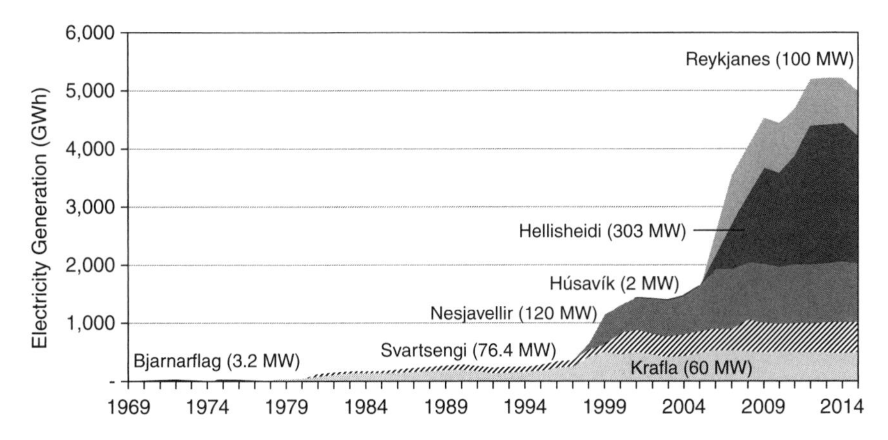

Figure 4-5 Electricity Generation of Geothermal Power Plants (GWh).
CREDIT: Orkustofnun, OS-2016, T003-01. Note: Plant installed capacities are indicated as (MW).

major growth occurred in Icelandic geothermal energy use, starting in 1998 (Figure 4-5).

Alongside replenished fisheries, long-awaited economic growth once again was evident with a global economic recovery, export increases, and new investments in the aluminum sector (Landsvirkjun, 2009; Sedlabanki Islands, 2005). In 1994, the European Economic Area Agreement (EEA) was brought into force in Iceland. This partnership provided a path toward increased trade within the European Union (EU). As a consequence, Iceland was required to account for EU directives. The Act on Environmental Impact Studies (1993) and Electricity Act (2003) would be two legislative outcomes of this partnership. The same EEA mandate would later be tested as foreign ownership of domestic energy resources sparked public concern.

For the Icelandic energy industry that had served a mostly saturated and insular market, new contracts with foreign companies in heavy industry changed the outlook. Beginning in 1995, contracts were signed to enlarge the ISAL smelter in Straumsvík, to construct a new 60,000-ton aluminum plant for Columbia Ventures, and to enlarge a ferro-silicon plant that had been installed in the late 1970s (Interview, 2012; Palsdottir, 2005).[14]

Environmental priorities also figured more prominently in the public dialogue. Public focus on the environmental impacts of hydropower projects and

14. By this time, the industry convention for aluminum plant production had increased to a minimum of 360,000 tons. In energy terms, this translated to a necessary, power plant capacity of roughly 600 MW. Rather than construct a sizeable plant and 'grow into' capacity, Landsvirkjun pursued incremental development of several smaller power plants with capacities of 50–150 MW to meet emerging needs.

heavy industry became much more pronounced, as international non-governmental organizations, like Greenpeace and the World Wildlife Fund, engaged alongside emergent, national organizations, such as the Nature Conservation Association and Saving Iceland, to educate and advocate for the protection of Iceland's wilderness.

Drawing on ideas about sustainable development (World Commission on Environment and Development [WCED], 1987), the Ministry of Industry, Energy, and Tourism, and the Ministry of the Environment launched the Master Plan process in 1999 to review prospective geothermal energy and hydropower projects, with criteria for energy efficiency, economics, environmental impacts, and social/cultural value.[15] By 2010, two phases of participatory review were completed, with input from a mix of committees and experts (Rammaaaetlun, n.d.), and a third phase underway. As of 2017, geothermal projects have been ranked more favorably among the choices (see Box 4-1).

The Master Plan's integrated approach to energy planning has been a major and multifaceted effort in Iceland. The process has received some criticism (1) for using poor-quality information to determine energy, tourism, and environmental values, and (2) for a lack of consideration given to alternative scenarios (Logadottir, 2012). The first, two phases also offered no explicit guidance on a desirable pace of project adoption or patterns of ownership. Nonetheless, the process is seen as a way to engage society in difficult choices about the tradeoffs associated with energy, the environment, economics, and societal values.

As international environmental developments continued to evolve, the ratification of the Kyoto Protocol in Iceland in 2002 occurred with an unusual allowance. Iceland had already eliminated most fossil fuels from its heating and electricity systems prior to 1990—the Kyoto benchmark year—so Iceland's base value for emissions was set at 110% of the 1990 amount (Valfells et al., 2004, citing UNFCCC, COP3). Importantly, Iceland's GHG emissions are largely tied to foreign companies' manufacturing activity on domestic soil, so the country was also granted a separate addendum arrangement, 14/CP.7 of the Framework Convention on Climate Change, in which heavy industry emissions were to be delineated on a separate line from the country's total if best technology and environmental practices were employed (United Nations Framework Convention [UNFCCC], 2001; Valfells et al., 2004).

As time passed, a significant shift in public attitudes toward hydropower projects signaled a pivot for geothermal energy. Public discontent over hydropower, Iceland's second most used power source, manifested itself in large

15. The Plan used scoping similar to that of a Strategic Environmental Assessment in that it evaluated prospective energy options at a level beyond a single project. However, it also differed by not covering cumulative or higher order effects (Thorhallsdottir, 2007*a*, 2007*b*).

Box 4-1

Master Planning

In the first stage (1999–2003), 43 projects were reviewed: 19 hydropower and 24 geothermal (Steingrimsson et al., 2007). Based on a weighted scorecard ranking system, geothermal projects were largely found to be much more favorable. The second phase of the Master Plan (2004–10) evaluated all geothermal projects in high-temperature fields in addition to new or revised hydropower projects. Findings indicated that, among the 32,356 GWh/year of projects that were reviewed, the most acceptable were 9,170 GWh/year of geothermal projects and 2,741 GWh/year of hydropower (Logadottir, 2012, citing Rammaaaetlun). The third phase (2013–2017) focused on evaluating options that could not be adequately categorized during the second phase. This included new energy options such as wind power (Rammaaaetlum, n.d.). Certain hydropower projects that were on a wait list in Phase 2 were moved to a protected category. Final governmental decisions, based on the third stage are still pending at the time of this writing. A fourth stage is also due to commence.

---

demonstrations and debates between 2000 and 2006. This was associated with the Karahnjukavirkjun hydropower project to be constructed by Landsvirkjun, which would provide power for aluminum producer Alcoa. Proponents of the project argued that it would bring jobs and improve the area's economy, whereas opponents contended it would destroy the pristine wilderness.[16] The project is now operational, and the widespread preoperational opposition to the project revealed that the least costly power option would no longer be easily installed going forward in Iceland.

On the global front, the 2005 Stern Report on the economics of low carbon action drew special attention to Iceland's status as the largest producer of (primary) aluminum in the world on a per capita basis, noting:

> The near-future looks set to see a continuing sharp increase in aluminum production in Iceland . . . making Iceland the largest aluminum producer

16. This Eastern Iceland project had been rejected by the Icelandic Planning Authority on grounds that the environmental impacts were significant, but it was nonetheless approved later by the Ministry of the Environment (Interview, 2012). Construction began in 2003 on five dams (690 MW, six turbines, rated output 115 MW) designed to produce approximately 4,600 GWh annually to power an Alcoa aluminum smelter plant (Landsvirkjun, n.d.). Based on interviews and reporting, the project was highly controversial and encountered considerable opposition from both domestic and international groups (Interviews, 2011–12; Johannesson, de Roo, and Robaey, 2011; Petursson, 2008).

in western Europe. . . . Emissions of CO_2 from electricity production per capita in Iceland are the lowest in the OECD. . .. Iceland is also taking action to reduce emissions of fluorinated compounds associated with aluminum smelting. Expectations of future globalization action to mitigate GHG emissions are already acting as a key driver in attracting investment of energy-intensive sectors away from high GHG energy suppliers and towards countries with renewable energy sources. (2005)

If Stern's observations are juxtaposed with points of the public debate over Karahnjukavirkjun, a critical environmental tradeoff is obvious, pitting reductions of global GHGs through clean Icelandic energy (assuming no other increases occur) against the conservation of Iceland's resources. This trade-off continues today.

In this period of flux, the energy playing field also changed in institutional terms. The Orkustofnun was split into two organizations. Iceland Geosurvey (ISOR) is now responsible for science and technical consulting. The remaining Orkustofnun serves as the independent regulator for the electricity market, monitoring and licensing geothermal resources, as well as advising the government on energy issues, among its other functions. The Icelandic power market was also fully opened to competition in 2006 with third-party access to transmission and distribution.[17] Public utilities remain among the major players,[18] with the wholesale market dominated now by 100% state-owned Landsvirkjun, which generates about two-thirds of the electricity (Landsvirkjun, n.d.*a*; Olafsson et al., 2011).

17. This began with the Electricity Act, No. 65/2003 (based on EU Dir No. 96/92 and Dir 54/EC). Additional Acts included one establishing grid operator Landsnet, No. 75/2004; and Competition Act No. 44/2005. Some rules have since been modified; for example, Act 58/2008, which required CHP plants to maintain separate accounts for power and heating (Steinsdottir et al., 2009).

18. Landsnet is the public grid operator formed as a spinoff of Landsvirkjun's holdings. There are six distribution companies, all but one owned by either the State or municipalities (Olafsson et al., 2011). RARIK, the Icelandic State Electricity Company, is the largest (Logadottir and Lee, 2012).

Major companies operating in geothermal power include:

- Landsvirkjun, owned by the state, produces 2000 MW_e electricity, of that 63.2 MW_e is geothermal.
- Reykjavik Energy, owned by the municipalities of Reykjavik, Akranes, and Borgarbyggd, produces and distributes water and power from two co-generation plants (423 MW_e and 430 MW_t) and 750 MW_t from a number of geothermal areas.
- HS Orka, formed from the privatization and splitting of Hitaveita Sudurnesja, generates heat and power from 150 MW_t and 176.4 MW_e (Landsvirkjun, n.d.; Logadottir, 2012; Olafsson et al., 2011).

An effort to privatize state-owned energy company HS Orka brought questions about the governance of natural resources back to the forefront.[19] Through a sequence of acquisitions, Canadian-owned Magma Energy (via its Swedish subsidiary) held 98.5% of HS Orka at one point in the privatization process, leading to a public outcry over foreign control of Icelandic resources. This issue triggered a governmental review of HS Orka's acquisition process with the possibility of reclaiming ownership (an act that would have had international ramifications for the EEA agreement that was in place). The review found that Magma Energy was in full compliance with Icelandic law. Ownership has since been diluted with increased ownership by Icelandic pension funds at 33.4% and Magma Energy holdings at 66.6% (HS Orka HF, n.d).

On other fronts, advances in science and technology also shaped Iceland's energy trajectory, as the Iceland Deep Drilling Project (IDDP) enabled greater utilization of deeper geothermal sources. A consortium of Icelandic companies and the Orkustofnun initiated the IDDP in 2000, using advanced drilling technology and new fluid management techniques (Valfells et al., 2004) to test the economic feasibility of hydrothermal systems of geothermal resources at depths of 4 to 5 kilometers and 400–500°C (supercritical conditions). In 2009, the project stopped at molten rock at a 2,100 meter depth—a feat done only once before (see the later section "Innovations and Adaptations" for details). In 2017, IDDP 2 was drilled to a depth of 4,659 meters reaching supercritical fluid at 427°C (Iceland Deep Drilling Project [IDDP], 2017).

In recent years, Landsvirkjun has completed feasibility assessments for a subsea cable between Iceland and the United Kingdom as carbon reduction targets and favorable economics make the option more viable. Dubbed IceLink, the project is among the EU's projects of common interest. It would include the longest subsea cable in the world (at least 600 miles) with a 800–120 MW high-voltage direct current (HVDC) transmission link offering a bi-directional flow and able to deliver a volume of more than 5 TWh of electricity per year (Landsvirkjun, 2016). If this were to move forward, Iceland's energy market options could broaden considerably.

IceLink is not the only energy option for the Nordic nation. Iceland is also involved in the North Atlantic Energy Network, a multicountry project that investigates the feasibility of linking isolated energy systems with strong renewable energy potential in the North Atlantic to large markets in the UK and European continent through an electric grid (Orkustofnun, 2016; Orkustofnun et al., 2016).

Changes are also evident in jobs and industry terms, with an outflow of craftsmen during the banking crisis of 2008–11, and a more recent influx of craftsmen from other countries (Communications with S. Bjornsson, 2016).

19. HS Orka was formed from a decoupling of Hitaveita Sudurnesja (HS) entities.

Alongside these changes, there has been a considerable increase in the number of tourists to Iceland in recent years, spurring a construction boom and further influx of people to support the industries (Communications with S. Bjornsson, 2017). These dynamics not only alter the structure of the economy, but also potential political priorities. Only time will show their full relevance to the geothermal path.

Overall, large-scale investment by heavy industry, which began in the latter half of the 1990s, spurred growth in Iceland's geothermal-based power generation, increasing geothermal energy's share of the national power mix from 10% in 1998 to 29% in 2014 (IEA, n.d.). By this stage, space heating from geothermal energy was already quite advanced, so little changed: the period ended with a share of coverage at roughly 90% of the population. Considered in terms of the total primary energy supply, geothermal energy's share rose from 52% to 68% between 1998 and 2014 (Orkustofnun, n.d.).

EXPLAINING CHANGE

Iceland's hybrid form of sectoral intervention and expansionary model of readiness are outlined below.

Mode of Intervention

Iceland's geothermal transition reflected a widely dispersed and hybrid combination of three shifts that consisted of both bottom-up and top-down forms of intervention. In space heating, early and concerted sectoral effort was evident primarily in actions by the national government, industry, and localities during the 1970s. Municipalities sought the expansion of the infrastructure (emergent change). The national government provided loans for exploration and drilling, while the Orkustofnun completed studies of the regional resources and assisted with site-specific solutions (encouraged-deployed change). Publicly-owned industrial actors Landsvirkjun, Reykjavik Energy, and Hitaveita Sudurnesja (now privately-owned HS Orka) implemented the extension of the infrastructure and industry.

A second track of energy system change was tied to heavy industry's use of power. Here, the national government's long-term promotional efforts through various public agents (encouraged change), together with industry efforts (emergent-mandated-deployed change), drove development that reflected a top-down-led, hybridized form of transition.

A third track, which centered on combined heat and power, was led by industry, reflecting bottom-up change. Here, process and technical adaptations were made (emergent change), creating CHP plants that tapped geothermal energy for power generation more efficiently.

Comparing the transition since 1970 with the one since 1900, differences can be seen. The earlier transition had a different kind of bottom-up leadership, in which entrepreneurial homeowners and farmers brought about changes in heating (emergent change) similar to local mobilizers of the Danish wind transition that will be seen in Chapter 7.

Model of Readiness

Iceland's geothermal transition primarily reflects the expansionary model of readiness in that indigenous energy resources and expertise were harnessed through an existing geothermal industry and pathway. Energy utilization was also scaled with a build-out of infrastructure. Markets that were more local in nature, particularly for power, became more integrated over time.

INNOVATIONS AND ADAPTATIONS

Four primary areas reflect innovation or critical adaptations in the Icelandic transition: *carbon mineralization and renewable methanol*, *resource parks*, *fish farming and de-icing*, and *exploration-drilling-site optimization*.

Carbon Mineralization and Renewable Methanol

Two important and relevant innovations in the carbon playing field are notable in Iceland's geothermal development. These include the mineralization of carbon and the use of carbon dioxide (CO_2) in producing renewable methanol.

For the former, recent tests by geothermal energy company, Reykjavik Energy, and an international team of researchers working in Iceland have shown that CO_2 can be pumped into volcanic basalt and solidify. Reports of this breakthrough indicate that 95% of the CO_2 emissions were converted to solid carbonate material in less than 2 years. In contrast to expectations that the process would take hundreds or thousands of years in geologic reservoirs, this innovation may provide a more practical and long-term way to store CO_2 emissions (Kintisch, 2016; Matter et al., 2016).

Specific to renewable methanol, Carbon Recycling International (CRI) has implemented a process for producing methanol using CO_2 instead of carbon monoxide (CO) and electrolyzing water to generate hydrogen. This differs from conventional production of methanol in that it is done without fossil fuel and uses carbon dioxide from the neighboring geothermal plant. With the CRI approach, oxygen is the sole chemical released. The methanol is sold for blending with gasoline and is used in the production of biodiesel. The adapted process is less sensitive to change in feedstock prices and is now carried out at a demonstration level (CRI, n.d.). Renewable methanol reflects the concept of resource parks that is covered next.

Resource Park

A resource park is a community ecosystem in which streams are sustainably utilized with little waste (Albertsson and Jonsson, 2010). Iceland's Blue Lagoon is an example of this business innovation in industrial symbiosis in which waste from one firm serves as feedstock for another. Formed by the discharge of effluent brine fluid from the Svartsengi power plant into a lava field, superheated water powers turbines to produce electricity and a heat exchange process delivers hot water for municipal water systems before the water is channeled to the Lagoon. This man-made lagoon drew early swimmers who claimed health improvements tied to the silica-rich water. Scientific studies later completed between 1992 and 1996 indicated the healing capabilities of the Lagoon (Gudmundsdottir, Brynjolfsdóttir, and Albertsson, 2010). This work would serve as the basis for the launch of a skin care line of products in 1995 and the opening of a health clinic in 2005. Spillovers in microbial R&D continue to amplify the reach of this resource (Albertsson and Jonsson, 2010*a*, 2010*b*; Blue Lagoon, n.d.; Gudmundsdottir et al., 2010). The example of a resource park is now studied as a model for transforming areas like Cornell University into a zero carbon campus (Friedlander, 2016).

Fish Farming and De-icing

Geothermal resources are also used for fish farming in Iceland.[20] Young salmon and trout have been found to grow more rapidly in water temperatures of 10–12°C relative to the average groundwater temperature of 4°C that is typical for Iceland (Orkustofnun, n.d.). With this insight, warm water from geothermal boreholes has been used for decades for salmon and trout breeding practices. Warm effluent water from the aluminum plant at Straumsvik has also been used for this purpose.

De-icing and snow melting are other key applications of geothermal energy, used for roads, parking areas, and walkways in areas with geothermal district heating (Loftsdottir and Thorarinsdottir, 2006; Orkustofnun, 2012). This process employs systems that are installed under target areas and that channel geothermal water from space heating. As of 2003, an estimated 740,000 square meters were covered by snow-melting systems. In terms of annual energy use

20. Older applications, such as fish drying and greenhouse utilization, also continue.

for deicing and snow melting, more than half was derived from spent water used in houses (Ragnarsson, 2003).

Exploration, Drilling, and Site Optimization

To locate geothermal resources that are suitable for economical exploitation, exploration has improved with enhancements in prediction, drilling methods, and special drilling rigs. Specific to the low-temperature geothermal resources, advances were made in exploration using structural geology methods, resistivity surveys, and shallow temperature gradient holes, which have enabled discoveries of low-temperature reservoirs (<130°C) that have limited surface-level indications (Communications with S. Bjornsson, 2012). Improved geology methods have assisted with the prediction of fracture system geometry and the dynamics of potential fluid flows, allowing a more fitted stimulation of a geothermal reservoir (Phillip et al., 2007). Progress has also occurred in the development of down-hole pumps that facilitate increases in production by a factor of 10 (possibly even an order of magnitude larger) relative to the free-flow from wells (Communications with S. Bjornsson, 2012; Axelsson, Gunnlaugsson, Jonasson, and Olafsson, 2010, citing Axelsson and Gunnlaugsson, 2000).

For high-temperature resources, progress has been made with exploration and drilling methods. Exploration techniques, specifically with microearthquakes, geochemical thermometers, and transient electromagnetic as well as magnetotelluric resistivity have aided in the identification of targeted drilling areas in high-temperature fields (Communications with S. Bjornsson, 2012). The use of transient electromagnetic resistivity methods, for instance, allows the analysis of subsurface resistivity structures by sending an artificially induced alternating current into the earth (Hersir and Bjornsson, 1991).

Changes with directional drilling adapted from the oil and gas industry have also allowed for improvements in geothermal development. The technique enables multiple holes to be drilled from the same drill pad, which enhances the ability to manage the direction of drill holes that would intersect with faults and fractures (Communications with S. Bjornsson, 2012; Interviews, 2011–12). This technique has improved the success ratio of wells, where the average yield per drilled well is now approximately 5 $MW_{e,}$ with some wells yielding as much as 20 MW_e (Communications with S. Bjornsson, 2012). More extensive data to evaluate the increased efficiency are not readily available; nevertheless, researchers and industry widely recognize that these techniques substantially reduce drilling risk—a critical cost attribute in geothermal development profiles (Interviews, 2011).

Deep drilling for geothermal energy reflects a breakthrough area that has emerged largely over the past decade in conjunction with the IDDP project. With this project, research at depths of 4 to 5 kilometers and super-critical conditions of 400–600°C presents unusual opportunity for technological and procedural learning (Fridleifsson, 2010; Fridleifsson and Elders, 2007). The first phase of the project struck magma in 2009, at a depth of 2,100 meters, showing that superheated steam of 452°C had a power capacity of 36 MW_e, with CO_2 levels lower than those in the atmosphere (Fridleifsson, 2015). This reflects the first operable Magma-EGS system.[21] A milestone was met in the second phase with IDDP2 by: drilling deeply, extracting drill cores, measuring the temperature at 427°C with fluid pressure of 340 bars, and determining that the rocks appear to be permeable (IDDP, 2017). Continuing work, if successful, may offer a future for new geothermal operations, with fewer wells needed for power output, a higher thermodynamic efficiency of the power cycle, a smaller environmental footprint, and greater sustainability (Fridleifsson, 2015).

Beyond exploration and drilling, adaptations have been made to improve the harnessing of geothermal resources, many of which are site-specific. An example noted earlier was in the development of heat exchange methods and plant design, which allowed the utilization of high-temperature fluid for the Svartsengi plant. In that case, the high salinity of the fluid and large amounts of minerals precluded direct use. The integration of a heat exchanger to warm a secondary fluid provided a means to overcome many of these issues. Additional improvements with scaling and corrosion inhibitors, return water reinjection, and system efficiency enhancements addressed site-specific challenges such as scaling and corrosion, sea water incursion, and overexploitation, among others (Axelsson et al., 2010). Gains have also been made with pipe insulation in district heating systems.

One example that is worth emphasizing involves process redesign and repowering. The Svartsengi plant uses high-temperature geothermal energy to produce heat and electricity. The geothermal fluid is rich in minerals and has a high salinity level (roughly two-thirds that of seawater), which can be corrosive. In the 1970s, Orkustofnun scientists developed a dual flash plant design that heated groundwater through a heat exchanger at the site (Albertsson, Thorolfsson, and Jonsson, 2010; HS Orka, 2012; Interview, 2012). In subsequent adaptations, the plant was optimized with repowering (i.e., replacement of older equipment with newer), utilizing the plant's waste stream to generate an additional 8.4 MW of power (Kaplan and Schochet, 2005. With the changes, the plant's installed power output more than doubled (Kaplan and Schochet, 2005).

21. For a discussion of EGS, see the "Technology Primer" in the Appendix.

KEY DRIVERS AND BARRIERS

Major determinants of Iceland's energy transformation are outlined here. These reflect drivers and barriers of significance that were identified in case analysis and historical record review, as well as in interviews and data trends.

Reduction of Oil Imports/Improvement of Balance of Payments

Iceland adopted an explicit transition strategy to displace oil in the 1970s. Geothermal energy was already a fairly common energy source for heating in some Icelandic regions, so public actors and energy companies focused mostly on expanding infrastructure to eliminate imported fuels from space heating.

The Icelandic Orkustofnun evaluated savings from the use of geothermal energy in lieu of imported heating oil for the period 1970–2014. Annual savings (Figure 4-6) ranged from 1% to more than 7% of gross domestic product (GDP) per year between 1970 and 2014, with substantial costs avoided during periods of high oil prices. Cumulative savings were estimated at roughly 2,680 billion ISK or $20.7 billion (Ketilsson et al., 2015), assuming a 2% interest rate.

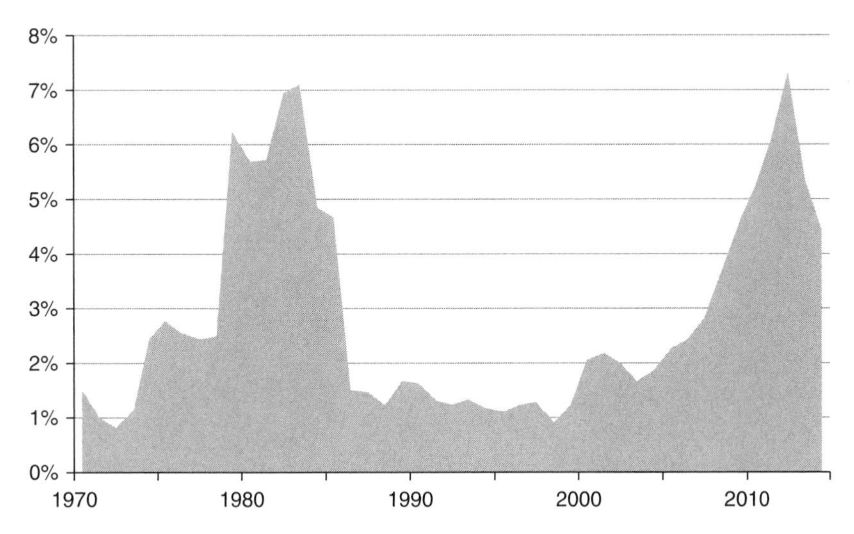

Figure 4-6 Avoided cost from use of geo-thermal energy in lieu of oil, % of gross domestic product (GDP).
SOURCE: Ketilsson et al. (2015).

Growth in Energy-Intensive Industry

To varying degrees, the Icelandic national government and domestic energy companies have pursued an energy-intensive industrial strategy designed to attract foreign investment in a manner that utilizes Iceland's hydro- and geothermal power. One estimate of economic gains associated with this suggests that energy-intensive industry contributed on average to about 0.5% of the annual GDP between 1969 and 1997 (Valfells et al., 2004).

Figure 4-7 shows the rapid growth of electricity consumption by the power-intensive industry since 1990. In line with major investments in aluminum smelting that commenced around 1997, the widening gap between the ordinary consumption of citizens and that of industry is starkly apparent.

While ordinary consumption has remained mostly flat, the growth in energy-intensive industry has encouraged a corresponding build-out of Icelandic installed generation. For example, the Century Aluminum smelter, which opened in 1998, has more than quadrupled its capacity, now producing 260,000 tons of aluminum per year. The Alcoa smelter, which came online in 2007, added capacity to produce 350,000 tons of aluminum per year. Even the first aluminum smelter that opened in 1970 (now owned by Rio Tinto Alcan) produces 180,000 tons of aluminum per year—6 times more than when it started.

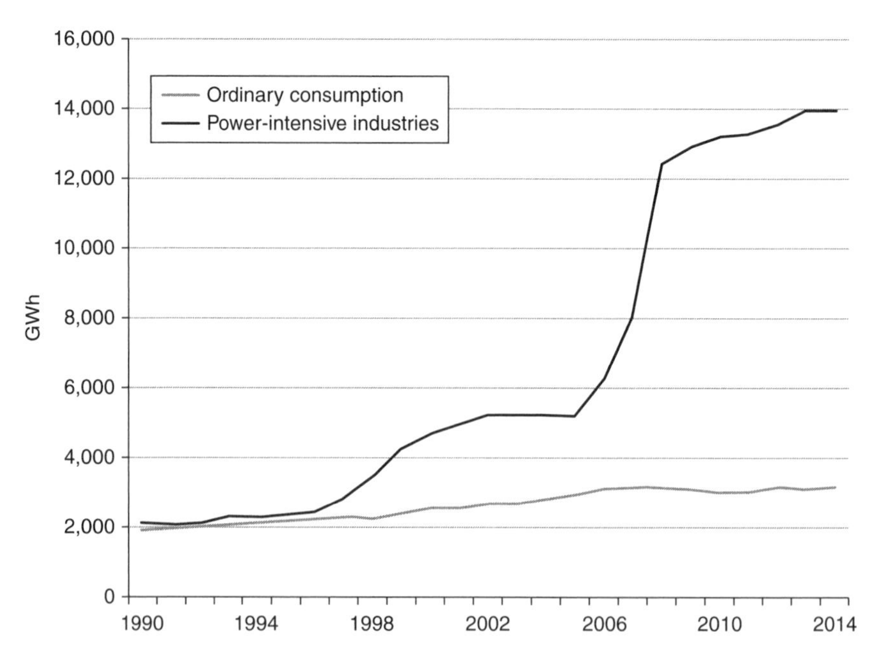

Figure 4-7 Gross power consumption in Iceland (GWh).
SOURCE: Data from Statistics Iceland (n.d.).

Figure 4-8 Exports by commodity (value as a share of total exports).
SOURCE: Compiled from Statistics Iceland data (n.d.), referencing Hagskinna. SI Classification.
NOTE: Value is free on board, which means the price on board by whatever means of transport. Some rounding applies.

In other areas of heavy industry, the Elkem ferrosilicon plant, which opened in 1979, currently produces double its original capacity, now at 120,000 tons of 75% ferrosilicon. Such growth is not inconsequential. Industry consumed roughly 81% of gross power in Iceland in 2014.[22]

Rising Socioeconomic Development

Tied to the preceding discussion of growth in heavy industry is the continued progress of socioeconomic development. Simply said, Iceland underwent a substantial change in its socioeconomic status as its energy path evolved. Regardless of whether one views energy demand as driving growth or resulting from it, geothermal energy is now the most-used fuel in Iceland's total primary energy supply.

Looking at Iceland's economic structure (Figure 4-8), a substantial shift is evident over the past century, moving away from agriculture and toward manufacturing. The restructuring of the economy was rooted in a broadly recognized effort to diversify and expand the domestic industry focus (Einarsson, 1991;

22. Due to the small size of the power sector, new power plants are generally not pursued today until power purchases are guaranteed (Ministry of the Environment, 2010).

Table 4-1. Iceland Profile - Human Development Index

	Life Expectancy at Birth	Expected Years of Schooling	Mean-Years of Schooling	GNI per Capita (2011, ppp $)	HDI Value
1980	76.6	12.6	7.5	24,077	0.756
2014	82.6	19.0	10.6	35,182	0.899
Change	8%	51%	41%	46%	19%

SOURCE: Compiled from UNDP data (n.d.), citing Human Development Report (2015).

Ingimarsson, 1996; Landsvirkjun, 2009). In line with diversification, the GDP grew by a factor of 4.5 since 1970 as primary energy increased by a factor of 6.6 (IEA, n.d.).

Another major shift is seen in improvements in the general quality of living, as measured by the Human Development Index (HDI).[23] Life expectancy, education, and the standard of living all increased, with substantial changes evident in the years of schooling and gross national income per capita (Table 4-1). Relative to peer countries, the average for OECD countries, and the world for 2000 and 2010, Iceland grew and exceeded the other classes (Figure 4-9).

Environmental Concerns

Environmental concerns served as both drivers of and barriers to the expansion of geothermal energy in Iceland. On the positive side, smoke-free air was a desired benefit in the move toward geothermal energy and away from oil and coal in heating. Now deemed a noticeably "smokeless city," Reykjavik has come a long way from its more polluted state in the 1940s (Arnorsson, 1975; Orkustofnun, 2010; author's observations; Figure 4-10).

Turning toward environmental challenges or barriers to geothermal adoption, geothermal waste water must be disposed of in a manner that doesn't pollute groundwater.[24] The 303 MW_e Hellisheidi plant, for example, utilizes roughly 600 kg/s of steam, with the remaining geothermal fluid of at least 1,000 kg/s waste water plus roughly 400 kg/s of condensed water requiring

23. The HDI is a proxy metric for the quality of living.

24. Flash plants use roughly 20–40% of the mass of geothermal fluid, so the rest of the fluid plus the condensed water from the turbine must be managed through reinjection (Communications with S. Bjornsson, 2012).

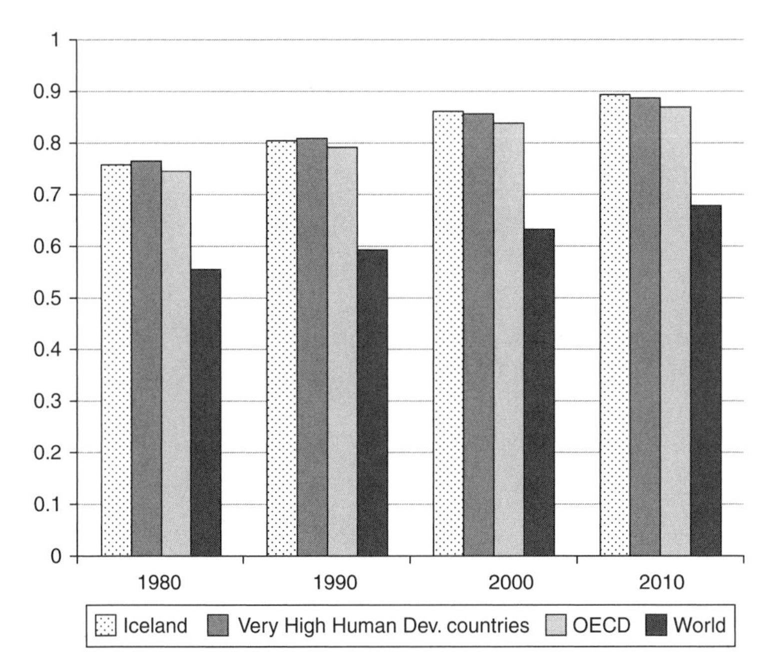

Figure 4-9 Comparative trends in the Human Development Index.
SOURCE: Compiled from UNDP data (n.d.).

Figure 4-10 Reykjavik in the 1930s with a cloud of smoke attributed to coal heating.
CREDIT: Reykjavik Energy.

reinjection (Communications with S. Bjornsson, 2012). Notably, reinjection at this site has induced seismic activity, producing earthquakes up to a magnitude of 4 and creating anxiety in the village of Hveragerdi (Communications with S. Bjornsson, 2012). Reinjection and earthquakes are environmental issues that require effective address for continuing development.

Considering the environment on a different level, views of nature preservation or conservation and climate intersect in an unusual way in Iceland. While citizens and environmental groups have articulated the desire to protect Iceland's resources over several decades, deepening industrialization since the late 1990s has engendered a wider discourse (Interviews, 2011; Saving Iceland, n.d.). Meanwhile, increased international attention to climate change and GHG reduction targets have spurred energy-intensive industries to seek sites like Iceland, where they can relocate their operations and leverage cleaner energy.

Focusing on master planning, the process has shown geothermal projects to be "more acceptable" than alternatives. Deep-seated societal discontent with recent hydropower developments appears to have solidified this point.

Geological Conditions, Time to Explore, Stepwise Adoption, Uncertainty

The technology profile of a geothermal resource has some attributes that may be seen as unfavorable relative to hydropower, the likely fuel substitute in Iceland. Geothermal energy is generally less efficient than hydropower generation (Euroelectric, 2003), more uncertain in terms of the amount of the actual resource that is available, and its sustainability hinges on a stepwise approach. As former Orkustofnun scientist Stefan Arnorsson (1975) noted, geothermal utilization requires "more elaborate technical and costly research in the early phase of economic evaluation," including geological assessments and drilling.[25]

Substitutes and Financing

Specific to Icelandic power plant investments, hydropower traditionally appears to have had lower investment costs than geothermal energy (see discussion of

25. Hydropower also has its own set of what may be called unfavorable characteristics, including the potential to eliminate locational biodiversity through flooding.

costs in the following section on developments). While much of the cost and pricing information in Iceland has been historically protected by law and/or the confidentiality of long-term contracts, a recent estimate indicates that geothermal energy costs are $2,500/kW versus hydropower at $2,200–2,300/kW (Steering Committee for Comprehensive Energy Policy, 2011, citing Mannvit, 2011). Combining insular market economics with the drilling uncertainty of geothermal energy resources and the investment profiles of hydropower and geothermal energy (i.e., no fuel costs), one finds that Icelandic hydropower has traditionally been viewed more favorably in project terms for large-scale projects. This, however, does not factor for new developments with the Master Plan, policy changes, or larger shifts in the market, such as what might be achieved with a subsea cable to Europe.

With respect to financing, geothermal projects are inherently encumbered by the uncertainty of their risk profile as well as by the front-loaded nature of investment costs. According to a report by Islandsbanki, an Icelandic bank that specializes in geothermal finance, debt financing can typically cover about 60% of financing for research, development, and drilling. The remaining 40% must be covered by equity (2010). For the Icelandic energy companies that are publicly as well as privately owned, sources of public funds were heavily constrained after the recent banking crisis (2008–11). When combined with a weakened Icelandic currency, such conditions for geothermal energy–based companies represent substantial challenges for raising necessary equity in an already complicated risk and financing landscape (Islandsbanki, 2010).

Monopoly of the National Power Company Landsvirkjun

The monopoly held by Landsvirkjun has been raised as another barrier to the adoption of geothermal energy in the Icelandic power sector (Interviews, 2011–12). Essentially, Landsvirkjun, through its governmental charter and early-mover status, is said to have had first claim on the power market until the market was opened to competition with the 2003 Act. Since that time, Landsvirkjun and Reykjavik Energy remain publicly owned, yet other market entrants can now compete.

It is worth noting that early charters for companies, like HS Orka's predecessor, were structured with mandates focused on heating/hot water services. A company's ability to sell electricity (even if produced efficiently in a CHP capacity) was arguably limited in *de jure* terms until Parliament expanded the charter.

DEVELOPMENTS IN COST, SOCIETAL ACCEPTANCE, AND INDUSTRY

Costs

Costs for Icelandic geothermal energy reflect favorable terms in a number of ways. Development costs of Icelandic geothermal energy are among the lowest, ranking along with those of Indonesia at $2.5–2.6 per MW, compared to $3.5–$15 per MW for countries like China, the United States, and Germany (Islandsbanki, 2010). Specific to heating within Iceland, geothermal costs were competitive in the 1970s and remain so today. By contrast, costs for electricity derived from geothermal energy in Iceland were not competitive in the 1970s, but are now (Interviews, 2011). Recent estimates for energy prices in Icelandic residential heating indicate that geothermal heating is the most competitive option as other fuel sources are subsidized to provide equity in cost coverage (Figure 4-11).

Considering costs over time, substantial differences can be seen between geothermal energy and oil for heating. Geothermal space heating costs remained very low for the period versus the radical flux in oil costs (Figure 4-12). Given the favorability of geothermal energy, the fairly steady rise in its use for home heating is not surprising.

Specific to electricity, power prices are protected and mostly confidential, yet geothermal power appears to have become competitive over the period studied (Interviews, 2011–12). Landsvirkjun's CEO Hordur Arnarson has also

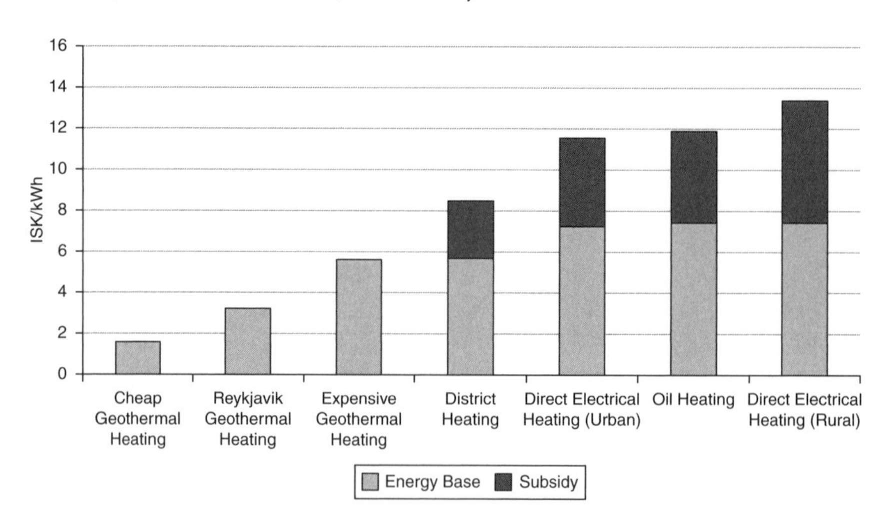

Figure 4-11 Comparison of energy charges for home heating in Iceland, September 2015 (ISK/kWh, $1 = 127.29 ISK, Central Bank of Iceland, September 15, 2015).
SOURCE: Data from Orkustofnun (n.d.).

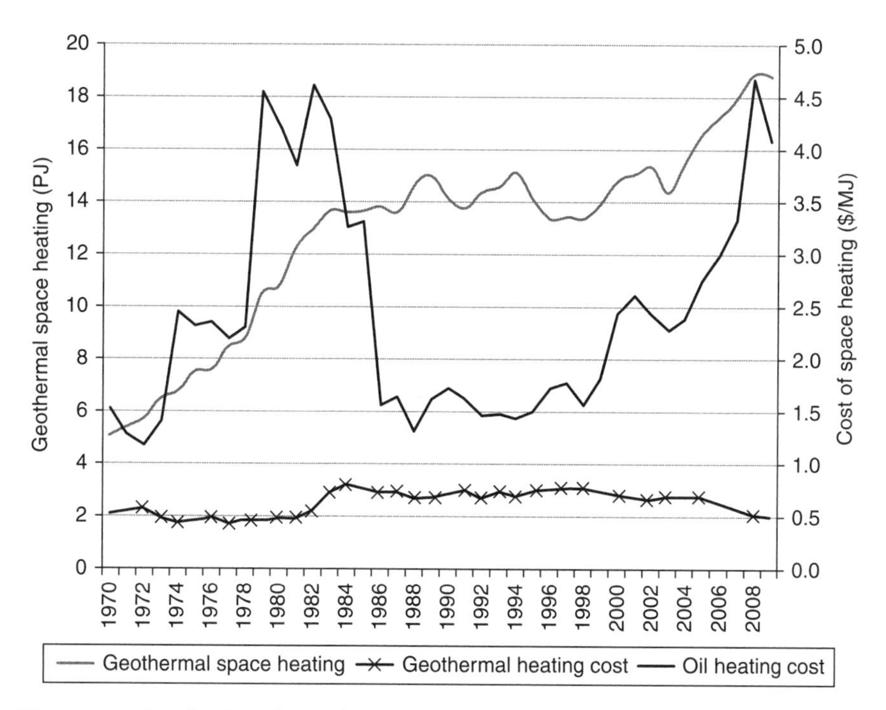

Figure 4-12 Supply of geothermal space heating, costs of geothermal heating and oil
SOURCE: Haraldsson (2014), adapted from Haraldsson et al. (2010). Based on annual average consumer price index and ISK/$ rate in February 2014.

highlighted the unreasonably low nature of some of the company's power contracts in 2011 (Landsvirkjun, 2011; Logadottir and Lee, 2012), emphasizing a business vision for Landsvirkjun to raise power prices in Iceland in proportion to European price increases.

Turning to the question of who paid for the transition to geothermal energy in Iceland, the short answer is that the adoption of geothermal heating during the 1970s and 1980s was covered by the government and municipalities broadly in conjunction with work of the Orkustofnun and the energy companies. This implies that national and municipality revenue plus customer utility bills financed this portion of the transition. The incremental adoption costs of geothermal power in CHP, as well as the robust scale-up with heavy industry, were, in turn, covered by the specific companies and their customers.

A number of nuances should, however, be noted. Following record inflation in the 1970s and early 1980s, together with the collapse of oil prices in 1986, the national government absorbed some district heating debt for municipalities. Additional costs were covered in the restructuring of utilities. Such costs are separate from those deferred by the Energy Fund that was set up in the 1960s to provide grants and loans with favorable terms for exploration and drilling for geothermal resources (Bjornsson, S., 2010; Interview, 2012). If a project led to failure (i.e., cold or dry wells), the Fund absorbed a portion of the drilling costs. With some degree of competition, including the participation of large public companies, now occurring in the Icelandic energy field, costs should generally be passed on to consumers.

Historically, heavy industry prices were linked to aluminum profits or aluminum prices, subjecting publicly owned Landsvirkjun and, by extension, Icelanders to the vagaries of the metals market. Landsvirkjun appears to be making some changes in this regard. In 2010, the company announced a new contract with Rio Tinto Alcan, its second largest client, in which pricing was benchmarked to the U.S. Consumer Price Index, rather than to aluminum prices (Reitun, 2010). The CEO's acknowledgment of some pricing details may foreshadow increased transparency to come.

Societal Acceptance

Broadly, social acceptance of geothermal energy appears to have increased if one considers geothermal project favorability in the Master Plan. Early objections by neighbors over the visual impacts of pipes and field structures in geothermal projects appear to have been addressed by adaptations over time that produce less overt layouts. Compared to the visual effects of coal emission clouds that used to linger over Reykjavik, geothermal energy seems to be faring well.

There are some concerns related to sulfuric emissions. The release of H_2S produces the scent of rotten eggs and may cause some respiratory problems.[26] In the capital area, increased concentrations of H_2S from the Hellisheidi plant have led to demands that H_2S be reinjected. Reykjavik Energy is now sequestering two thirds of the H_2S from Hellisheidi and plans for complete injection of H_2S along with the CO_2. The H_2S reacts with iron in the rock, creating FeS, a golden compound called the 'gold of fools' (Communications with S. Bjornsson, 2017).

Ensuring that effluents from the geothermal fields do not contaminate groundwater is another societal concern. This is addressed by reinjecting the effluents into the ground, but this process can induce seismicity. The issue

26. A recent study of Icelandic adults exposed to low levels of geothermal gas releases found asthma-like symptoms that may be associated with the H_2S (Carlsen, Zoega, Valdimarsdottir, Gislason, and Hrafnkelsson, 2012). This subject is still rather new. Interestingly, health changes were noted by Icelandic doctors as far back as the 1950s and 1960s, indicating a reduction in common colds associated with reduced use of coal furnaces (Jonasson and Thordarson, 2007).

has recently grown among concerns. Given that Iceland is located in one of the most seismically active regions in the world, earthquakes must be recognized for their common occurrence, as well as for the danger associated with inducing them.

Overall, societal acceptance appears to be undergoing redefinition. The intense discontent associated with hydropower expansion in recent years appears to leave geothermal energy positioned in a more favorable light. However, continued discussion about the need for powering additional heavy industry by yet untapped hydropower or geothermal energy remains open within Icelandic society (Interviews, 2011–12).

Industrial Development

The industrial development associated with geothermal energy has taken highly varied forms. Iceland's district heating services have a long and proven track record. Reykjavik Energy (including its predecessor) is one of the oldest and largest district heating services in the world. Such services emerged in the early 20th century and have undergone numerous changes, including market consolidation and expansion as well as infrastructural modernization.

Utilization of geothermal energy in recreation and agriculture is also prominent. Geothermal energy use in swimming pools accounts for roughly 138 of the 169 swimming centers in the country (Orkustofnun, n.d.). The energy and its effluent fluids or minerals in some cases are also used to produce the Blue Lagoon, the Myvatn Nature Baths, and Nautholsvík, a geothermally heated beach. In agriculture, an estimated 120,000 square meters of fields are also warmed by geothermal water in order to harvest produce earlier in the season (Orkustofnun, n.d; Bjornsson, S., 2010; Loftsdottir and Thorarinsdottir, 2006).

Private-sector firms, academia, and international research entities recognize Iceland's considerable expertise in geothermal resource management. In 2015, Iceland's GeoSurvey was awarded four grants totaling $4.5 million from the EU Framework Programme at a success rate of 80% compared to typical application success rates of 15–30% (Richter, 2016). The World Bank and Iceland also partner to assist African countries in evaluating and harnessing their indigenous geothermal energy (Iceland Review, 2012). The UNU Geothermal Training Programme also holds an unparalleled track record of producing geothermal experts in Africa, Central America, and Asia. Courses from the school are now being offered in locations like Kenya, and more than half of the instructors are graduates of training programs in Iceland.

CONCLUSION

Iceland is a country with an unusual abundance of low carbon energy, both tapped and untapped. Despite this seemingly-optimal, energy situation, Icelanders wrestle (as citizens may do in other countries) with choices about which energy source to use, what market adaptations could best serve the people, and how environmental tradeoffs at the local versus global levels should be reconciled.

As Iceland's economy evolved to become more diversified, its energy-related policy has also changed to reflect an increasing footprint of international agreements and stakeholder engagement. Iceland now is positioned to expand its renewable energy utilization with advanced technological connectivity to other markets. Going forward, choices about greater international market integration remain open. In the meantime, continuing breakthroughs in deep drilling and carbon storage could radically advance the country's low carbon leadership.

5

French Nuclear Energy

Concentrated Power

No coal, no oil, no gas, no choice . . . .

—French maxim (mid-1970s)[1]

INTRODUCTION

Nuclear energy is one of the most significant sources of low carbon energy in use in the power sector today.[2] In 2013, nuclear energy represented roughly 11% of the global electricity supply, with growth projected to occur in China, India, and Russia (International Atomic Energy Agency [IAEA], n.d.*a*; NEA, n.d.). As a stable source of electricity, nuclear energy can be a stand-alone, base-load form of electricity or complement more variable forms of low carbon energy, like wind and solar power.[3] Among the energy technologies considered here, nuclear energy is complex not only for the science behind it, but also for its societal, environmental, and economic dimensions.

This chapter explores the rapid rise of French nuclear energy in the civilian power sector. It considers what a national energy strategy looks like under

1. Palfreman (n.d.). This maxim may be a shortened version of Prime Minister Messmer's announcement of the Nuclear Program on March 4, 1974.

2. Nuclear plants generally do not produce carbon emissions in their operations. Indirect emissions are produced in uranium mining and enrichment, as well as during the construction of nuclear plants.

3. Base-load plants are operated nearly all the time and are able to continuously generate energy, typically at low cost relative to other plants. For more discussion, see the Appendix.

conditions of high concern about energy supply security when limited domestic energy resources appear to exist. The case reveals that centralized planning with complex and equally centralized technology can be quite conducive to rapid change. However, continued public acceptance, especially for nuclear energy, matters in the durability of such a pathway.

PROFILE FOR FRENCH NUCLEAR ENERGY

France is a traditional and currently global leader in nuclear energy, ranking the highest among countries for its share of domestic electricity derived from nuclear power at 76% of total electricity in 2015 (IAEA, n.d.*b*). France is highly ranked for the size of its nuclear reactor fleet and amount of nuclear generation, second only to the United States. In 2016, this nation of 67 million people and economy of $2.7 trillion had 58 nuclear power reactors (CIA, n.d.; IAEA, n.d.*b*). Due to the level of nuclear energy in its power mix, France has some of the lowest carbon emissions per person for electricity (IEA, 2016*a*). France is also one of the largest net exporters of electricity in Europe, with 61.7 TWh exported (Réseau de Transport d'électricité [RTE], 2016), producing roughly $3.3 billion in annual revenue (World Nuclear Association [WNA], n.d). This European country has the largest reprocessing capacity for spent fuel, with roughly 17% of its electricity powered from recycled fuel (WNA, n.d.). Nuclear technology reflects a significant export in France and includes reactors as well as fuel products and services (WNA, n.d.). In 2015, French nuclear generation represented 17% of the world's total (IAEA, n.d.*b*).

BACKGROUND

When one considers France's energy transition, its long history in atomic energy is an important backdrop, dating at least as far back as the late 19th century. In 1896, French physicist Henri Becquerel discovered natural radioactivity while studying the fluorescence of uranium salts (Leclercq, 1986). Two years later, French physicists Marie and Pierre Curie discovered polonium and radium, and, in the course of their work, Marie Curie named the phenomenon of study "radioactivity" (Leclercq, 1986; Marcus, 2010). The Curies' contributions to atomic energy continued with their daughter Irene and son-in-law-Frederic Joliot-Curie's discovery of artificial (induced) radioactivity in 1934 (Goldschmidt, 1962; Marcus, 2010). In 1938, researchers Irene Curie-Joliot, Hans Halban, Lew Kowarski, and Pavel Savic advanced understanding of the occurrence of nuclear fission in tandem with research by Otto Hahn, Fritz Strassman, and

Lise Meitner in Germany (Goldschmidt, 1962; Marcus, 2010; Communications with B. Barre, 2016).[4] By 1939, teams in France, Germany, and the United States demonstrated neutron production during fission (Goldschmidt, 1962; Marcus, 2010). Building on such discoveries, Enrico Fermi's team at the University of Chicago produced the first self-sustaining nuclear reaction in 1942 (Marcus, 2010), followed three years later by the production and use of nuclear bombs by the US to end World War II.

Following the War, the French government set out to create a modern and independent nation, as it rebuilt its infrastructure and economy (Bernstein, 2006; Bernstein and Rioux, 2015; Dormois, 2004; Hecht, 2009; Rioux, 1989). The French Atomic Energy Commission (Comisariat a l'energie atomique or CEA) established in October 1945, was part of this effort. The CEA was tasked with a mission to conduct research and design nuclear reactor technology, to oversee radiation protection standards, and to manage associated fuel supply development (Saumon and Puiseux, 1977). In addition, Électricité de France (EDF) was also formed from the nationalized power sector to provide reliable and abundant electricity for the French people (Hecht, 1991). EDF and the CEA would jointly manage the nascent French nuclear program, which included civilian and military dimensions (Hecht, 1991, 2001, 2009; Schneider, 2009). An advisory body consisting of senior members from EDF, CEA, and the government, known as the PEON Commission (Commission for the Production of Electricity from Nuclear Energy), would later be formed to provide recommendations for decisions regarding civilian nuclear development (de Malleray, 2011; Interviews 2011–12).

In the early years of the nuclear program, the French pursued nuclear technology based on designs for natural uranium, gas-cooled reactors (UNGG) and fast breeder reactors (FBR) (Jasper, 1988; Leclercq, 1986; Thais, 2014). This was due in part to the limited availability of enriched uranium and to interest in producing plutonium for weapons use. [5] Enriched uranium was accessible at the time primarily through the United States or Soviet Union (Jasper, 1988). Circumstances, however, changed with international advances in light water reactors (LWRs) and the successful construction of a domestic uranium enrichment plant in 1969 (International Directory of Company Histories,

4. For a more in depth discussion of work by Joliot-Curie and Savic, see Savic (1989) and Goldschmidt (1989). Radioactivity is discussed in the Appendix "Technology Primer".

5. Fast breeder reactors can generate power and repeatedly recycle their own fuel, so may be used to optimize energy output, minimize waste, or produce new plutonium (Karam, n.d.; Pearce, 2012). Their ability to produce plutonium, which can also be used in weapons, raises security concerns. For more discussion of reactor technology, see "Technology Primer" in the Appendix.

1998; Thais, 2014).[6] Domestic attention turned to what was deemed to be more competitive LWR technology, with a decision to license the pressurized water reactor (PWR) design from Westinghouse for the Fessenheim 1 plant (Golay, Saragossi, and Willefert, 1977; Hecht, 2009; Interviews, 2011–12; Surrey and Huggett, 1976).[7] Government planning at the time considered construction of one or two reactors per year (Thais, 2014).

MODERN TRANSITION: 1970–THE PRESENT

Looking at France in 1970, one would see a postwar boom known as the "30 Glorieuses" with substantial growth in the standard of living, domestic consumption, and population (Fourastie, 1979; Interview, 2012; Price, 2005). In this period, French power production had shifted from near exclusive reliance on coal to include shares of oil, gas, hydropower, and nuclear power (Golay et al., 1977; Leclercq, 1986). Within the mix, nuclear energy reflected 4% of electricity output in 1970 (Figure 5-1) and oil represented 68%.

Acceleration of the Transition: 1970–1987

In the period preceding the 1973 oil crisis, foreign oil dependence was high, yet changes were being developed. Imports equaled roughly 75% of France's total energy consumption (IAEA, 2016), meanwhile the PEON Commission recommended that, in addition to the Fessenheim 1 reactor, another 21 GW_e of light water reactor (LWR) capacity and 1.2 GW_e of fast breeder reactor capacity should be added (Golay et al., 1977).[8]

The oil shock, however, catalyzed the country to mobilize an untraditional energy strategy. Within weeks of OPEC's decision to curtail the oil supply, the French government instructed EDF to increase nuclear plant construction (International Directory of Company Histories, 1998). On March 5, 1974, a decision was settled at an inter-ministerial meeting to move forward with a series of contracts for nuclear power plants (NPPs). Nuclear developer Framatome

6. The presence of a domestic fuel enrichment plant meant that France could consider light water reactor (LWR) technology, like PWRs, without depending on the United States or Soviet Union for the necessary fuel supply. The earlier natural uranium gas graphite reactors did not require enrichment. See Appendix "Technology Primer" for a discussion of reactor technology.

7. With LWRs, gas graphite reactors were eliminated on the basis of price considerations (Leclercq, 1986). High temperature and breeder reactors were deemed to be unready for commercial adoption (Leclercq, 1986). Canadian heavy water and advanced gas reactors were also seen as problematic, in connection to anticipated developments (Leclercq, 1986).

8. Before the oil shock of 1973, the provision for nuclear capacity additions was 1.5 GWe per year (Communications with B. Barre, 2016).

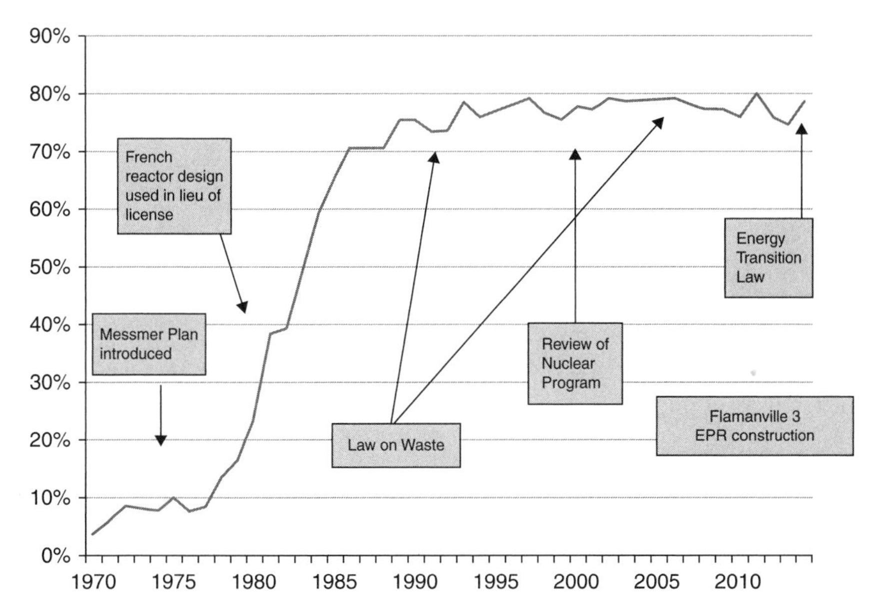

Figure 5-1 Share of total electricity from nuclear energy.
SOURCE: Slope based on IEA (n.d.).

was awarded a contract by EDF for sixteen 900 MW_e PWRs in an initial series (Leclercq, 1986). Two years later, a 1,300 MW_e series was also launched (Leclercq, 1986).

In what could easily be called the most comprehensive nuclear program to date globally, French state actors, EDF, Framatome, and the CEA initiated an accelerated roll-out of an advanced nuclear program. EDF was the architect and manager of the NPPs. Framatome was the primary manufacturer and developer. The CEA managed nuclear research and development (R&D), oversaw the fuel cycle, and advised on nuclear technology. By the beginning of the 21st century, this program would produce a nuclear fleet of 58 PWRs (Table 5-1 and Figure 5-2).[9]

In tandem with the planned expansion of the nuclear fleet, a complete, industrial-sized nuclear fuel cycle was developed, encompassing both the front-end and back-end of the process (see Appendix "Technology Primer" for a discussion of the fuel cycle). This was done by integrating and expanding capabilities, including reprocessing and fuel fabrication that to date had been used for the nuclear defense program (House, 2008; IAEA, n.d.*a*). In 1976, COGEMA was established as a subsidiary of the CEA with the mandate to create an industrial group which would service all stages of the nuclear fuel cycle for civilian and military program needs (Schneider, 2008*b*). With its formation, COGEMA was granted oversight authority for the industrial

9. Two experimental fast breeder reactors were also launched during the period of this study—the Phenix in 1973 and the Superphenix in 1986 (IAEA, n.d.*b*).

Table 5-1. PRESSURIZED WATER REACTORS INSTALLED IN FRANCE FOLLOWING THE LAUNCH OF THE *MESSMER PLAN*

Number of Units	Type	Capacity per Unit (approx)	Commissioned
34	PWR (CP0, CP1, and CP2)	900 MW_e	1977–1987
20	PWR (P4 and P'4)	1,300 MW_e	1984–1993
4	PWR (N4)	1,450 MW_e	1996–2002

SOURCE: IAEA, n.d.*b*; IEA, 2004*a*.

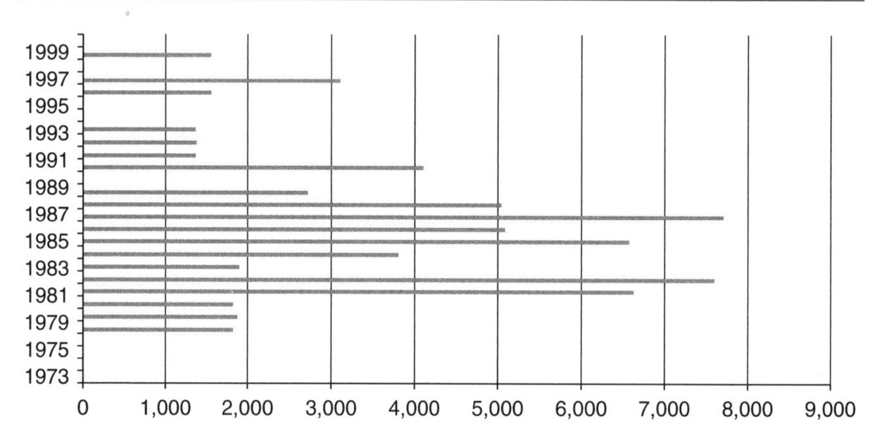

Figure 5-2 Additions in gross electricity capacity from new pressurized water reactors (PWRs) following the launch of the Messmer Plan (MW).
SOURCE: IAEA-PRIS data (n.d.*b*).

uranium fuel and reprocessing facilities that had been previously overseen by the CEA (Schneider, 2008*b*).

The decision to adopt the French nuclear program, dubbed the *Messmer Plan*, was guided by a number of factors. A strong network of scientists played a prominent role, as did concerns about security and independence, projections of growth in energy consumption, and the nation's industrial aspirations (Faro, 2013; Interviews, 2011–12).

Broadly, there was a sense that France had limited indigenous energy resources to use. On March 4, 1974, war hero-turned Prime Minister Messmer publicly announced the Plan on television (Topçu, 2008). Based on overprojections of French energy needs,[10] Messmer indicated that 40 reactors would be ordered by 1980 (Surrey and Huggett, 1976).

10. Estimates point to different outlooks. According to Grubler (2009, citing G-M-T, 2000), EDF estimated electricity demand of 1,000 TWh for the year 2000 to be covered by 80 PWRs and 20 FBRs, whereas Surrey and Huggett indicated that EDF forecasted a need for 200 reactors on 50 sites by 2000 (1976). The PEON Commission in 1973 projected demand of 750 TWh for 2000. The actual demand was 430 TWh. A tapering off of demand growth explains at least some of the difference.

The announcement of the *Messmer Plan* was made without the executive branch consulting with Parliament or the public (Faro, 2013; Interviews, 2011–12; Nelkin and Pollak, 1980, 1982). This was not unusual at the time, since large infrastructure projects including the Roissy airport and the freeway network were also implemented in a similar way (Communications with B. Barre, 2016). For nuclear-related expenses, Parliament voted on the budget each year in large part without objection (Communications with B. Barre, 2016).

A common perspective then and among countless still today (Interviews, 2011) is that nuclear energy was the obvious and only solution to meet France's power needs (Interviews, 2011; Topçu, 2008). For decades, France's nuclear energy program had been in place, demonstrating potential at a smaller scale. Given the risks that fuel disruptions could create emergency-like conditions, one could say that the government also had to act quickly to minimize the effects on French progress in industrialization and socioeconomic development.

Technical complexities of nuclear energy also made public discussion unwieldy (Interviews, 2011–12), so the government relied on the experts from the PEON Commission, EDF, Framatome, and the CEA. Giscard d'Estaing, French president from 1974 to 1981, spoke about this pivotal period later in 2011, indicating that the choice was validated scientifically and politically (de Malleray, 2011). According to him, the subject was debated with the best engineers involved and in relevant fora. Moreover, the decision remained largely unchallenged by successive governments since there was no perceived alternative (de Malleray, 2011).

The decision to move aggressively toward nuclear energy was not, however, uncontested. Even before the *Messmer Plan* was announced, a reported 10,000–15,000 people had protested the siting of the Bugey 1 nuclear plant in 1971, with demonstrations continuing across many cities (Surrey and Huggett, 1976; Nelkin and Pollak, 1982).[11] The launch of the *Messmer Plan* spurred additional protests. In 1975, the "Appeal of the 400" petition was signed by 400 scientists, then extended to include several thousand more names. Among its petitioners were employees working for the French Scientific Research Center (CNRS), College de France, CEA, and EDF, challenging the "all-nuclear" nature of the *Messmer Plan*, the lack of public information, and absence of independent control (Surrey and Huggett, 1976; Topçu, 2007, 2008).[12] Between 1975 and 1977, a reported 175,000 people also turned out in a series of 10 demonstrations

11. Two, opposition groups were Scientists for Information on Nuclear Energy (GSIEN) (Topçu, 2007, 2008) and the Committee Against Atomic Pollution and Survive and Live, an ecological group formed by mathematicians (Topçu, 2008). For a discussion of other forms of anti-nuclear activity in this period, see Surrey and Huggett (1976).

12. *Alternatives to Nuclear*, published by scientists with the Institut Economique et Juridique de l'Energie et Grenoble, also criticized the *Messmer Plan*, but on the basis of unsound economics (Surrey and Huggett, 1976).

(Kitschelt, 1986; Nelkin and Pollak, 1982). This number reportedly included many demonstrators from outside France (Interviews, 2011). In 1977, one protest gathering included more than 100,000 people at Creys-Malville to oppose the Superphenix FBR being built, with violence occurring and one fatality (Nelkin and Pollak, 1982).[13] Despite signs of anti-nuclear dissatisfaction, polls after 1977 consistently showed a vast majority supporting the nuclear program, except in periods following nuclear accidents (Fourquet and Pratviel, 2013; Communications with B. Barre, 2016).

Other concerns came to the foreground in the late 1970s. The 1979 accident at the Three Mile Island nuclear plant in Pennsylvania in the United States spurred concerns about safety as the second oil shock aggravated an already weakened global economy. Political parties in France campaigned for parliamentary elections, with the Socialists appealing to anti-nuclear concerns (Kitschelt, 1986; Locher, 2011). During this time, a parliamentary discussion and vote on energy and domestic nuclear policy occurred.[14] Under the leadership of President Mitterand, the newly elected Socialists briefly interrupted, but then continued the rollout of the nuclear program, ultimately supporting an expansion of domestic and international nuclear plant sales (Collins, 2003; Kitschelt, 1986; Marshall, 1982; Schneider, 2008*b*; Topçu, 2007).

By the early 1980s, a temporary freeze in French nuclear energy deployment was evident. Some contend that economic growth was less than expected, and an overcapacity of power existed (Interviews, 2011–12). Others maintain that the Socialists reneged on their pre-election promises (Locher, 2011; Topçu, 2007) because of a deal struck with the trade union French Democratic Confederation of Labor (CFDT) that marginalized the anti-nuclear platform (Schneider, 2008*b*). In either case, the momentum of the national anti-nuclear protests receded, but local struggles would continue and amplify, primarily around issues of waste storage in the ensuing years (Interview, 2012; Kitschelt, 1986; Topçu, 2007, 2008).

Questions about reliance on nuclear power were reopened in 1986 with the Chernobyl nuclear accident in the Soviet Union. The French Ministry of Agriculture reportedly issued a statement saying that the French territory was exempt from radioactive fallout associated with the disaster due to France's distance from the accident (Schneider, 2008*b*). Pierre Pellerin, the head of the

13. Other forms of protest were evident. The home of the EDF president was bombed, and EDF engineers' wives received small coffins by mail (Interview, 2011–12). Later, rockets were fired on the unfinished Superphenix plant (Marshall, 1982; Besson, 2005; Surrey and Huggett, 1976).

14. Many have argued that this process lacked both adequate procedures for informing citizens and a genuine debate about both current and past decisions that had already been made (Fagnani and Moatti, 1984, citing Bourjol and Lamer, 1982).

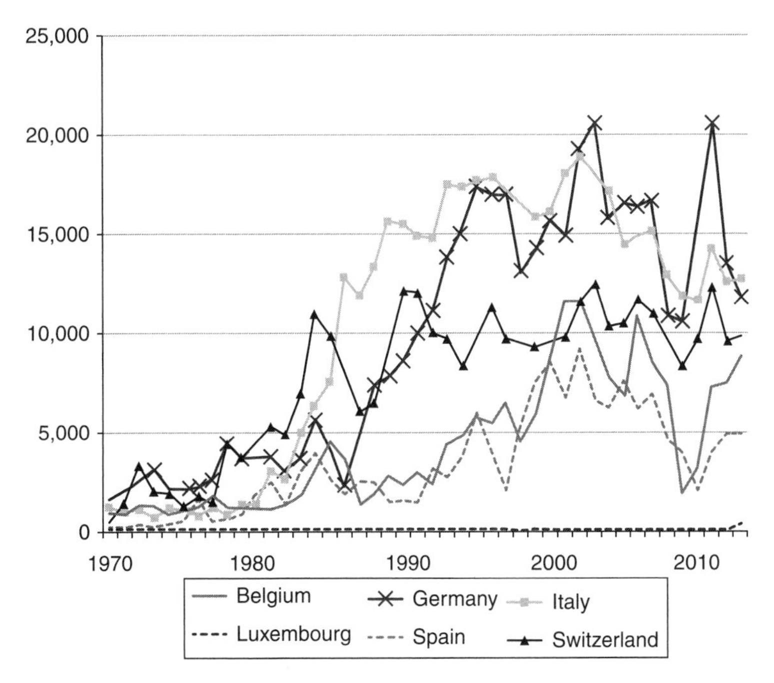

Figure 5-3 French electricity exports by country (GWh).
SOURCE: IEA data (n.d.).

French radiation protection agency, also sent a notice indicating that precautionary measures would be justified at levels that were 10,000 to 100,000 times higher than those evident at the time. Alongside the public actions, bottom-up efforts were carried out by citizens who were scientists or linked to nongovernmental organizations (NGOs). These efforts included the collection of radiation data around clusters of nuclear installations (Lehtonen, 2010*a*).[15] The actions of this period would be revisited in 2006, when Mr. Pellerin was indicted for aggravated deceit and later vindicated (Vanlerberghe, 2012).

By 1986, 37 new PWRs and 1 FBR had been brought online (IAEA, 2016). That same year, 70% of France's electricity was derived from nuclear energy, compared to 4% in 1970 (Figure 5-1). France also began exporting increased amounts of electricity to neighboring countries (Figure 5-3).

To address its now overcapacity of power, France began the practice of *load-following* with some of its 900 MW_e PWRs (WNA, n.d.). This practice entailed technical refinements to the reactors and modifications in the management of the grid which were unconventional for the nuclear energy sector. The

15. The NGOs ACRO in Normandy and CRIIRAD in Rhones-Alpes developed independent metrics to verify radiation levels in nuclear regions (Topçu, 2008).

refinements allowed nuclear plants to be used more flexibly, when accounting for reduced demand (see discussion under the "Innovations and Adaptations" section).

New Questions and Engagement: 1987–1999

As the French nuclear program progressed with increasing volumes of spent fuel, concerns about nuclear waste management became more pronounced. Scientific experts at the time saw irreversible disposal in geological repositories as the only approach, yet protests turned violent, elevating the topic to the public agenda (Interviews, 2011; Lehtonen, 2010*b*). Some citizens questioned the criteria for site selection, particularly when a chosen rural area would principally bring benefit only to the urban demand centers (Interviews, 2011; Lehtonen, 2010*b*, citing Barthe et al., 2010). The government responded by appointing a parliamentary commission. At the request of the commission, Christian Bataille met with stakeholders and identified reversibility as an important consideration (Interviews, 2011–2012; Lehtonen, 2010*b*, citing various). A direct outcome of the Bataille's engagement was the creation of the 1991 Waste Act, a law which postponed national decision-making on waste management for 15 years (Lehtonen, 2010*b*). The Act opened up the possibility of alternatives other than geological repositories with a new research program launched to consider (1) partitioning and transmutation,[16] (2) deep geological storage, and (3) interim storage (Lehtonen, 2010*b*; IEA, 2000*a*).

A solicitation was made, seeking expressions of interest from communities to host an underground laboratory for nuclear waste R&D. In the course of the following years, a site would be chosen in Bure (Box 5-1).

With the maturing of the French nuclear program, additional changes could be seen in the handling of information and in the forms public engagement. Authorities began the practice of announcing technical issues with nuclear plants through online platforms (Topçu, 2008). Regulations were also tightened in La Hague, for instance, where the NGO ACRO measured radiation releases (Topçu, 2008). When a public controversy arose over possible links between leukemia and nuclear energy production or fuel cycle management, the government responded by promoting public discussion (Topçu, 2008). A commission was established and found that the link was unlikely (Institut de Radioprotection et de Suret Nucleaire [IRSN], n.d.).

16. Partitioning and transmutation (P&T) entail the separation of minor actinides and long-lived nuclides from high-level waste, followed by the conversion of segregated elements into other nuclides to minimize the hazard and volume of long-lived waste (Knebel, 2009; Tsinghua University, 2010). P&T decreases the lifetime of the waste (Communications with B. Barre, 2016).

Box 5-1

Nuclear Waste Research and Site Selection

Assessments for waste siting between 1987 and 1990 encountered local opposition, prompting the French government to institute a 1-year moratorium on site evaluations (Lehtonen, 2015). In 1993, the French government encouraged regions to volunteer to host an underground research laboratory (Lehtonen, 2015). The following year, the government commenced assessments at four locations—Gard, Meuse, Vienne, and Haute-Marne—from among the 30 regions that indicated interest. In 1996, the candidacy of Meuse and Haute-Marne sites was combined as one region—Bure—and, in 1998, it was chosen as the site for the laboratory. The other two regions were eliminated largely on the basis of scientific reasons and local opposition. Construction of the underground laboratory in Bure began in 2000.

In 2005, the national waste management agency ANDRA concluded that the Bure site was "perfectly appropriate" to host a deep geological repository. During the period from 2005 to 2006, the National Commission on Public Debate organized country-wide discussions on waste management, which informed laws that were established in 2006 (Lehtonen, 2015). In 2009, a proposal was put forward by ANDRA to create a 250 square kilometer geological disposal facility, named Cigeo. The size was reduced to an area of 30 square kilometers, following consultations with geologists and local stakeholders (Lehtonen, 2015, citing various; OECD-NEA, 2014). The proposal was approved in 2010, and, in 2013, a national public debate highlighted interest in inserting an industrial, pilot phase between commissioning and normal operations. In May 2014, ANDRA published its plan to incorporate the requested pilot phase for Cigeo with a modified timetable (Lehtonen, 2015). Plans for Cigeo include the reversible geological disposal of both high-level and long-lived intermediate level radioactive waste that is produced by the French nuclear reactors.

In 1995, another formal step was taken with the passage of the Act of Michael Barnier. Named after the European Commissioner for the Internal Market and Services, this Act indicated that all infrastructural projects (nuclear ones in particular) required "democratic debate" (Interview, 2012). In line with this, the National Commission on Public Debate (CNDP) was created and charged with guaranteeing an airing of public perspectives for all infrastructure projects with the potential for substantial socioeconomic or environmental impacts (Lehtonen, 2010*a*).

Specific to research and industry, an important joint venture was established between Framatome of France and Siemens of Germany to produce a next-in-class Generation III reactor, the European Pressurized Reactor (EPR). This collaboration would serve as the basis for "new builds" of the following decade.

As the century came to a close in 1999, attention was drawn to French nuclear safety testing when a Level 2 event occurred at the French Le Blayais nuclear power plant.[17] A combination of high tide and winds produced flooding that breached sea walls, resulting in a partial loss of the plant's power supply (Interview, 2011; Mattei et al., 2001). Circuit failure for two units led to an automatic shutdown. Diesel generators based at a higher level commenced and maintained power until the primary power supply was restored (Interview, 2011; Mattei, Vial, Liemrsdorf, and Turschmann, 2001). This incident spurred an industry-wide re-examination of the safety mechanisms associated with multiple equipment failures (Mattei et al., 2001)—a theme that would be revisited following the events in Fukushima, Japan, in 2011.

The year 1999 was mixed in terms of gains and losses for the French nuclear industry. The 58th PWR of the Messmer Plan (and final N4 unit, Civaux 2) was connected to the grid (IAEA, 2016). Meanwhile, the Green Party's emergence as part of the new Government Coalition precipitated the shutdown of the Superphenix FBR and fast neutron reactor development (Communications with B. Barre, 2016). The closure of the Superphenix was a non-negotiable condition of the Green Party's participation (Communications with B. Barre, 2016).

During this period, parliamentary debate on energy also reaffirmed three fundamentals of French energy policy: namely the security of supply, environmental respect, and appropriate attention to radioactive waste management (WNA, n.d.). Public discussions on nuclear energy further acknowledged that natural gas had become increasingly competitive with technology developments in combined cycle gas turbines (IEA, 2000*a*). However, natural gas did not hold an economic advantage over nuclear power and was subject to volatile pricing (IEA, 2000*a*; WNA, n.d.). IEA/NEA reviewers confirmed the competitiveness of nuclear energy for France relative to gas and coal (IEA, 2000*a*, citing IEA 1998). This same analysis also found that France was the only IEA country where nuclear power was cheaper than gas as a base-load option (IEA, 2000*a*, citing IEA 1998).

New Challenges and Directions: 2000–The Present

If one were taking stock of France's energy profile at the turn of the century, he or she would find France had the second-largest electricity market among IEA

17. Nuclear incident rankings are discussed in the Appendix.

European countries, after top-ranked Germany (IEA, 2000*a*). France was also the largest electricity exporter among the IEA countries (IEA, 2000*a*).

Specific to France's power sector, a wholesale electricity market was put in place in 2001, based on European Union (EU) Directives 96/92/EC and 2003/54/EC (IEA, 2010*a*). In line with liberalization, the Commission for Energy Regulation (CRE) was set up with devolved oversight powers to ensure the smooth and efficient management of practices related to competition (Laws of February 10, 2000 and January 3, 2003; IEA, 2010*a*). EDF was also converted to a limited liability company in 2004, with the Act of August 9, 2004 stipulating that the French State would continue to hold at least 70% of the capital and voting rights (IEA, 2010*a*).[18] Over the course of the decade, accounting and legal unbundling of the transmission and distribution system operators for the grid was completed, with ownership of the system operators remaining mostly in the hands of EDF. The change aligned with international trends in privatization (see Chapter 3). Liberalization of the French electricity market was completed in 2007. Nonetheless, the power sector remained quite concentrated publicly, with EDF accounting for 88% of the supply (IEA, 2010*a*).

As liberalization progressed, the existing nuclear fleet of reactors began to undergo license renewals in order to extend operational life spans for plants to 40 years. The 900 MW_e units all cleared their extension reviews in 2002 (WNA, n.d.), and, in 2006, the 1300 MW_e series was cleared, contingent upon minor modifications between 2005 and 2014 (WNA, n.d.). EDF also indicated that it was evaluating the prospect of 60-year life spans for the NPPs (WNA, n.d.).

Public ownership in the nuclear industry was reorganized under newly-formed AREVA by consolidating Framatome, COGEMA, ANP, Technicatome, AREVA T&D, and FCI (IEA, 2004*a*). This initiative integrated nuclear reactor manufacturing, development, and services with the nuclear fuel cycle. In 2003, the partnership of now AREVA and Siemens won a contract to build the first third-generation EPR in Finland (IEA, 2004*a*). EDF would later lead in the installation of an EPR in France after a decree officially named Flamanville the project site in 2007 (Interview, 2012).

In the summer of 2003, a vulnerability of the nuclear infrastructure was highlighted as Europe experienced an extreme heat wave. The majority of nuclear plants were cooled with water from hotter-than-usual rivers, and 19 French NPPs required a ramping down of operations. Energy shortages occurred

18. EDF (France's main electricity company and manager of its nuclear power facilities) was a vertically integrated power company and the largest power company in Europe (IEA, 2000*a*). Since 1969, government oversight of EDF has been carried out through multiyear agreements (*contrats de plan, contrats d'enterprise*) (IEA, 2000*a*). With each consecutive agreement, EDF gained greater autonomy, based on its attainment of performance goals like price reductions (IEA, 2000*a*).

during this period. Although these shortages could not be wholly attributed to the French nuclear power scale-back, nonetheless, thousands of heat-related deaths occurred that summer in France and Italy (Kenward, 2011).

The following winter, thousands of people gathered in Paris to oppose the development of new EPR projects (ABC News, 2004). Such EPRs would be the first, new nuclear builds in Western Europe since 1991 (Schneider and Froggatt, 2012). Opposition would continue, with groups challenging the transport of nuclear waste, raising concerns about seismic impacts, and attempting to block the operation of the aging Fessenheim plant (Euronews, 2012; Interviews, 2011–12; Reuters, 2014). Against a backdrop of continued protests, a series of public debates on energy was led by the CNDP in 2004–05, followed by still more debates on the construction of the EPR, the siting of a high-voltage line for the EPR, and waste management.[19]

In 2005, an energy law was passed setting targets for renewable energy technology, carbon dioxide emissions, and efficiency (IEA, 2010*a*). Renewables were set to equal 10% of total primary energy and 21% of gross electricity by 2010 (IEA, 2010*a*). CO_2 emission reductions were targeted for reductions from the base year of 1990 by 75% in 2050 (IEA, 2010*a*). Leaving choices about nuclear energy open, the law emphasized energy security, the diversification of energy imports, and energy savings (IEA, 2010*a*; Interview, 2012).

Following a review of waste management research and public debate (Box 5-1), two laws were passed in 2006 centered on safety, transparency, and the management of radioactive waste. Continuing research for long-term geological disposal, transmutation, and interim storage was mandated (Lehtonen, 2010*b*) with reversible geological disposal stipulated as the reference point (Lehtonen, 2010*b*, citing ANDRA, 2010).[20] An independent safety regulatory agency, the National Agency for Nuclear Safety (ASN), was also created to replace the General Directorate for Nuclear Safety and Radiation Protection.

In the area of research, construction began at the Cadarache scientific research center in 2007 for the International Thermonuclear Experimental Reactor (ITER) complex. With the ITER, 35 countries would partner to develop

19. While these debates reflected a step toward transparency and participation, the debates (particularly those on the EPR and waste) have been criticized for being devoid of any impact on decision-making. According to some, the events were used to legitimize predetermined decisions (Lehtonen, 2010*a*, citing Lhomme, 2006; Schneider, 2008*b*). Nonetheless, critics acknowledge that the CNDP was successful in resisting pressures from vested interests. The agency gained credibility for altering the rhetoric and for withstanding pressure on how the public discourse should be managed (Lehtonen, 2010*b*, citing GC, 2006 and Chateauraynaud et al., 2005).

20. The new laws provided clarification, distinguishing recoverability from reversibility, with the former term corresponding to the technical capacity to recover waste, whereas the latter term refers to the prospect of altering and/or reversing decisions (Lehtonen, 2010*b*).

the world's largest magnetic fusion device (ITER, n.d.). In conjunction with related work on ASTRID and the Gen IV, the project reflects key international engagement in the area of nuclear energy.[21] If successful, ITER would be the first, commercial-scale fusion power facility (Bouveret et al., 2013).

Looking beyond international research partnerships, the French nuclear industry was regularly promoted by former President Sarkozy for opportunities in trade and technology development. Between 2007 and 2012, Sarkozy signed memoranda initiating talks on nuclear technology trade with Libya, China, and the United Arab Emirates, among other countries (Bouveret et al., 2013).

In 2008, an energy and climate package was adopted that included EU-based targets for GHG reductions, efficiency, and RETs (IEA, 2010*a*).[22] French climate plans were also updated to strengthen actions, in addition to incorporating measures that reflected insights from public environmental roundtable discussions (e.g., Grenelle de l'Environnment) (IEA, 2010*a*). With the passage of the Grenelle I and II Acts in 2009–10, the French energy sector was encouraged to diversify with a variety of renewables through feed-in tariffs (FIT), tendering schemes, income tax credits, and tax exemptions (IEA, 2010*a*). Energy and environmental policy that developed during this period reflected a major shift from earlier years in how French priorities were discussed and implemented.

On March 11, 2011, the Fukushima accident in Japan renewed questions about nuclear power. As with Chernobyl, radiation measurements were made in France. Communications issued by the French government demonstrated marked improvement over those made during the earlier disaster (Interview, 2012). Later that year, the French agency on nuclear radiation and safety, the Institut de Radioprotection et de Surete Nucleaire, released a 500-page report that surveyed improvement measures proposed by operators. The report indicated that each French nuclear power facility needed an additional safety layer for cooling and power functionality, such as with an independent and externally positioned diesel generator (Boselli, 2011; Communications with B. Barre, 2016). The head of the IRSN also noted that the Bugey, Fessenheim, and Civaux plants could be strengthened against seismicity, while the Fessenheim, Chinon, Craus, Saint-Laurent, and Tricastin plants should have additional measures put in place for flooding (Boselli, 2011; Communications with B. Barre, 2016).

21. ASTRID is an industrial demonstration project of a sodium-cooled FBR (~600 MW_e) (Devictor, 2015).

22. According to this, France must meet a 14% reduction target in GHGs in sectors outside of the EU-Emission Trading System and increase its share of RETs in total final energy consumption to 23% by 2020 with a 10% target in the transport sector (IEA, 2010*a*).

In early 2012, the French Cour des Comptes, a court of audit and accountability, reported on the cost and strategy of the French nuclear industry and fleet. Among its findings, the court indicated that planned replacement of the current French fleet could not be adequately accomplished, based on existing assumptions about time and cost (Cour des Comptes, 2012). Lifetime extensions of the nuclear fleet or a new energy strategy were required (Cour des Comptes, 2012).

More recently, the French energy playing field has been under new scrutiny, with President Hollande's pledge to shift the French energy mix from 75% nuclear-based electricity to 50%. Following a period of public debate, a law was passed in 2015 codifying the objectives to be met by 2025 with a focus on increased renewable energy, efficiency, and savings (WNA, n.d.). The law caps the total for installed nuclear capacity at the existing level, requiring that old plants must be taken offline as new ones are brought online.

Specific to the nuclear industry, EDF and AREVA have been besieged by financial pressures in recent years in tandem with diminished political support, liberalization, and shifts in the investment environment. Moreover, performance in the construction of two first-in-class EPRs has experienced substantial time and cost overruns (Schneider and Froggatt, 2015).[23] To address some of the economic issues with consolidation and restructuring, EDF and AREVA have agreed on a deal in which EDF will acquire a majority ownership position in AREVA's reactor business.[24] The two companies also recently developed a new model of EPR—EPR NM—which has simplified design and lowered expected costs (WNA, n.d.). EDF indicates that the revised design will be the model for fleet replacements in the 2020s (WNA, n.d.). EDF, AREVA, and CEA also announced a new partnership in March 2016: the French Nuclear Platform. This aims to improve the efficacy of member performance with medium- and long-term strategies for industry (WNA, n.d.).

All told, the French power sector underwent a striking change over the past four and a half decades. The share of nuclear power rose from 4% of total electricity in 1970 to 65% in 15 years, finishing out the period of this study at 76%

23. Issues in fabrication and staffing are among the challenges (Interviews, 2012; Marignac, 2015*a* and 2015*b*), with both projects overshooting deadlines by a number of years and with current cost overages on the order of 2–3 times greater than originally projected. The Flamanville 3 plant in France was originally estimated to cost €3.3 billion for a completion date in 2012. It is now estimated to have first power in 2018 at a cost of €10.5 billion (WNA, n.d.). The Olkiluoto plant in Finland was planned to begin commercial operation by mid-2009 at €3.2 billion. As of 2016, it is not yet operational and its reported cost is €8.5 billion (WNA, n.d.). In 2015, the option for a second EPR to be constructed in Finland was cancelled.

24. See the section "Industrial Development" for details.

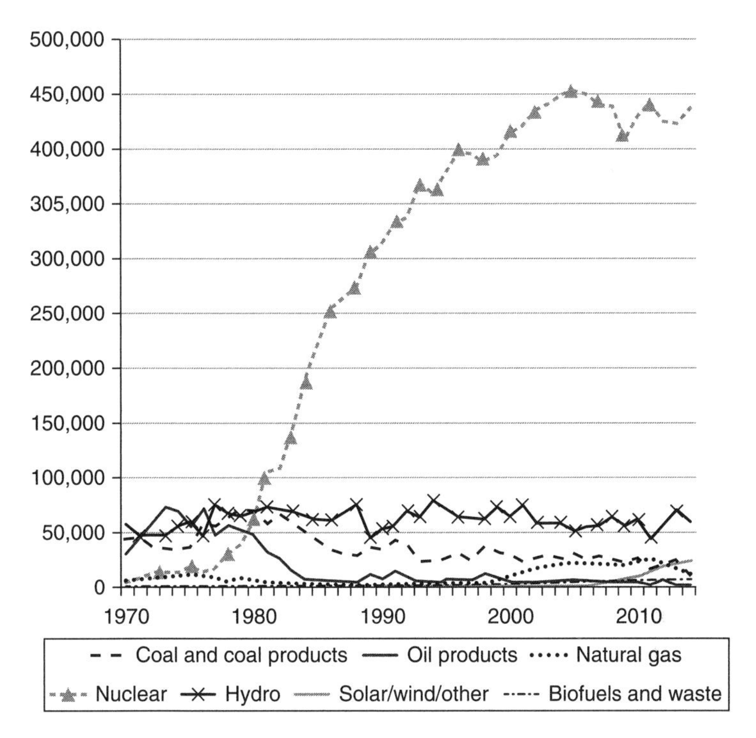

Figure 5-4 French power generation by fuel type (GWh).
SOURCE: IEA data (n.d.).

(Figure 5-1 and 5-4). Targeted policies that enabled this transition were often not explicit but rather embodied in the actions of state agents that expanded infrastructure and adapted practices to meet changing conditions. The following section highlights dimensions of the change.

EXPLAINING CHANGE

The French nuclear transition can be described as a robust and front-loaded shift for the period under consideration. The types of sectoral intervention and readiness are described next.

Mode of Intervention

In terms of the mode of intervention, as described in Chapter 3, the French nuclear change was driven and deployed from the top. Public actors, namely the CEA, EDF, and AREVA (i.e., Framatome and COGEMA), managed the transition. Two of the three actors were partly privatized over the period. The policy mix largely consisted of deployed actions with increased use of mandates. Civil society brought scrutiny to the changes and to discussion about waste management, but did not appear to be a principal driver of the nuclear transition.

Model of Readiness

A combination of the expansionary and reconstitutive models of readiness, as outlined in Chapter 3, characterizes the French nuclear transition. The transition is expansionary since many of the constituent elements of the model were extended from an existing nuclear pathway. Repurposing was also evident with the integration and realignment of institutions and resources.

France's modern nuclear transition was based on expertise and institutions that were developed for the defense industry and research organizations' use of nuclear power. With the transition, the nuclear sector was expanded to include civilian power generation. Industry and infrastructure for the fuel cycle were retooled from the defense–research apparatus and then enlarged. Alongside industry growth, markets for the fuel supply and reactors increased over time. In terms of the fuel supply, resources were indigenously developed in the early years of the transition with uranium and mixed oxide (MOX).[25] Over time, uranium would be increasingly sourced from external markets, whereas MOX remained indigenously derived.

INNOVATIONS AND ADAPTATIONS

Three primary types of innovation and adaptation played a role in France's energy transition: *reactor technology, development of an industrial-scale fuel cycle, and power market/grid system changes.*

Reactor Technology

Design standardization is a hallmark of the French nuclear program, achieving unusual uniformity across the 900, 1300, and 1450 reactor series. In addition to fostering economies of scale, standardization enabled gains in learning and knowledge transfer across operators and reactor models (Interviews, 2012). This uniformity provided greater predictability in the scale-up and utilization of the nuclear fleet, yet it also brought inherent risk if a generic technical problem

25. Mixed oxide is a combination of uranium and plutonium. For more discussion of the fuel and fuel cycle, see the "Technology Primer" in the Appendix.

were to appear across a model series. Combined with management practices, this standardization brought structure to user (operator) innovations.

Two, other key changes occurred in reactor design: load-following and the use of MOX fuel (see Appendix "Primer"). Specific to load-following, the practice was introduced to provide management flexibility with overcapacity of supply. Less efficient and less absorptive "gray" control rods were added to the fuel assembly of some PWRs of the 900 MW_e design to adjust the reactor and control the power distribution, thereby allowing sustained variation in power output (Interview, 2012; WNA, n.d.). Today, new plant designs incorporate these types of features (WNA, n.d.), but at the time they were novel. In addition to load-following, roughly 21 of the 900 MW_e PWRs were also adapted to operate with an enriched uranium-MOX blend that included 30% MOX (AREVA, 2016; Commissariat a l'Energie Atomique [CEA], 2010). For this to occur, plant modifications required more control rod clusters (Interview, 2012). Two plants are also fueled with re-enriched, reprocessed uranium (Communications with B. Barre, 2016).

Additional reactor changes occurred in the shift from the Westinghouse license to the French design, the growth in plant size, and adaptations relating to safety and efficiency, among others. Most noticeable among these was the change in size from the 900 MW_e to the 1650 MW_e reactor that is currently being built with Flamanville 3.

Development of an Industrial-Scale Fuel Cycle

The development of a fully integrated and industrial-scale fuel cycle represented a major step in the French energy transition. Based on a closed cycle with reprocessing of spent fuel and recycling of plutonium and reprocessed uranium (see Appendix "Primer"), the fuel cycle leveraged facilities and expertise that had previously served only defense purposes (Albright et al., 1992; Interviews, 2011–12). The Marcoule reprocessing plant, for instance, was adapted to separate plutonium from spent fuel rods for both military and civilian program needs (Marsh, 1985). Such adaptations enabled France to become a commercial reprocessing hub for international clients in the nuclear sector as least as far back as the 1980s (Schneider and Marignac, 2008).[26] Today, France's nuclear fuel market serves domestic needs and the needs of international clients in areas including conversion, enrichment, fuel fabrication, and reprocessing

26. The Hague Reprocessing Plant (UP3) that was opened in 1989 by state-owned COGEMA was almost entirely "pre-financed" by foreign clients (Schneider and Marignac, 2008). Germany and Japan collectively covered about 84% of service contracts for the UP3's first 10 years of service (Schneider and Marignac, 2008).

(IEA, 2005 Interviews, 2011–12). Underpinning such advances were process redesign and institutional modifications. The AREVA Group currently oversees the front- and back-end of the fuel cycle by providing engineering and services that include prospecting and mining, conversion, enrichment, and fuel fabrication, plus reprocessing and packaging (IEA, 2010*a*). ANDRA manages the radioactive waste management and disposal operations (IEA, 2010*a*).

Power Market and Grid System Changes

Similar to what is seen in some of the other cases, the liberalization of the power market in France had some bearing on the energy system. The turn to competition, increase in new entrants, and decrease in centralized control were some of the hallmarks. State-owned energy company EDF unbundled a number of its functions, spinning off grid oversight functions to the Transmission System Operator Reseau de Transport d'Electricite (RTE).[27] Competition with liberalization brought the opportunity for consumers to purchase power from multiple providers, meaning that the EDF no longer held a full monopoly over the portfolio of options. In conjunction with load-following and technology adaptations mentioned earlier, altered practices in the power market and grid management were also necessary, beginning in the 1980s.

KEY DRIVERS AND BARRIERS

Major determinants to France's energy transformation are outlined here. These reflect drivers and barriers of significance that were identified in case analysis and historical record review, as well as in interviews and data trends.

Reduction of Oil Import Dependence and the Deleterious Impacts on the Balance of Payments/Self-Sufficiency

The *Messmer Plan* was launched in the wake of the first oil shock, and it has been actively supported or, at minimum, passively allowed to continue under roughly 16 prime ministers and 6 presidents to date (Gouvernement, n.d.). This was driven at least in part by an aim to reduce oil imports and their deleterious effects on the balance of payments in a country that appeared to possess limited natural energy resources and had undergone major postwar industrialization.

27. RTE was formed as an internal division of EDF in 2000 with independent finance, management, and accounts (RTE, n.d.).

While import dependence and the change in the energy mix are covered elsewhere in this book, the following elaborates on the import aspects of the balance of payments.

Figures 5-5 and 5-6 illustrate import costs of petroleum in nominal dollars for France and as a share of all French import costs. While both indicators are affected by influences beyond the scope of this study, they do reveal a number of relevant insights. First, import costs of petroleum were mostly flat for much of the 1990s and then sharply increased in the past decade, not unlike the flux in international oil pricing. Second, the share of total, French import costs for petroleum was high during the period of the two oil shocks, yet has been substantially lower since then. This underscores that (although oil import costs remain a consideration for the balance of payments), the nuclear program has allowed France to minimize its vulnerability to oil price flux in the power sector.

Concentrated Decision-Making, Limited Public Participation, and Centralized Planning

Historically, political and economic bases of power in France have been highly centralized in comparison to most member countries of the Organization for Economic Cooperation and Development (IEA, 2010*a*). In line with this,

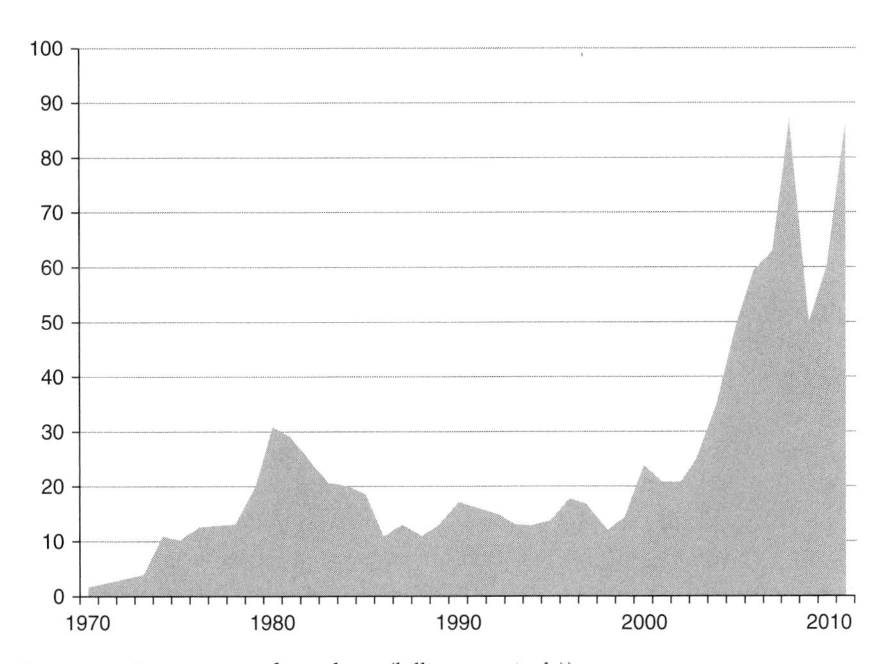

Figure 5-5 Import costs of petroleum (billion, nominal $).
SOURCE: UN Comtrade data, SITC Rev 1, Petroleum and Petroleum Products (n.d.).

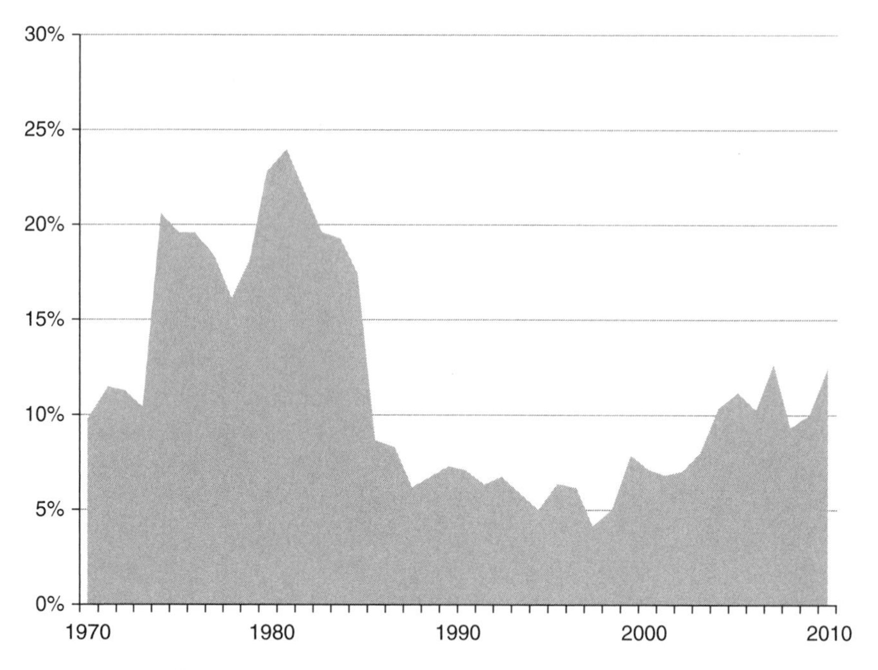

Figure 5-6 Petroleum as % total import costs (trade balance).
SOURCE: UN Comtrade data, SITC Rev 1, Petroleum and Petroleum Products (n.d.).

people who examine the French nuclear energy trajectory may point to the limited number of institutions and actors making decisions as a factor that enabled the rapid scale-up of the French nuclear program (Finon and Starpoli, 2001; Grubler, 2010; Jasper, 1992).

For the period of this study, the government worked with and through the CEA, EDF, Framatome/COGEMA/AREVA, and other organizations, like Alstom, to deploy the nuclear program in conjunction with advice from the PEON Commission in the early years. More recently, the Nuclear Policy Council has supported the effort. Bridging these institutions was a small circle of expert managers, often drawn from the Corp des Mines and Corps des Ponts (Grubler, 2010; Interviews 2011–2012; Schneider, 2009). This network and the PEON Commission both appear to have been central nodes for policy development, implementation, and continuity (Interviews, 2011–12).

French planning has also been characterized as highly centralized (Golay et al., 1977). For the period 1946–2006, the Commissariat General du Plan (CGP) was the principal institution behind public investment strategy with its use of 5-year plans that distilled objectives to guide social and economic development (Lindberg, 1977). One example of this structured prioritization was the development of France's TGV high-speed rail for public transportation.

In more recent years, the 5-year plans have ceased to exist, and the CGP has since been replaced by the Strategic Analysis Center, which serves in a similar capacity (Interview, 2012). However, old planning practices continue in a new form. Today, the government requires Pluri-annual Investment Plans (PPIs) by companies like EDF and AREVA to evaluate investment options, ensuring they are in line with desired future development goals (IEA, 2010*a*).[28] For electricity (and, by extension, nuclear energy), the government may open a tendering auction if the goals of energy security, for example, are not met (IEA, 2010*a*).

Robust Energy Policy Backed by Political Support

Throughout the period studied, nuclear energy policy in France has been robust and principally embodied in the actions of state agents who are supported by strong political backing (Interviews, 2011–12). While the Socialist Party in the late 1970s questioned the nuclear plan during the campaign season, party members under Socialist President Mitterand ultimately backed the nuclear program's continuation. The Socialist-Green parties of the late 1990s also challenged aspects of the program, which led to the closure of the Super Phenix FBR in 1997/1998. These mid-course corrections, however, were comparatively small relative to the overall program rollout.

It is important to also remember that an overlap exists between French military and civilian nuclear programs (Bouveret et al., 2013; Hecht, 2009). It is possible that some R&D expenditure for the civilian nuclear program could be covered by the military program, akin to a subsidy, although no evidence was seen over the course of this study. A primary nexus also may be found in the CEA, which was initially created to oversee both lines of development and which continues to do so today (CEA, n.d.; Schneider, 2009). While analysis of the military intricacies of French nuclear power goes well beyond the scope of this research, it is critical to underscore that senior-level decision-making for the programs often accounted for the duality of purpose. Choices, for instance, on reprocessing of spent fuel, use of FBRs, and production of plutonium intersected both civilian and military domains. Limitations from international agreements on exports and transport of nuclear fuel also mattered. In more concrete terms, the creation of the industrial fuel cycle under COGEMA gained an early advantage by absorbing infrastructure, materials, personnel, and related expertise from the military program. To ignore the

28. This is currently to be done by all EU countries, as indicated in the EU Electricity Directive.

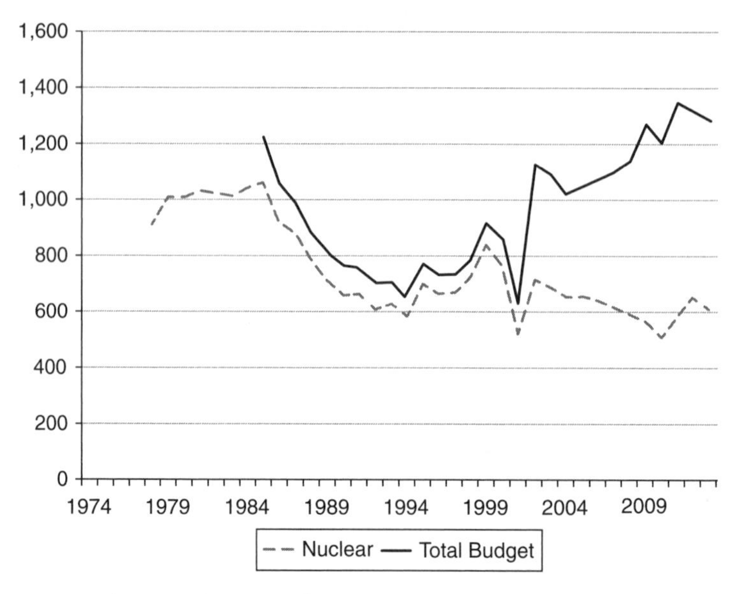

Figure 5-7 French RD&D (2014 $ million, ppp).
SOURCE: IEA (n.d.). Data for 1974–78 exist but were not collected.

military dimension would deny dimensions like synergies, spillover effects, and feedback loops.[29]

Another way to consider support is to look at R&D spending. Figure 5-7 shows the dominance of nuclear energy in overall expenditure until roughly the point in time at which liberalization occurred. This highlights long-term continuity of policy in action. There are, however, recent signs of discontinuity in policy, beginning under former President Hollande. How the nuclear industry adapts under the 2015 law, which caps nuclear capacity as it stands, is a key question. How also will the French energy path evolve differently under President Macron's administration remains to be seen.

National Independence and Leadership in Science and Technology

One cannot discuss French nuclear development without seeing independence and leadership in science, technology, and international politics as important drivers. Independence can be traced at least as far back as early energy policy

29. The nuclear industries in other countries, including the United States, have also benefitted from defense contracts and other forms of crosscutting support.

adopted around the time of World War I. Moreover, experiences with the Suez Crisis and in President de Gaulle's choice to remove France from NATO are emblematic of a national independence streak that also motivated the adoption of nuclear energy and shielded the industry from some criticism (Interviews, 2011–12; Sovacool and Valentine, 2012).

Complementing the interest in independence was France's historical leadership in physics, which provided important cultural traction for the nuclear program. Breakthroughs by French scientists Becquerel, the Curie-Joliot family of scientists, as well as Goldschmidt, among others, were familiar reference points for the public and political decision-makers' view of contemporary nuclear developments. Many French viewed the nation's postwar nuclear program as a source of national pride and a pursuit mirroring that of the U.S. Manhattan Project or the Space Race by the United States and Soviet Union (Hecht, 2009; Interviews, 2011–12). If these elements are considered in conjunction with co-benefits of the expert network described earlier, they present an explanation of why at least some people were willing to defer to experts (Sovacool and Valentine, 2012) and why a technologically complex course of action was seen as reasonable.

Public Support and Opposition

Public perception of nuclear energy in France is a phenomenon that continues to be studied (Chausssade, 1990; Fourquet and Pratviel, 2013; Hecht, 2001; Jasper, 1988; Kitschelt, 1986; Nelkin and Pollak, 1980, 1982; Topçu, 2007, 2008). Opposition assumed numerous forms on the basis of safety as well as discontent over the process, transparency, regional equity, uncertainty with respect to waste, and the extent of local engagement in decision-making (Interviews, 2011–12). Despite protests and other signs of discord, surveys indicate that long-term support also existed (Fourquet and Pratviel, 2013). According to some who favor the nuclear program, reporting by the press contributed incorrectly at times to heightened opposition by providing facts without adequate context (Interviews, 2011–12). Interestingly, the presence of radical opposition during different periods may have had the unintended consequence of insulating the program (Sovacool and Valentine, 2012).

Financing

The way the nuclear program was financed could be a discrete, companion study to the current one. According to some interviewees, the civilian nuclear energy program was not subsidized (2011). For companies like EDF, which carried substantial program-related debt and began to see a profit in 1985 from sales of electricity exports (Jeffery, 1986), self-financing was an impediment to the adoption of new nuclear plants (Boulin and Boiteux, 2000; Interviews, 2011). According to one source, public regulations also did not allow EDF to raise its electricity tariff in order to balance its accounts (Jeffrey, 1986, citing *Business Week*), which would have complicated the economic sustainability of the nuclear program's rollout. EDF secured external (i.e., non-state) funds; however, some point out that the company's favorable status as a state-backed enterprise would have improved the conditions of the loans (Interviews, 2011–12).

Returning to the subject of subsidies, it is reasonable to say that the military program was paid for by the state and, by extension, its taxpayers. Gains in research, expertise, and equipment which overlapped the defense and civilian nuclear programs likely reduced some of the financing challenges for both. Nonetheless, EDF and AREVA have also paid taxes and dividends to the state for their nuclear activities and have absorbed related costs (Communications with B. Barre, 2016; Jeffery, 1986; Stoffaes, 2016), thereby leaving a highly mixed support picture.

Expert Pool

A continuing challenge for nuclear development is the need for a sustained pool of experts. While France's schools are well known for training top science and technology experts, and the nuclear program was deemed by many to be the French equivalent to the US-Soviet Space Race, the contemporary pool of available nuclear experts has, at different points in time, been seen as thinning (Interviews, 2011–12). In line with this, the French government has instituted a program to educate French and international students in nuclear energy (Interviews, 2011–12).

Control of the Nuclear Fuel Supply

Finally, control of the nuclear fuel supply served as both a barrier to and driver of the nuclear transition. Until the 1970s, autonomy and flexibility in nuclear power decision-making was encumbered by incomplete management of the nuclear fuel cycle. Restrictions on and the control of fissile material sourced from the United States and Canada, for example, drove France to develop its own enrichment technology (Interview, 2012). Until the Eurodif plant came online in the late 1970s, enriched uranium was acquired from the United States or the USSR (Lindberg, 1977).

Table 5-2. CHANGE IN PROJECT COSTS/MW

Time (Plant)	Reactor Cost/MW
1978 (Fessenheim)	€1.07 ($1.4) billion$_{2010}$/MW
2000 (Chooz 1 and 2)	€2.06 ($2.6) billion $_{2010}$/MW
2002 (Civaux)	€1.37 ($1.7) billion $_{2010}$/MW
Average for 58 plants	€1.25 ($1.6) billion $_{2010}$/MW

SOURCE: Cour des Comptes (2012).

NOTE: 2010 average exchange rate for $/Euro = 0.785 (IRS, 2016).

DEVELOPMENTS IN COST, SOCIETAL ACCEPTANCE, AND INDUSTRY

Costs

Cost information about the French nuclear program became available in the late 1990s, with a number of studies commissioned for Prime Minister Jospin and Parliament (Bataille and Gailley, 1999; Charpin et al., 2000; Girard et al., 2000; see Grubler, 2010, for a discussion of these). Findings indicated that nuclear energy was competitive for France (Bataille and Gailley, 1999) and that the country should maintain its use of fuel reprocessing (Grubler, 2010, citing Charpin et al., 2000).[30]

In 2011–12, the French Cour des Comptes assessed the costs and the status of the French nuclear program (2012). Its findings, published in January 2012, found the cost of the program to be €121 billion (€2010 or $154 billion) for facility costs for 58 PWR reactors, excluding the Superphenix FBR.[31] This amounted to a full economic cost of €59.5 ($75.8) per MWh, the least cost source of electricity in France except for hydropower (Communications with B. Barre, 2016).

The report also indicated that initial construction costs (including engineering) had nearly doubled between 1978 and 2000, whereas the average cost for the plants across the 58 PWRs was roughly 17% more than the original cost of the first plant (Table 5-2). According to the report, the increase was due primarily to safety requirements (Cour des Comptes, 2012).

30. Cost of electricity analysis by the IEA around the same period (1998) found that new nuclear energy, based on discount rates of 5% and 10% and prevailing market assumptions, was competitive for France.

31. Within this, €96 billion ($122 billion) equates to: (1) the construction costs of initial investment between 1973 and 2002 (i.e., overnight costs) of €83 billion ($106 billion), plus (2) interest during construction, estimated at €13 billion ($17 billion) (Cour des Comptes, 2012). Note: the 2010 average exchange rate for $/€ = 0.785 (IRS, 2016).

A related view of costs may be seen in construction times for nuclear plants. A study of France and the United States by Mark Cooper revealed a considerable increase in the time to completion for nuclear plants in both countries, a finding that is consistent with the results of other studies, such as that of MIT's *Future of Nuclear Power* (Massachusetts Institute of Technology, 2003, 2009).

Arnulf Grubler (2009, 2010) and Charles Komanoff (2010) also identified a cost escalation in the French PWRs, although at a higher level than that found in the Cour des Compte study.[32] When viewed together, the cost increases underscore a limit to the basic learning curve expectations that costs should decrease with cumulative increases in an energy technology over time (Grubler, 2010). With such negative learning, Grubler points to a concept raised by Amory Lovins (1986), and referencing the Bupp-Derian-Komanoff-Taylor (BDKT) hypothesis, which states that technology scale-up can introduce greater system complexity (i.e., load-following, fuel cycle, etc.), thereby leading to cost increases. Here, regulatory encumbrances should not be underestimated (Communications with B. Barre, 2016).

If one were to compare energy research, development, and demonstration spending on nuclear technology for France, the United Kingdom, the United States, and Japan (a number of OECD countries with nuclear power) from 1974 to 2014, one would find that France spent significantly less than did the United States or Japan on nuclear energy for much, if not all, of the period (Figure 5-8). This suggests that France's low-cost, nuclear-backed electricity was not secured at an extraordinary expense in relative terms (or was well managed across civilian and defense purposes).

Today, electricity prices for consumers are low (Commisariat General au Developpment Durable [CGDD], 2015; RTE, 2016). These are, however, expected to rise to account for safety upgrades for NPPs (Schneider and Frogatt, 2016; WNA, n.d.) and for the new EPR that is expected to become operational in the near future.

Turning to the question of who paid, the answer is complicated by the mix of information that is available. EDF managers maintain that EDF was not subsidized (Interviews, 2011) and that it carried significant debt during program rollout, not earning a profit until the mid-1980s. According to one report, half of France's nuclear program cost, excluding construction cost interest, was self-financed by EDF; 8% was invested and discounted by the state in 1981; and the remaining 42% was covered by commercial loans (WNA, n.d.). There

32. The distinction with the Cour de Compte study can be partly attributed to the differences in data. Grubler and Komanoff's costs assessments (as well as Cooper's) are based primarily on Grubler's data, which were developed with educated estimates from the late-1990s governmental studies (2010).

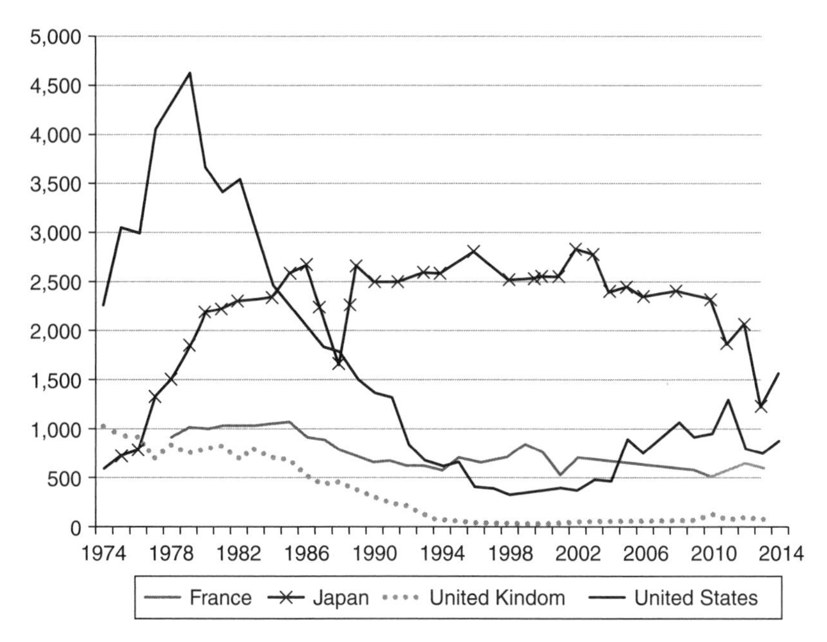

Figure 5-8 Comparative RD&D on nuclear (millions $ 2014, ppp).
SOURCE: IEA (n.d.). Data for France 1974–77 and France/UK 2014 exist but were not collected.

are, as discussed earlier, also some indications that the commercial loans that the EDF secured may have been on highly favorable terms, given its governmentally backed position (Grubler, 2010; Interviews, 2011–12). At some point, there were also indications that a portion of the debt was written off, but the limit on available details leaves the door open for further study. Essentially, rate payers (i.e., electricity consumers) covered a credible portion of the costs for the nuclear energy program. Yet taxpayers also paid, primarily through indirect costs of state-owned enterprises and expenditures for military uses that impacted the civilian program.[33]

Societal Acceptance

French society's view of nuclear energy as a centerpiece of its national energy system and military program is a complex subject with many angles. Gabrielle Hecht's study of the French nuclear development highlighted how the valuing

33. Cost information on the overlap between the civilian and defense-based nuclear programs is not readily available.

national identity as well as excellence in science and technology contributed to positive views of nuclear development (2009). Others, like former President Giscard d'Estaing, also indicated that there was "no choice" (de Malleray, 2011). By contrast, Georges Vendryes, former Adviser to the CEA Administrator General, illuminates the unusual nature of the decision-making in this way:

> [for] forty years the big decisions concerning the development of the French nuclear program are taken by a very restricted group of personalities that occupy key positions in the government or in the top administration of EDF, CEA and the few companies involved in the program. The approach remains unchanged in spite of the change of ministers thanks to the permanence of these personalities that occupy the same position generally for some ten years. (Vendryes, 1986)

A combination of these three perspectives likely captures the nuances of the early transition.

There are, however, signs that a deliberative turn began in 1990 toward a greater degree of engagement and transparency. Examples can be found in the public reporting of nuclear program costs and waste management, as well as in incident alerts. Public debates on policy agendas also occur with more frequency today. Whether they inform decision-making or foster acceptance remains a subject for continued study. Related research on the influence of key groups, including trade unions, scientists, and NGOs, would more fully extend knowledge of this topic.

Industrial Development

The French nuclear industry encompasses electricity, reactors, the fuel supply, and related services. Each has progressed under the transition.

France has for years had some of the least cost power in Europe due in part to plant extensions and likely gains from dual-purpose civilian–military use (CGDD, 2015; Interviews, 2011–12; RTE, 2016; Schneider and Frogatt, 2016, 2015). In the past decade, France also exported up to 70 billion kWh (70 TWh) on an annual net basis (WNA, n.d.). In 2010, the IEA encouraged France to seriously consider expanding its export role in the context of the EU electricity market going forward (IEA, 2010*a*). Given that EU states like Germany have nuclear phase-out plans under way, an opportunity exists for increasing France's electricity exports at the EU level.

Over the years, France also actively engaged in developing and selling its reactor technology. The 900 MW_e reactor design, for example, has been sold in a variety of export markets, including two units to Iran, two units to South Africa

(Koeberg), two units to South Korea (Ulchin), three units to Belgium (Tihange and Doel), and four units to China (Daya Bay and Ling Ao) (Schneider, 2009; WNA, n.d.).

EDF is the largest nuclear plant operator in the world (EDF, n.d.; Schneider and Frogatt, 2015). As of 2014, it had 73 reactors in France and the United Kingdom with a net nuclear capacity of 72.9 GW—the largest in Europe (EDF, n.d.). In recognition of a "green transition," EDF was the first large corporation to issue green bonds (in Euros) in 2013, and it has been in the process of building the lowest carbon-generating fleet among the world's top 10 energy companies through its use of nuclear generation and renewables (EDF, n.d.). With amenable conditions, EDF plans to renovate the fleet with additional safety measures and extend the operating life beyond 40 years at a cost estimated at €55 2013 (EDF, n.d.). As of the end of 2015, EDF had $40.9 billion in debt (Schneider and Forgatt, 2016).

The AREVA group has worked on reactor design and construction, as well as on the nuclear fuel cycle. It is the sole company with a foothold in every part of the fuel cycle (WNA, n.d.). It has operated in 100 countries, with 47,000 employees, and an annual operating revenue of $8 billion (Bouveret et al., 2013). Recently, its world market shares included about 20–30% uranium mining, uranium conversion, and uranium enrichment; and 30–35% for low enriched uranium fuel fabrication (Schneider, 2009). It also has been the largest builder of nuclear power reactors, holding a 20–25% share of the global market for NPP construction and services (Schneider, 2009). As of 2016, AREVA had 102 PWRs supplied or under construction worldwide (AREVA, 2016). It has dominated the back-end fuel cycle, controlling 70–75% of spent fuel reprocessing and 65–70% of MOX fabrication (Schneider, 2009). In 2015, AREVA's reported debt was €6.3 billion (~$6.9 billion) (AREVA, 2016). With current changes under way, its market profile can be expected to evolve. In June 2016, AREVA announced that it would sell a majority stake of AREVA NP to EDF for €2.5 billion ($2.8 billion) and undergo corporate restructuring, establishing a new company that would focus on the nuclear fuel cycle (Schneider and Frogatt, 2016; WNA, n.d.). The French government aims to inject €5 billion into AREVA by the first quarter of 2017 (Schneider and Frogatt, 2016). This appears to recognize the industrial implications of recent policies, but it is not clear if this will align with EU rules.

Overall, the French nuclear industry is experiencing pressures on many sides, including debt, new international market players, policy shifts, and an aging workforce. The industry, which carried some of the early program debt, now is retooling its finance strategy and market approach to effectively perform in a more privatized playing field. The French industry has an abundance of operational knowledge, and, if it adapts well to current pressures, it is positioned to continue as a leader in nuclear technology, fuel, waste management, and low carbon electricity (Interviews, 2011–12) in a global industry that is also undergoing redefinition.

CONCLUSION

French nuclear development for the period since 1970 is a classic case of an energy transition being largely deployed by public actors. Placing essentially full political backing behind its nuclear energy development, the French government and experts mobilized to implement a robust and rapid shift, drawing upon historical momentum and knowledge. Traditions in science and technology, a system of governance with limited public engagement, apt policy support, and the experience of strong public actors all played important roles.

Unique attributes of nuclear technology add to the nuanced nature of this case. Cost increases, rather than decreases, were seen with new projects. Technological complexity, standardization, and the centralized nature of projects also led to a different kind of learning and innovation—rather than user innovations driving incremental change, mostly formal R&D, institutionalized partnerships, and internationally recognized accidents in nuclear energy shaped learning and change.

Ultimately, a small group of French decision-makers and experts rolled out an extraordinary energy transition. The strength of the future, nuclear energy path in France may well ride on the heels of incumbent company adaptations, societal acceptance, integration with renewables, and sunk costs in a playing field in which "newer" options are on the rise.

6

Brazilian Biofuels

Distilling Solutions

Ethyl alcohol is the fuel of the future.

—Henry Ford (*September 20, 1925*)[1]

INTRODUCTION

Worldwide, transportation accounts for roughly a quarter of the total final energy demand and a similar share of energy-based carbon dioxide emissions (IEA, 2016*f*). The transport sector has the most homogenous of fuel mixes, with petroleum-based products accounting for roughly 95% of the overall final share (Kahn Ribeiro et al., 2012). Biofuels and other options, like electric vehicles, have the potential to displace a notable portion of petroleum and CO_2 emissions in the transport sector. Global use of ethanol, the most widely used among biofuels, has grown significantly in recent years. Between 2000 and 2010 alone, ethanol utilization increased 350% worldwide, with trade increasing by a factor of 5 and usage equaling 74 billion liters in 2010 (Valdes, 2011).

This chapter examines the underlying roots of the biofuels transition in Brazil. Two micro-shifts—one that is government-led and a second that is industry-led—are evaluated, demonstrating how a new, energy market and industry can develop at a national scale through the retooling of existing

1. Henry Ford's comment in the *New York Times*, September 20, 1925, was referenced by Kovarik (1998). Note: The Ford Model A car (1896) used pure ethanol (Banco Nacional de Desenvolvimento Economico e Social [BNDES] and Centro de Gestao de Estudos Estrategicos [CGEE], 2008).

industries and infrastructure. Insights on policy inflections, market longevity, and dual-use technology are also covered.

PROFILE FOR BRAZILIAN BIOFUELS

Brazil is the historical leader in biofuels and the only country to substantially alter its automotive fuel mix with ethanol, shifting from 1% in 1970 to 34% in 2014 (see the section entitled "Modern Transition" later in this chapter). Ranked sixth globally for its population of roughly 206 million people and eighth for its economy of $3.1 trillion in mid-2016 (CIA, n.d.), Brazil has been a leading pioneer in the production and export of ethanol, its principal biofuel. In 2015, Brazilian ethanol equaled 28% of the global supply (Renewable Fuel Association [RFA], 2016). The country is known for having the lowest production costs of ethanol (Goldemberg, 2008; Shapouri, and Salassi, 2006; Valor International, 2014). Brazil also has a unique distribution network of more than 35,000 fuel stations supplying the renewable fuel (Agência Nacional do Petróleo, Gás Natural e Biocombustíveis, 2008).

Today, one can find the largest fleet of flex fuel vehicles in Brazil. As of mid-2015, cumulative sales of light-duty, flex fuel vehicles were estimated at 25.5 million units (Rato, 2015). These vehicles can be powered by any mix of ethanol and gasoline, so Brazilian drivers are able to choose their preferred fuel mix—gasohol (a gasoline-ethanol blend), pure ethanol, or others—based on preferences that may include environmental, cost, or other aims. In 2014, Brazilian drivers consumed roughly 13,008 thousand tons of oil equivalent (TOE) of ethanol, an increase by a factor of 132 since 1970 (MME, 2016).

BACKGROUND

As far back as 19th century inventors, one can observe international developments in biofuels and related engine technology. In 1826, for instance, Samuel Morey developed an engine that was powered by ethanol and turpentine (Fuel Testers, n.d). Among his contemporaries, Nicholas Otto utilized ethyl alcohol in an early version of the internal combustion engine (see Kovarik, 1998), and Rudolph Diesel used peanut oil to fuel his first engine (Pacific Biodiesel, n.d).

Specific to Brazil, early signs of ethanol development were evident at the beginning of the 20th century (Figure 6-1). In 1903, the First National Congress on Industrial Applications of Alcohol recommended that infrastructure be developed to promote alcohol production and use. By 1931, a 5% ethanol blending requirement was already in place for imported oil, alongside guidelines for transport and commercialization. Two years later, the federal government established an oversight body, the Institute of Sugar and Alcohol (IAA).

Box 6-1

Early R&D in the Brazilian Sugar Industry

The Brazilian sugar industry had fairly advanced research and development (R&D) centers in place well before the modern biofuels transition emerged. The Cooperative of Sugarcane, Sugar, and Alcohol Producers in the State of São Paulo (Copersucar), for instance, was created in 1959 to drive development in sugar, sugarcane, and ethanol technologies by addressing the needs of the larger, southern sugar industry actors that were not met by the Institute of Sugar and Alcohol (IAA). Copersucar's research arm, the Center for Sugarcane Technology (CTC), was established 10 years later to conduct studies on plant breeding. The research was extended to include all agricultural and industrial activity. In 1971, the federal government created a public-sector counterpart to the CTC, the National Program for the Improvement of Sugarcane (Planalsucar) under the IAA, to conduct R&D on genetic improvement of sugarcane varieties, with experimental stations throughout the main sugar and ethanol-producing states. While some competition may have existed between the R&D entities in seed breeding, the two entities differed in their level of focus, with the CTC concentrated mainly on production in the state of São Paulo and Planalsucar focused on breeding areas throughout the country.

Figure 6-1 Vehicle used for alcohol testing in early 20th-century Brazil.
CREDIT: Instituto Nacional de Tecnologia archives.

The IAA was created to regulate the sugar industry, promote alcohol fuel, provide technical assistance, and advocate for the small to medium-sized sugar industry actors in the Northeast of Brazil. While a comprehensive, national market for ethanol would not appear for decades, the IAA assumed a leading role for the sugar-ethanol industry by managing the planning and setting of supply targets, prices, and allowable export targets.

During the 1950s–70s, engineer Urbano Stumpf pioneered efforts to test ethanol as a fuel for internal combustion engines (ICEs) at the Aerospace Technical Center (CTA). His research and related activity with the Fuel and Mining Experimental Station would provide a basis for the production of *neat vehicles* (i.e., vehicles that would run on hydrous ethanol/100% ethanol/E100) enabling the launch of Brazil's modern biofuels transition.[2]

MODERN TRANSITION: 1970–THE PRESENT

Brazil's modern shift to biofuels (Figure 6-2) represents two distinct transformations: one linked to the *ProAlcool* Program of the 1970s (the National Ethanol Program, discussed in detail below) and a second tied to innovations with flex fuel vehicles that emerged in the 1990s. To understand these developments, one must begin first with the sugar industry.

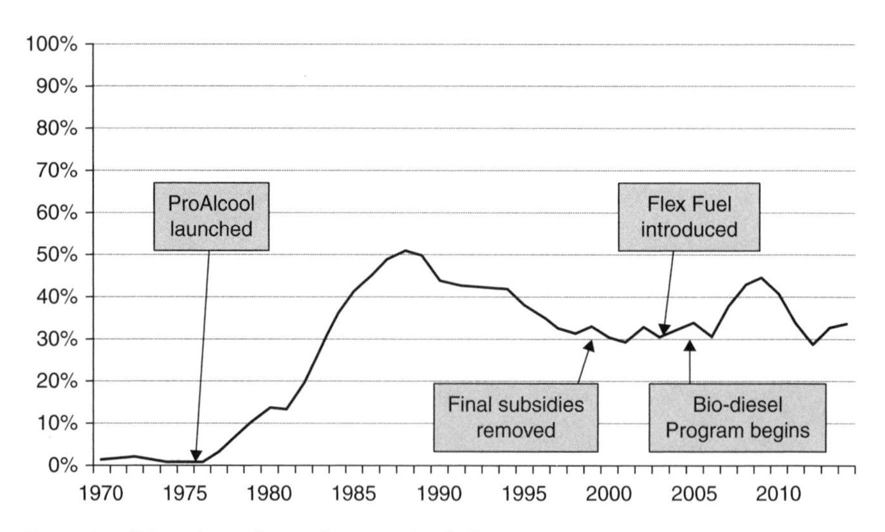

Figure 6-2 Ethanol as a share of automotive fuels.
SOURCE: MME data (2016).
NOTE: Auto fuels are defined, here, as gasoline plus ethanol.

2. For a discussion of vehicle and fuel technology, see the "Technology Primer" in the Appendix.

Launch of *ProAlcool*

Before the first oil shock of 1973, interest already existed in developing a Brazilian market for ethanol. The domestic sugar industry, in particular, considered ways to overcome stagnation following a period of heavy reinvestment. A decline in sugar prices combined with limitations imposed by the International Sugar Agreement left this industry with underutilized capacity. Here, creating a domestic (sugar-based) ethanol market was an obvious choice for a tightly-organized industry facing limited export opportunities, especially given the historical foundations of ethanol use and public mandates in Brazil. To revitalize, the Brazilian sugar industry—namely, the Cooperative of Producers of Sugarcane, Sugar, and Alcohol in the State of São Paulo (Copersucar) and the IAA—together with the Ministry of Industry and Commerce put forward a plan promoting increased production of ethanol for use in transport. By introducing a separate market, this stimulus plan was designed to insulate the sugar industry against future flux in sugar prices, while bringing additional gains to the capital goods and automotive sectors. The plan, however, remained largely untouched until the oil shock of 1973.

With the oil shock, petroleum prices quadrupled from $2.90 to $11.65 per barrel between late 1973 and early 1974 (Corbett, 2013). Like many countries, Brazil was caught off guard, importing roughly 80% of its oil at the time (Ministry of Mines [MME], 2016). The national oil import bill increased by a factor of 6 from $566 million to $3.2 billion for Brazil in the period between 1972 and 1974 (UN Comtrade, n.d., SITC Rev 1, Petroleum and Petroleum Products).

If the oil shock and under-utilization of the sugar industry were not enough, extreme volatility also hit the global sugar market, with sugar prices dropping by almost a factor of 3 between 1974 and 1976. Brazilian President Geisel responded by launching the National Ethanol Program (Programa Nacional do Alcool or *ProAlcool*) in November 1975 (Decree 76.593). Its aims were to increase the domestic production of ethanol in order to more effectively utilize spare sugar capacity, to reduce oil imports, to bring some control to foreign exchange savings, and to improve regional income disparities (Decree 76593; da Silva et al, 1978; CGEE and BNDES, 2008; Goldemberg, unpublished). Ethanol had already been used as a gasoline additive for more than 40 years in Brazil at blends of 1–6% in fuels. With the program launch, the government introduced a bundle of policies to radically alter automobile fuel use and sourcing through price controls, incentives, loans, and direct deployment.

On the supply side, the government established targets to achieve new ethanol production levels. Working from a baseline production of 0.6 billion liters in 1975/76, targets were set for 3 billion liters in 1980 and 10.7 billion liters for 1985 (Goldemberg, unpublished). Favorable public financing was also put in place, including credit guarantees and loans, to expand ethanol production at interest rates that, in real terms, were below that of inflation (Barzelay, 1986; Ministry of Industry and Commerce [MIC], 1981; Walter, Rosillo-Calle, Dolzan, Piacente, and Borges da Cunha, 2007). Additional measures set ethanol prices for producers 5% higher than for a comparable unit of sugar to make fuel production more attractive relative to that of food. *ProAlcool* provided R&D grants to strengthen ethanol productivity (see also Box 6-1). The Program also set fuel economy standards for ethanol-based engines, incrementally increased blending requirements, and protected the emerging industry with import restrictions on ethanol (Goldemberg, 2009; MIC, 1981; Lehtonen, 2007; Sandalow, 2006). Going further, the Federal Government mandated that domestic fuel stations offer ethanol at all locations, prompting adaptations in distribution and refining by the state-owned energy company, Petrobras. By 1977, the ethanol production target for 1980 was met well ahead of schedule (Goldemberg, unpublished).

In terms of demand-side policies (Barzelay, 1986; MIC, 1981; Moreira and Goldemberg, 1999; Goldemberg, 1994 and unpublished; BNDES and CGEE, 2008), the government required that Petrobras become a guaranteed buyer of specified amounts of ethanol each year. Information campaigns also promoted the link between ethanol use and national industrial strength. Meanwhile, retail prices of ethanol were set below that of gasoline (i.e., 59% of gasoline with a guarantee not to exceed 65%). The state also purchased neat cars (i.e., ethanol only or E100 vehicles) for demonstration in a fleet of taxis. Furthermore, favorable financing and vehicle registration fee rebates were made available for the purchase of such vehicles.

Ethanol adoption began to increase with the blending of anhydrous ethanol into gasoline.[3] By adopting this type of ethanol in spark-ignited, internal combustion engines, no real adaptations were required for the automotive fleet. Ethanol blends rose from about 10% to about 20% by the end of the 1970s and would remain at this higher range, except for a period of retrenchment in the 1990s (Figure 6-3).

Measures were also put in place to convert the existing fleet of cars to run on E100/hydrous ethanol. Instituted by the Secretary of Industrial Technology within the Ministry of Industry and Commerce, this plan was administered through Technological Control Centers that worked in conjunction with 13 institutes/universities.

As the Brazilian energy system began to reach a new level of equilibrium in 1979, another oil shock occurred. At that time, petroleum products accounted for 42% of energy consumption in Brazil, with roughly 85% of the oil supply

3. For a discussion of fuel types and technical requirements, see the "Innovations and Adaptations" section and "Technology Primer" in the Appendix.

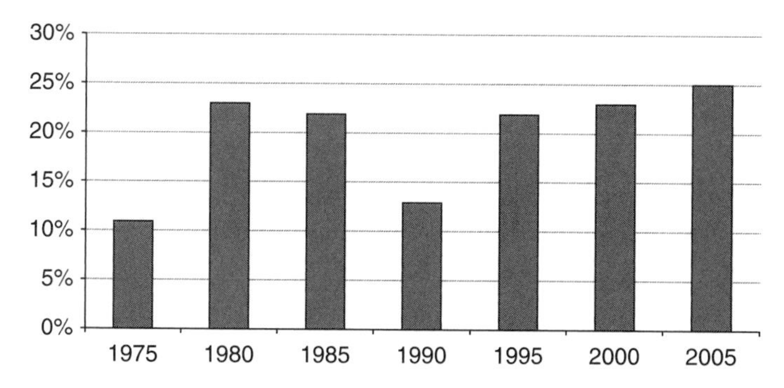

Figure 6-3 Average share of anhydrous ethanol in gasoline-ethanol fuel blends. Adapted from Walter and Dolzan (2009), referencing F. O. Lichts (2006).

imported—approximately 1 million barrels a day (World Bank, 1981*a*, 1981*b*).[4] Up to that point, the focus of *ProAlcool* had been on fuel blending and adapting the existing fleet of vehicles to ethanol-only, neat vehicles.[5] With this new disruption, *ProAlcool*'s strategic focus shifted from adaptations to full mobilization of a market. Arrangements were made with auto manufacturers—VW, GM, Fiat, Mercedes Benz, and Toyota—to produce a fleet of neat vehicles dedicated to hydrous ethanol. To sweeten the deal, favorable financing was provided to auto manufacturers for adjusting production lines. Urbano Stumpf's earlier research on E100 engine utilization became the blueprint for the auto industry.[6] In line with the agreement between the Brazilian government and auto manufacturers, neat ethanol vehicles were commercially introduced in 1979, beginning with the Fiat 147 model (Revista Veja, 1979). These neat vehicles contained modified engines with increased compression ratios, altered fuel injection, more corrosion-resistant materials, auxiliary cold-start systems, and colder spark plugs to disperse heat. Such changes in the auto fleet produced an immediate and robust demand for E100 fuel.

A second, strategic adjustment in the *ProAlcool* Program centered on moving from the utilization of untapped capacity and modified mills to building new infrastructure, including roughly 350 distilleries. Building

4. In 1980, Brazil spent $10.3 billion on petroleum imports in nominal/current dollar terms (World Bank, 1981*a*, 1981*b*).

5. By mid-June 1980, 80,000 cars of the roughly 7.7 million in the fleet had been converted to run on E100, with a focus on government and other public vehicles (World Bank, 1981*a*, 1981*b*).

6. By many accounts, it was Stumpf's vehicle experiments that convinced President Geisel to adopt the ethanol program (Hammond, 1997; Joseph Jr., 2010; Interviews, 2010; Goldemberg, unpublished; BNDES and CGEE, 2009; Walter, 2009).

on the existing sevenfold increase in ethanol production between 1975 and 1980, a new target was also set to increase ethanol production from approximately 3.4 billion liters in 1980 to 10.7 billion in 1985 (World Bank, 1981*b*). The revised fuel targets would bring hydrous ethanol more prominently into the plan and required the production of vehicles that could run on E100. Measures were also implemented to establish a national distribution network for managing hydrous ethanol in all service stations, with supply guaranteed by controls. This was only the beginning.

Complicated Times: The Early 1980s

Entering the 1980s, Brazil's biofuels transition could be observed in an aggressive shift spurred by government, essentially creating and guaranteeing a market with support for distribution. Automobile drivers responded by adopting neat or neat-converted vehicles. The automotive and sugar industries adapted their inner workings and scaled relevant vehicle and fuel production. Biofuels system players—namely automobile drivers and mechanics, ethanol mill owners, and farmers—underwent rapid learning, which complemented the more targeted R&D efforts of auto manufacturers, the sugar industry, academia, and government programs (see later sections for additional R&D detail).

Meanwhile, other developments complicated the playing field. The Brazilian economy experienced high inflation and foreign debt due in part to the economic shocks of the 1970s. During the 1980s, the World Bank provided a $250 million loan to Brazil to support the continued advance of the *ProAlcool* Program amidst the economic flux (World Bank, n.d.*c*., 1981*a*, *b*).

In 1984–85, the national government also peacefully transitioned from a military regime to the first civilian regime in 20 years. Amid the political realignment, public investment and subsidies were reduced. When world oil prices collapsed and world sugar prices increased, the pull of these markets and variability of sugar use placed ethanol at an extreme disadvantage commercially compared to gasoline and sugar.

By 1985, the number of cars designed for ethanol use had risen from 0% to 85% of all cars sold in Brazil (Figure 6-4).[7] This share would not be seen again until 2015. Ethanol as a share of automotive fuel (namely ethanol and gasoline) reached 41% that year, increasing from <1% in 1970 (Figures 6-2 and 6-5). At this juncture in time, disruptive change was evident across the automotive fleet, fueling infrastructure, and fuel supply.

7. This does not include aftermarket conversions of vehicles to the neat features.

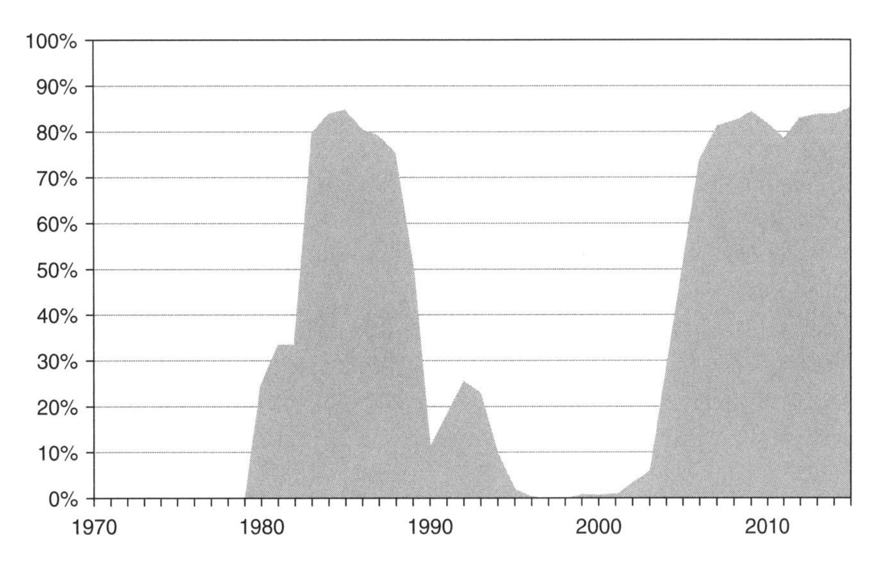

Figure 6-4 Share of newly registered automobiles designed for ethanol use in Brazil (neat and flex fuel vehicles).
SOURCE: Associação Nacional dos Fabricantes de Veículos Automotores (ANFAVEA) data (2017).

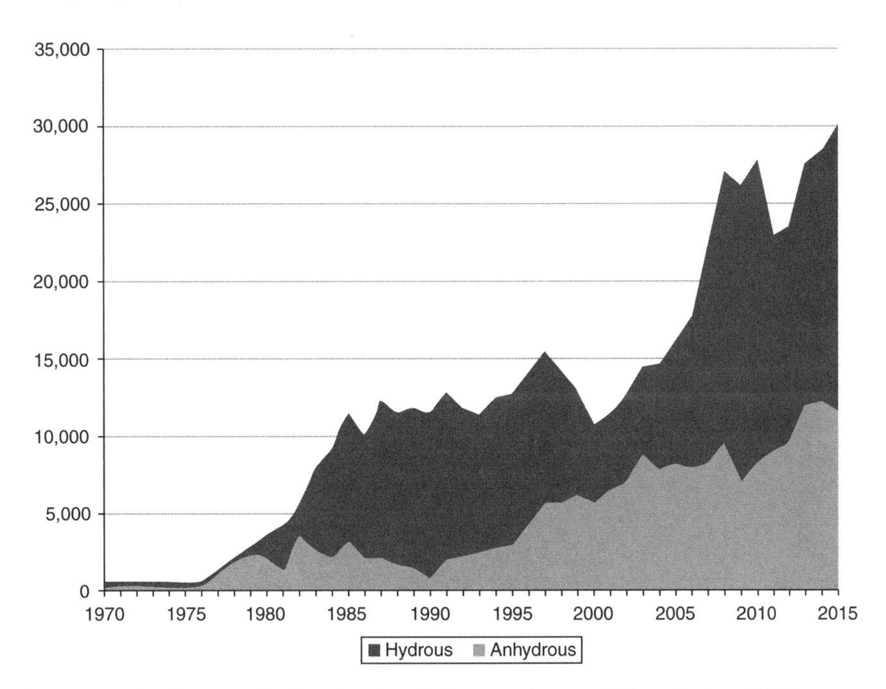

Figure 6-5 Production of ethanol in Brazil (thousand tons of oil equivalent/TOE).
SOURCE: MME, data (2016).
NOTE: This does not account for vehicles converted from internal combustion to neat technology.

Era of Retrenchment: The Late 1980s and 1990s

In marked contrast to the mobilization of *ProAlcool*, the late 1980s and the 1990s reflected a major downturn in biofuels and associated vehicle adoption. As details will show, however, key progress would continue less visibly in R&D and other industrial advances.

Beginning in the late 1980s and following through to the early 1990s, world sugar prices recovered, oil prices dropped, and poor weather conditions negatively affected sugarcane harvests. In line with these conditions, ethanol production declined, and with the convergence of these events, the still-emerging ethanol market underwent a period in which the demand for ethanol exceeded the available supply. This status produced a need for imports between 1989 and 1995 to meet the domestic shortfall (Figure 6-6). In response, the government adopted measures to mitigate the situation by introducing a methanol-ethanol gasoline blend using imported methanol (Coelho and Guardabassi, 2014).

Under the pressure of hyperinflation, considerable attention in Brazil also turned to stabilizing a volatile economy. Reforms, such as that promoted under Federal Bill 8,723/1993, removed nearly all *ProAlcool* policies, other than blending requirements.

Amid these uncertain times, consumer confidence for ethanol eroded. Diminished interest could be attributed not only to the ethanol supply shortfall, but also to early-stage technology issues, shifting political-economic conditions, lower gasoline prices, and new policy support for less costly, non-E100 vehicles. Car manufacturers responded by refitting production lines back to conventional, gasoline-fueled vehicles. With this, the auto fleet reverted from neat vehicles specifically designed for E100 use to standard internal combustion engine vehicles that principally, if not wholly, used gasoline.

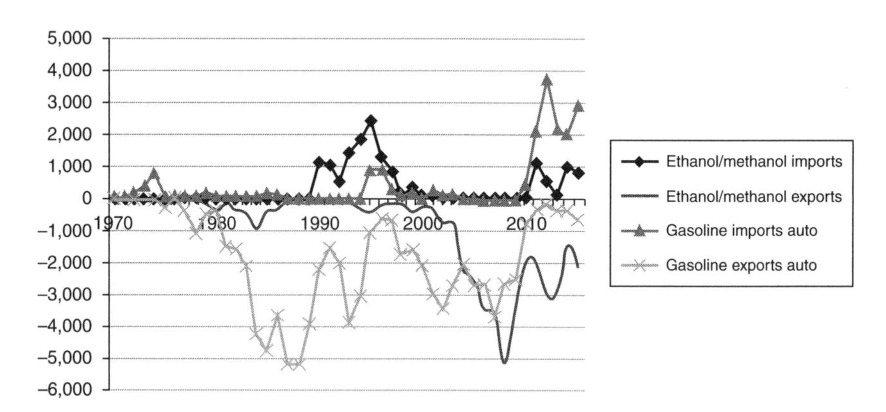

Figure 6-6 Exports and imports of ethanol/methanol and automotive gasoline *(thousand TOE)*.
SOURCE: MME (2016).

The ensuing years were ones of substantial consolidation and retooling for the biofuels business. The Federal Government redirected policies toward privatization, competition, globalization, and decentralization of the energy sector. One major initiative in 1997 was the partial privatization of Petrobras, currently a $81 billion company in terms of revenue (Nasdaq, 2017).[8] The initiative raised capital for foreign exchange, introduced competition, and partly deregulated the industry. Along related lines, the electricity market was liberalized,[9] and the sugar-ethanol industry was deregulated, in such a way that restrictions were largely removed.

The sugar-ethanol industry evolved as its management base professionalized. Mills modernized as international companies and foreign capital entered into the ethanol production chain. From 1999 on, Brazilian ethanol was created without government subsidies in the production or market chain (Dornelles, 2010). One rare, but favorable policy development for biofuels during this time was the introduction of a tax mechanism to support cleaner fuels. Professional associations, like the São Paulo Sugarcane Agroindustry Association (União da Agroindústria Canavieira de São Paulo [UNICA]) also emerged. Meanwhile, ethanol producers established the Brazilian Ethanol Exchange System (Bolsa Brasileira de Alcool [BBA]), a voluntary contractual model for sugarcane pricing.

Because the decree that established *ProAlcool* was revoked, key institutions like the IAA and Planalsucar were dismantled (IEA, 2004*b*/2004*c*). In conjunction with these changes, Copersucar, the São Paulo cooperative of sugar industry actors, continued many of the dismantled institutions' research activities.[10] Its R&D entity, CTC, partnered with the state of São Paulo's research foundation FAPESP to extend sugarcane research in genomics. These efforts drew investments and the development of start-up companies that would eventually lead to the establishment of agro-biotechnology R&D centers around Campinas, São Paulo (Arruda, 2011). The Ministry of Agriculture absorbed members of IAA and Planasucar's institutional teams and infrastructure. A newly-formed university network, known as Rede Interuniversitária para o Desenvolvimento do

8. The Brazilian government retains ownership of 54% of Petrobras' common shares with voting rights. The Fundo Soberano, a sovereign wealth fund, and BNDES, a federal-level public company linked to the Ministry of Development, Industry and Foreign Trade, each hold 5%, which gives the government 64% control. The remaining 36% of the shares are traded on the Bolsa de Valores, Mercadorias and Futuros de São Paulo (BOVESPA), as well as the New York Stock Exchange (Petrobras, n.d.).

9. See Box 3-1.

10. Entering the 1980s, Planalsucar is said to have lost researchers to Copersucar and individual mills. In the period that followed, Copersucar conducted parallel lines of research in crop diversification, uses for vinasse and bagasse, and biological control (Sorj and Wilkinson, 1993).

Setor Sucroenergético (RIDESA), also emerged in support of the sugar-energy sector's development.[11]

As the institutional landscape underwent dramatic realignment, two new policy agencies were formed: the National Energy Policy Council (Conselho Nacional de Politica Energetica [CNPE]) and the National Petroleum Agency (Agencia Nacional do Petroleo [ANP]) that, in 2005, would be renamed the National Agency for Petroleum, Natural Gas and Biofuels (USDA, 2010). CNPE is an executive branch agency charged with the formulation of energy policy and directives, issuing key programmatic directives associated with biofuels (Law 9478/1997; Decree 3520/2000). ANP implements national biofuels policy and oversees regulation as well as quality standards.

During the broader period of retrenchment, one could observe signs of decline at the infrastructural level. Neat vehicles, as a share of the new Brazilian car fleet, diminished from 12% to 1% between 1990 and 2000 (Associação Nacional dos Fabricantes de Veículos Automotores [ANFAVEA], 2011*a* and 2017). The share of total automotive fuel associated with ethanol also decreased from 43% to 31% (MME, 2016).

Despite the downturn and market stagnation, the sugar-ethanol industry's infrastructure and fuel distribution network sustained, albeit in leaner form. The compulsory blend of anhydrous ethanol was maintained, helping producers by preserving a share of ethanol demand. Less visible R&D progress also continued in niches of the automotive sector. In particular, engineers working for Fiat subsidiary, Magneti Marelli, and competitor, Bosch, began to adapt flex fuel technology from the United States to fit Brazilian market needs (see "Innovations and Adaptations"). These developments set the stage for a second transition.

Second Transition: 2000 to the Present

In what could be described as a technological "comeback" in Brazilian biofuels, flex fuel technology was commercially launched in March 2003 with VW's Gol 1.6 Total Flex. The Gol automobile adapted earlier flex fuel technology with knowledge accumulated from decades of ethanol use and R&D efforts in Brazil. R&D had focused on producing advanced flex fuel functionality that was suited for a broader automotive market. The key innovation involved the use of an existing sensor in the vehicles rather than a separately-dedicated probe. This breakthrough enabled costs to remain essentially at par with conventional,

11. Ridesa carried the work of Planalsucar forward, eventually absorbing the technology institutions and the coordination of experimentation stations (Ridesa, n.d).

gasoline-powered vehicles (see "Innovations and Adaptations"). In line with this advance, tax credits for neat vehicles were extended to flex fuel vehicles. The negligible cost impacts and improvements over neat vehicles for consumers strengthened the new vehicle's appeal for both consumers and producers. In short order, flex fuel technology was rapidly diffused, with other major auto manufacturers releasing their own models, including Chevrolet, Fiat, Ford, Honda, Kia, Mitsubishi, Nissan, Peugeot, Renault, and Toyota. Since then, flex fuel technology has evolved in areas including fuel efficiency, compression ratios, and engine power (ANFAVEA, n.d.; Interviews, 2010–2012). These technical technology changes were also later applied to motorcycles and buses (ANFAVEA, n.d.).

Drawing on earlier experience with the *ProAlcool*, and other programs, the Brazilian government established the Biodiesel Production and Use Program (*PNPB*) in 2004 to substitute diesel fuel with domestically-produced biofuels from vegetable oils, such as palm, soy, and castor (No. 11.097; Nogueira, 2005). In conjunction with environmental and energy supply security aims, the program focused on fuel diversification, rural development, and job creation to reduce regional income disparities. The *PNPB* set voluntary blending requirements for biodiesel at B2 (2% biodiesel), which subsequently shifted to mandatory rules with stepped increases (Resolucao No. 6/2009; Agencia Nacional do Petroleo [ANP], n.d). In March 2016, new legislation indicated that blends must be 8% by 2017, 9% by 2018, and 10% by 2019 (Law 13,263/2016; USDA, 2016). This rule also required testing of blend feasibility at 15%.

Standard testing is overseen by the Ministry of Science and Technology (MCT) with the participation of the national auto manufacturers association (ANFAVEA). The *PNPB* employs a public auction system that sets a volume for production quantities and an average sales price. These biodiesel auctions began in 2005, and it is through such auctions that the Petroleum National Agency (ANP) contracts for future production. To participate, biodiesel companies must hold a "social fuel seal" (SFS), indicating that they have contracts with family farmers. As of June 2016, 49 auctions had occurred (USDA, 2016).

Other than blending requirements, standards, and auctions, *PNPB* policies have consisted principally of tax incentives/exemptions, a producer subsidy for small family farmers in the poorest states, and a preferred seal to encourage use of feedstock from small farmers (USDA, 2011). If biodiesel fuel producers acquire allowable feedstock from small Brazilian farmers, they are eligible for federal tax reductions of up to 68–100%. If feedstock purchases are made from other groups, the maximum reduction is 31%.

In 2015, biodiesel production reached 3.9 billion liters, up 15% over 2014 (USDA, 2016). Biodiesel production has outpaced targets, but farmers, producers, and distributors have encountered challenges in terms of access to capital, seed optimization, weather, and a lack of support services in areas, including the Northeast. As of August 2016, 51 biorefinery plants were in operation to produce biodiesel (USDA, 2016).

New opportunities emerged for the biofuels industry with the reform of the electricity sector. In 2004, the wholesale (liberal) electricity market that was set up through the privatization of the 1990s was replaced with a new model that included power auctions (Law 10,848/2004 and Decree 5,081/2004). These power auctions allowed independent power producers, such as sugar mills, to participate in concession bidding for long-term power purchase agreements with energy distributors in a pool organized by the electricity regulatory authority (Agência Nacional de Energia Elétrica [ANEEL]).[12] This reform was adopted to foster competition, while addressing market failures. As a consequence, the reform opened a new line of business for sugar mills to compete with power generation from sugarcane waste (bagasse). Up to that point, bagasse had been used by sugar mills to self-sufficiently produce heat and electricity. In 2007, an auction was held specifically for renewable power producers supplying generation from biomass (i.e., plant material, like bagasse), small-scale hydro, and wind power. In 2010, nearly 6,000 MW of installed capacity associated with bagasse was in place, compared to 14,000 MW_e for the Itaipu hydropower dam, one of world's largest sources of power generation. For the 2009–10 season, 20,031 GWh of electricity was produced from sugarcane bagasse in Brazil (Coelho and Guardabassi, 2014).[13] In conjunction with this, 28.2% of the mills sold excess generation to the grid (Coelho and Guardabassi, 2014, citing EPE, 2011).

In related areas of development, the Ministry of Agriculture, Livestock and Supply (MAPA) outlined agro-ecological zoning in 2009. This revision delineated protected land and other areas that are not suitable for large-scale sugarcane farming. The zoning designates regions of ranching and agricultural land where sugarcane is not currently grown to inform infrastructure investment decisions, financing policies, tax regime changes, and potential socioeconomic certification (BNDES and CGEE, 2008, citing Strapasson, 2008). To classify prime areas for sugarcane planning, indicators on soil, climate, and other relevant scientific data are combined with information on land use, environmental legislation, and the like. Of the national

12. Regulation 2003 (1996) established the right for independent power producers and self-generators to operate. By paying transmission charges, they have "free access" to the interconnected grid and to distribution lines (GTZ, 2007).

13. These numbers are based on boilers and turbines up to 80 bar. Additional mills used 20–40 bar (Coelho and Guardabassi, 2014).

territory, 92.5% was deemed "off limits" for sugarcane cultivation (Coelho and Guardabassi, 2014).

Furthermore, the federal government and state of São Paulo enacted laws to phase out the use of sugarcane burning in manual harvest, thus spurring mechanization. Based on Federal Decree 2,661/1998, the practice of burning cane will no longer be possible in areas with a slope of less than 12% within a set timeline, where mechanized harvest is possible. The state of São Paulo implemented an even more aggressive plan that was then superseded in 2007 by an agreement between UNICA and the São Paulo state secretaries of the Environment and Agriculture and Supply to accelerate the schedule (Interviews, 2011). When the burning restriction is factored, manual harvesting of sugarcane becomes almost economically infeasible.

In 2010, the domestic consumption of ethanol equaled that of gasoline (ANP, n.d). That same year, the U.S. Environmental Protection Agency (EPA) also designated sugarcane-based ethanol an advanced biofuel that reduces GHG emissions by more than 50% relative to gasoline (PR Newswire, 2010; Schnepf and Yacobucci, 2013; U.S. Environmental Protection Agency [EPA], 2011).

In terms of international training and diplomacy related to biofuels, Brazil has taken a lead, particularly with industrializing countries. In conjunction with this, Brazil had signed more than 60 agreements on bilateral technical cooperation with countries from Central America, the Caribbean, Africa, and the United States. Brazil also put forward the idea of an International Biofuels Forum, which was officially launched in the United Nations in 2007 (United Nations, 2007; Whitehouse, 2007). The Forum and other international efforts by Brazil promote the creation of an international commodities market for biofuels, related jobs development, and more uniform standards.

Specific to vehicular technology developments, Honda launched the first commercial flex fuel motorcycle, the CG 150 Titan Mix, in 2009, followed by an on-and-off-road version later that same year. By 2011, four flex fuel motorcycle models were commercially available, with production equaling nearly 1 million units, increasing the existing market share to 57% (Abraciclo, n.d.; ANFAVEA, n.d. and 2017; UNICA, 2011).

As of August 2016, 383 first-generation ethanol plants and 3 cellulosic plants were operating with a total ethanol nameplate production capacity of 39.8 billion liters (USDA, 2016). R&D also continues at Brazilian labs, schools, and companies to advance a new generation of sugarcane and biofuels. The national laboratory for bioethanol, Laboratório Nacional de Ciência e Tecnologia do Bioetanol (CTBE), together with Alellyx-Canavialis (Monsanto), Dedini, CTC-BASF, Petrobras-Novozymes, UNICAMP, Amyris, Butamax (BP-DuPont), Syngenta, Shell-Cosan, and IPT reflect regional research hubs focusing on aspects of the science and technology. The research now includes spillovers into bioplastics.

Changes specific to biofuel infrastructure have also been in various stages of play within Brazil. Pipeline projects are under development (Sapp, 2013). Export terminals, barges, and rail are also under consideration (*Biofuels Digest*, 2011; Scandiffio, 2014). Gasohol is now sold at every fuel station. The majority of stations sell B7 7% biodiesel) since B100 is unavailable for the regular road transportation sector. Some stations also sell natural gas (mainly for use by taxi fleets) due to its lower operational cost compared to gasohol, but vehicles equipped to use natural gas require extra investment for vehicle refurbishment.

Importantly, Brazilian ethanol became more competitive with gasoline in the past decade as production continued to rise.[14] Consumer demand for flex fuel vehicles has also continued to grow (ANFAVEA, n.d. and 2017, see also Figure 6-4).

Despite the advances with biofuels, consolidation has occurred in recent years for what, in part, may be a surprising challenge—gasoline prices were kept artificially low in Brazil as a means to counteract inflation. This condition presented an uncompetitive playing field that was amplified by the drop in world oil prices to record lows alongside the growth of fracking that tapped new oil reserves. The National Bank for Social and Economic Development (BNDES) extended credit to the sector at roughly $2.8 billion per year on average between 2011 and 2014 (USDA, 2015); however, plant closures for ethanol and biodiesel occurred in about 15% of the sector, with a net decrease of 58 plants for ethanol and 8 for biodiesel (USDA, 2015).[15] These shifts partially offset the rapid industry growth between 2006 and 2011 in which 138 new ethanol plants were brought online (Valdes, 2011).

On balance, the period since 2000 provides a mixed picture into how a second energy transition can emerge and evolve. At the start of this period, pro-ethanol vehicles represented 1% of new car sales and, as of June 2016, flex fuel vehicles represented more than 90% of total, monthly vehicle sales (USDA, 2016). The Brazilian light duty vehicle fleet was recently estimated at 35.3 million units, with 72% being pure, hydrous ethanol or flex fuel vehicles (USDA, 2016). Ethanol as a share of auto fuels also grew from 31% in 2000 to 41% in 2013 (MME, 2016). Currently, about 11 auto manufacturers produce more than 70 flex fuel models in the Brazilian market at a price that is equivalent

14. It is worth noting that price may be used as a proxy for cost but does not fully reflect costs (Araújo, 2016). See Meyer et al. (2014) for a more in depth discussion of Brazilian biofuels price calculations.

15. For a discussion of loan programs, see Valdes (2011).

to conventional car models, bringing choice and better economics to the consumer.

EXPLAINING CHANGE

Brazil's biofuels transition reflected a combination of two energy transitions with an initial shift that was front-loaded and robust, followed later by a second, more tempered shift. The type of sectoral intervention and model of readiness are outlined next.

Mode of Intervention

Drawing upon the modes of intervention, as described in Chapter 3, early change for the Brazilian biofuels transition was led and induced by the government. The sugar industry lobbied for a stimulus plan (emergent change); however, the government catalyzed the shift with state-owned energy company, Petrobras, along with ministries and public agencies driving a reconfiguration of both the sugar and auto industries (deployed-mandated-encouraged change). The public was mostly engaged via financial support and blending requirements (mandated-encouraged change).

In the 1990s, little occurred overtly to advance the biofuels transition except for R&D that continued behind-the-scenes with auto manufacturers, the sugar industry, and academic labs. Given this, the second major shift is characterized as bottom-up development (emergent change), with a much more modest role played by the government in blending rules and tax credits (mandated-encouraged change). Consumer demand and learning also fostered some additional bottom-up momentum during the later period (emergent change).

Taking the above combination of sectoral influences into account over the entire period, a hybrid form of intervention is observed.

Model of Readiness

Looking next at readiness for change, Brazil's biofuels transition is the clearest example of the reconstitutive model of change, in which the infrastructure and industries were repurposed using indigenous expertise to create a new market. A biofuels pathway existed mostly in remote niches in 1970 and was negligible in numbers before the modern shift. Biofuel also has been indigenously derived, except in times when imports covered gaps, typically in years of limited harvest.

INNOVATIONS AND ADAPTATIONS

Four primary areas reflect innovation or critical adaptations in the Brazilian transition: *automotive technology, ethanol production, bioelectricity*, and *biodiesel*.

Automotive Technology

Specific to automotive technology, key developments in Brazil can be seen with flex fuel vehicle commercialization in 2003, gasoline/gasohol-powered vehicles converted to use hydrous alcohol, and neat vehicle commercialization in the late 1970s.

Flex Fuel

The design breakthrough in flex fuel technology was enabled by a Bosch patented optimization (Bastin et al., 2010) that revolutionized contemporary Brazilian biofuels development. Flex fuel advances were then furthered by Magneti Marelli, a manufacturer of high tech components for the auto industry. Essentially, the innovation entailed moving beyond the costly oxygen probe (fuel-line capacitive sensor) of an earlier design to utilize an existing sensor (oxygen lambda) found in the exhaust area of vehicles, thereby allowing the gasoline–alcohol ratio to be determined (Interviews, 2012; de Lima, 2006). Bosch engineers employed the readapted sensor with electronic injection software in prototypes for GM, VW, and Fiat (Interviews, 2012; Bastin et al., 2010). Magneti Marelli extended the design with an algorithm that calculated the fuel composition, thus increasing the accuracy of the system without raising the cost. The optimized technology allowed cost reductions from the unification of several differentiated parts, namely, ignition wire harnesses, injection valves, and fuel pumps (Bastin et al., 2010, citing Abreu and Ribeiro, 2006). The release of flex fuel technology in 2003 revolutionized the auto fleet, enabling consumers to apply a simple rule of thumb at the pump that held that ethanol would be economically advantageous if its price were 70% or less than that of gasoline.[16] Continued advances include vehicle capabilities to utilize tetrafuel (i.e., four types of fuel) and modifications for cold starts that eliminate the need for an auxiliary gasoline reservoir (Cenbio, 2008; Joseph, Jr. 2013).

Vehicle Conversion

Although ICE vehicles were not seen as requiring any modifications to run on E20 blends, a number of modifications were needed to convert ICE vehicles to run strictly on E100. Adaptations to E100 in the existing fleet commenced fairly early with *ProAlcool*. These included cylinder head modifications, the addition of

16. This difference accounts for the variance in heat value and octane index (Costa and Sodré, 2010).

a fuel preheating system and cold-start facilities, and the calibration of the carburetor, as well as changes in the ignition system and to the materials used in the fuel system (World Bank, 1981*a*) (see the "Technology Primer" section in the Appendix for details on the technology efficiency). Such changes were deemed necessary to attain the higher compression ratio of the neat/ethanol-only vehicles, to address ethanol's corrosive nature, and to allow gasoline to start the engine in cold weather via a small additional gasoline tank (World Bank, 1980).

Neat Vehicle Technology

Prior to the launch of *ProAlcool*, labs such as the CTA had been working with and testing ethanol and vehicle technology extensively for many years. Building on this work, the leaders of the auto industry entered into an agreement with the Government for the rapid roll-out of neat vehicles in 1979. Auto manufacturers refined and commercialized neat technology with financing for investment provided by the BNDES. To convert vehicles, automakers increased compression ratios and added automatic cold starting, dual carburetors, nickel coating, and more corrosion-resistant liners, thereby enabling the vehicles to run on E100 fuel (Bastin, Szloko, and Pinguelli Rosa, 2010, citing Figueiredo, 2006). The initially commercialized, neat models were released in 1979 and appeared in government fleets, such as that of TELESP in São Paulo (Gatti, 2010; Joseph, Jr., 2010) and taxis. They were later diffused into mainstream use.

Ethanol Production

Advances in ethanol production encompass developments from the planting and harvesting of sugarcane, to the extraction of sugar, and alcohol fermentation processes. The following paragraphs outline specific changes of note in resilient seed varieties; mechanization in harvesting; fertilizer adaptations, like vinasse; and logistical improvements in the timing, transport, storage, and processing of sugarcane ethanol. These changes play a role in the growth of the ethanol industry, reflecting a quadrupling of sugarcane output and growth in ethanol output by more than a factor of 7 between 1980/81 and 2014/15 (Figure 6-7).

Seeds

Sugarcane research and breeding programs enabled gains in seed varieties. With hybridization (i.e., crossing of seed varieties), breeding has modified the sucrose and fiber content, the amount of stalks, early maturity, resistance to pests, disease and flowering, regional particularities, and the adaptability and stability of variety, among possible traits (Creste, Pinto, Xavier, and Landell,

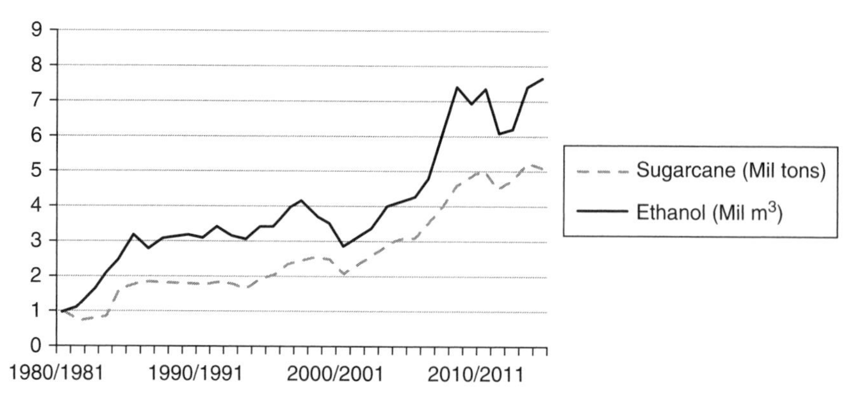

Figure 6-7 Brazilian sugar and ethanol production, 1970–2008.
Brazilian shifts in the production of ethanol and sugar (Base year = 1980–81 harvest).
SOURCE: UNICA (2016).

2010; Gazaffi et al., 2010, citing Matsouka et al., 1999*a*, 1999*b*).[17] Roughly six new commercial varieties per year have been obtained, furnishing more than 600 seed varieties in Brazil (BNDES and CGEE, 2008).

The major contributors to sugarcane seed development have been: Ridesa, the CTC, the Instituto Agronomico de Campinas (IAC), and CanaVialis (Cheavegatti-Gianotto et al., 2011; Creste et al., 2010; Souza and Sluys, 2010). Ridesa, the inter-university network that absorbed activity and infrastructure of Planalsucar in the early 1990s, is now a public–private partnership (PPP) consisting of 300 companies and nine federal universities engaged in the development of the sugar and alcohol sector (Ridesa, n.d.; Souza and Sluys, 2010). Ridesa holds the largest public collection of sugarcane genotypes in Brazil (Ridesa, n.d.; Souza and Sluys, 2010). By contrast, the former Copersucar R&D entity, CTC, manages what is likely to be one of the most important, private germplasm collections in the world, with restricted access to its collections (Creste et al., 2010; Souza and Sluys, 2010). CTC's leadership is an outgrowth of partnering by the CTC with the state of São Paulo research foundation (FAPESP) to extend sugarcane research in genomics (Arruda, 2011, citing Arruda, 2001). In contrast to the CTC, the IAC is the

17. For context, estimates indicate that the selection and launch of a new seed variety requires no fewer than 10 years and includes the testing of experimental clones, the evaluation of yield variation in different cultivation environments, and the monitoring of disease and pest outcomes (Dal-Bianco et al., 2012; BNDES and CGEE, 2008, citing Ridesa, 2008). On average, one new commercial seed variety may emerge from every 250,000 seedlings that are assessed in a breeding program's early testing and development stage (Cheavegatti-Gianotto et al., 2011).

oldest R&D institute in Latin America dedicated to the optimization of agriculture and applied fields. It holds a key academic-based seed collection. The Monsanto subsidiary, CanaVialis, is another major player, also engaged in seed varietal management and genetic improvement in Brazil. Like the CTC, it has a developing collection with restricted access (CanaVialis, n.d.; Creste et al., 2010). Overall, work currently by Ridesa and the CTC represents 95% of the seed varieties used in Brazil (Souza and Sluys, 2010).

Mechanization

The principal mechanization of sugarcane production between 1975 and 2005 involved the standardization of operations, the incorporation of new planting equipment, and procedural or technical training. This activity drew on technological development that occurred in Australia, Cuba, and the United States to optimize the recovery of plant stalks plus the economic elimination of straw in harvest practices (Magalhaes and Braunbeck, 2010).[18] Mechanization led to a reduction in production costs, which are currently the lowest worldwide (Magalhaes and Braunbeck, 2010). Actors driving this change included sugar mill technicians, regional manufacturers, and research institutes interested in cost reductions (Braunbeck and Magalhaes, 2010; Magalhaes and Braunbeck, 2010). Among principals mentioned in the research, Dedini, an agro-energy-infrastructure company that produces agricultural equipment as well as turnkey plants, was a key private-sector player in the modernization and mechanization of sugar-ethanol production during this period (Interviews, 2010–12).

When combined with practices like fertilization, mechanization was crucial for explaining why limited soil degradation occurred in Brazil after decades of sugarcane plantings on the same soil (Boddey, 1995; Cantarella and Rossetto, 2014; Rossetto, Cantarella, Dias, Landell, and Vitti, 2008). Such mechanization has been sought in recent years for environmental and economic reasons. In environmental terms, awareness has increasingly acknowledged the detrimental effects of emissions and soot from combustion associated with manual harvests. The Brazilian government has instituted measures to eliminate manual harvesting practices by 2020 (Goldemberg, 2010; Jank, 2009). Specific to economics, mechanization reduces manual jobs and increases plant residue that may be burned for bioelectricity. Figure 6-8 reflects an increase in mechanized harvesting from 34% to 84% in the state of São Paulo between harvest seasons 2006/7 and 2013/14.

18. The principles underpinning this are known as (1) the Soldier or Louisiana system, which harvests the whole stalks and lays them parallel to each other, and (2) the Push-Rake system, which cuts and loads whole stalks in a disorderly manner (Magalhaes and Braunbeck, 2010).

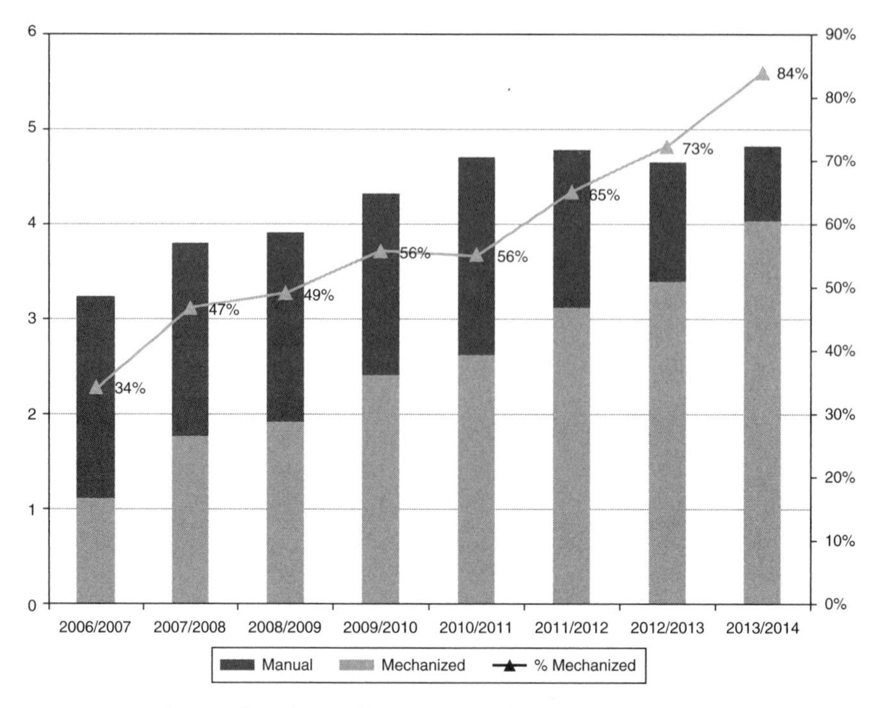

Figure 6-8 Evolution of mechanical harvesting in the state of São Paulo (million tons). Adapted from Goldemberg et al. (2014).

As a consequence of the shift in Brazil from manual harvesting to mechanization, as well as environmental interest in reducing carbon releases in farming, research at the CTBE national lab focuses on conservation tillage methods. These methods include no tillage practices (i.e., soil preparation with minimal mechanical disruption) and limited cultivation of the soil (Magalhaes and Braunbeck, 2010).

Fertilization: Vinasse

For decades, nutrient enhancement of soil through fertilization has been an area studied by Brazilian sugar producers, industry associations (Copersucar [CTC]), and government agencies (IAA). An example of Brazilian innovation in fertilizers entailed the science and practice of fertilizing with vinasse.

Vinasse is a liquid by-product of ethanol and *cachaça* (a common alcoholic beverage) production that is a rich source of potassium, sulfur, and other nutrients (Interview, 2010; Mutton et al., 2010). Its highly acidic quality means that it can act as an environmental toxin if used inappropriately. Environmental reports from the 1970s indicated that the sugar and alcohol industry contributed to environmental pollution due to the dumping of vinasse (Mutton et al., 2010). Large quantities of this substance degrade the quality of raw outputs and can cause an area to become fallow (i.e., useless for agriculture), primarily in response to high salinization and ion lixiviation of the soil (Mutton et al., 2010, citing Ferrerira, 1980). Early research in this area by Professor Nadir da Gloria of Usina da Pedra was extended by the Brazilian sugar industry and academic/government researchers to identify not only a proper dosage for sugarcane nutrition, but also process enhancements, which included spraying from trucks and automated feeding (Mutton et al., 2010; Mutton et al., 2010, citing Carvalho, 2007; Braunbeck and Neto, 2010). Advances in understanding vinasse nutrient–toxin levels led to altered fertilization practices that enhanced sugarcane yields, optimized waste management, reduced environmental impacts, and, if done with ferti-irrigation, reduced costs for fertilizer imports (Interviews, 2010–12; Mutton et al., 2010). Although there are many other commercial opportunities for vinasse, today it is principally used for fertilizing soils in the proximity of ethanol-producing mills in Brazil (Cheavegatti-Gianotto et al., 2011). Today, the gains are clear, and vinasse continues to be a subject of study.

Logistics and Information Management

Areas such as logistical planning as well as operation and information management also played an important role in improved ethanol production in Brazil. The global advent of electronic monitoring and navigation devices, together with digital communications and advanced management practices, enabled growers, producers, and distributors to incorporate substantial efficiencies across the Brazilian agro-industrial-ethanol chain.

One specific area related to ethanol planning entails the synchronized logistics of harvesting, loading, and transport systems, which are responsible for the milling rate of sugar mills (Braunbeck and Neto, 2010). A guiding principle is to keep the mills in constant operation with little idle time or excess stock because cane can spoil quickly (Braunbeck and Neto, 2010, citing Chiarinelli, 2008; Interviews, 2010–12). Consulting related to such forms of agricultural management and integrated information technology (IT) solutions for logistics began with the deregulation of the industry and mill optimization in the 1990s.

More recently, *precision agriculture* represents a new direction for biofuels development that leverages global positioning systems (GPS) in equipment and productivity mapping (Inamasu and Neto, 2010). To date, precision agriculture has been used primarily for soil correction. If adopted more widely, it may serve as a guide for crop management by identifying areas of low productivity and improving knowledge with quality-related parameters.

Gains with Sugarcane

Sugarcane crops are cultivated in the majority of Brazilian regions. They are the third largest crop domestically in terms of harvested area, with about 68,000 farms producing sugarcane (Valdes, 2011; also citing IBGE, 2010). Sectoral contributions of roughly $33 billion annually represent about 2.3% of the gross domestic product (GDP) and 15% of value added in agriculture (Valdes, 2011, citing IBGE, 2010).

Brazilian sugarcane productivity grew by 66% between 1975 and 2010 in terms of the tonnage of sugarcane produced per hectare, with an increase of 34% in the sugar yield per tonnage of hectare for the same period (Dal-Bianco et al., 2012, citing CONAB, 2011 and MAPA, 2009). These gains made Brazil the world's largest sugarcane producer, with roughly 632 million tons of sugarcane produced in the 2014–15 harvest season (UNICA, 2015). In that season, roughly 57% of the harvest was used to produce approximately 28 billion liters of ethanol, with about 43% of the harvest utilized in the production of 36 million tons of sugar (UNICA, 2015).

Bioelectricity

Adaptations with co-generation technology in the sugar-alcohol mills allowed biofuel residuals to be used as a feedstock for electricity (Arruda, 2011). This change enhanced mill self-sufficiency, enabling mills to sell surplus electricity to the grid (Arruda, 2011). Higher pressure boilers were needed for mills to do this (Arruda, 2011; BNDES and CGEE, 2008, citing Horta Nogueira, 2006). Regulatory reform, namely the introduction of renewable energy technology-based power auctions in 2007, and other technology improvements also enabled this development to advance. In 2008, the first biomass-only reserve auction was held by the government for 2,379 MW of power derived from sugarcane and napier grass (IEA, 2015*e*).[19] In that year, 4% of electricity was derived from biomass (Arruda, 2011, citing Jank, 2008).

Biodiesel

Although biodiesel has been researched at various points in time in Brazil, recent progress offers some promise for future progress. An example of such advance is in the development of a type of biodiesel product called H-bio. Petrobras's R&D center, Centro de Pesquisas Leopoldo Americo Miguez de Mello (CENPES), developed H-bio through a process of hydrogenation, specifically catalytic hydrogenation and a cracking of vegetable oils, rather than the more common method of trans-esterification (Green Car Congress, 2006). The change allows

19. As of 2009–10, biomass power was to be supplied through 15-year power purchase contracts at an average final price of BRL 58.84/MWh (IEA, 2015*e*; Interview, 2010).

vegetable oil from various source feedstocks to be mixed directly into a mineral oil in a refinery unit (Green Car Congress, 2006; Petrobras, 2009). The approach is designed to eliminate waste, increase fuel quality, and complement growing bioelectricity practices (Green Car Congress, 2006). H-bio aligns with existing requirements for diesel handling, optimizes diesel fuel processed at the refinery, allows flexibility in load processing, and may reduce the need for testing since the product mimics existing diesel products in use (Green Car Congress, 2006).

KEY DRIVERS AND BARRIERS

Major determinants to Brazil's energy transformation are outlined below. These reflect drivers and barriers of significance that were identified through the analysis of cases, historical records, interviews and data trends.

Reduction of Oil Import Dependence, Improvement of the Balance of Payments, Foreign Exchange Savings, and National Energy Security/Self-Sufficiency

In the early period of *ProAlcool*, greater self-sufficiency was a primary goal, aiming to reduce oil import dependence and improve the balance of payments or savings on foreign exchange. The program was launched with the stated aim of substituting imported oil with domestically produced ethanol, thus improving the foreign exchange position among other objectives. Figure 6-9 shows Brazil's heavy dependence on oil imports in the 1970s.

Such early reliance on energy imports is associated with petroleum import costs increasing by a factor of 41, from $285 million in 1970 to $11.7 billion in 1981 in nominal dollars (Figure 6-10). Viewed somewhat differently, the share of Brazil's total import costs derived from oil rose from roughly 10% to 50% between 1970 and the early 1980s (Figure 6-11).

Over the course of the period studied, domestic oil reserves increased, based on growth in exploration and development (Figure 6-12). In line with this, domestic oil production also increased substantially, enabling Brazil to become self-sufficient in oil on a net basis in 2006. This strengthening of Brazil's oil independence altered the transport fuel playing field, making the later-staged biofuels transition arguably more interesting as the energy independence rationale was less pressing.

Better Utilization of the Indigenous Sugar Industry

The desire to better utilize the spare capacity of the domestic sugar industry was another critical driver of biofuels development (Demetrius, 1990; Goldemberg,

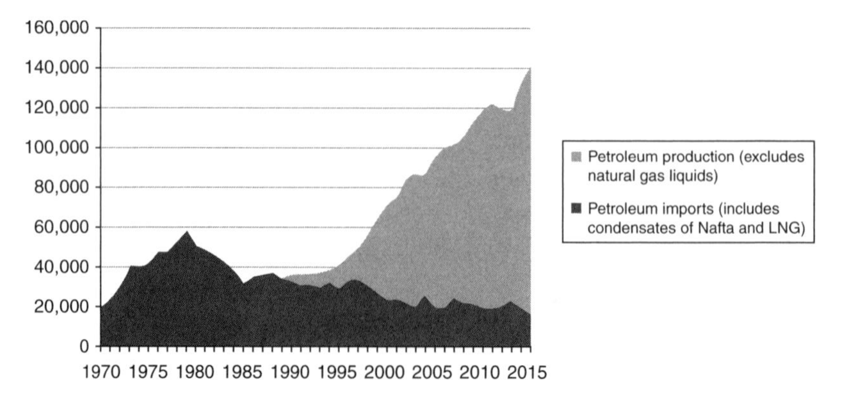

Figure 6-9 Petroleum imports and production (thousand TOE).
SOURCE: MME data (2016).

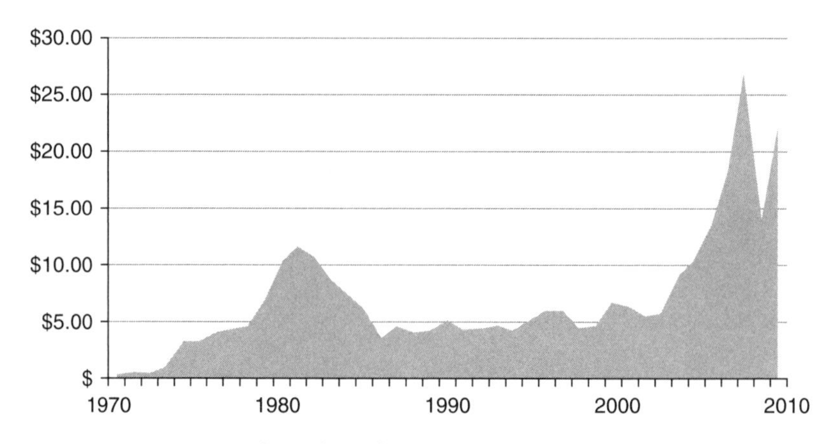

Figure 6-10 Import costs of petroleum for Brazil (billion, nominal $).
SOURCE: UN Comtrade data, SITC Rev 1, Petroleum/Petroleum products (n.d.).

unpublished; Interviews, 2010–11). This explanation broadly coincides with reported data on global sugar price flux and domestic sugar industry production, as well as with major investments in the sugar industry that preceded the oil shocks (Barzelay, 1986; Goldemberg, unpublished).

Regional Development, Jobs, Rural Development

The socioeconomic aims of biofuels growth—namely, jobs and regional development, particularly in rural areas—were also critical in the development story. While such priorities were noted in the *ProAlcool* decree, they also were criticized for not being met as well as other objectives (Barzelay, 1986; Demetrius, 1990). The rationale behind this aim in the early period was to reduce income disparities

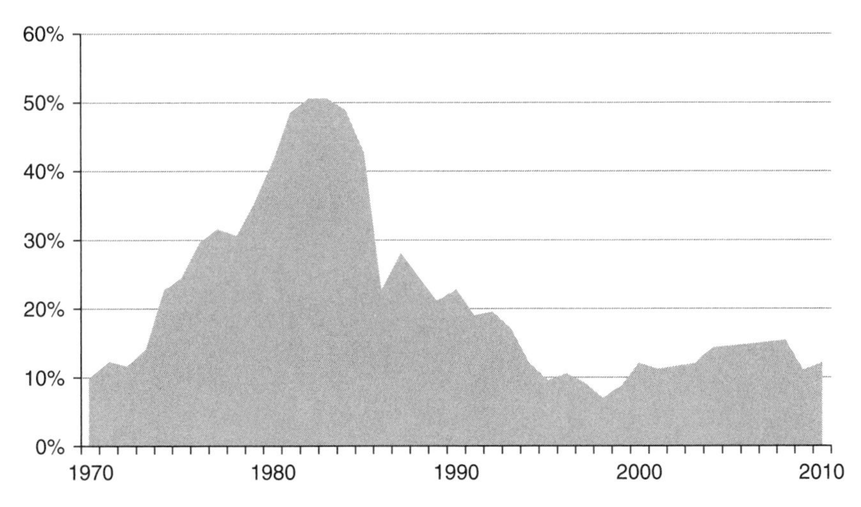

Figure 6-11 Petroleum as a share of total import costs.
SOURCE: UN Comtrade data, SITC Rev 1, Petroleum/Petroleum products (n.d.).

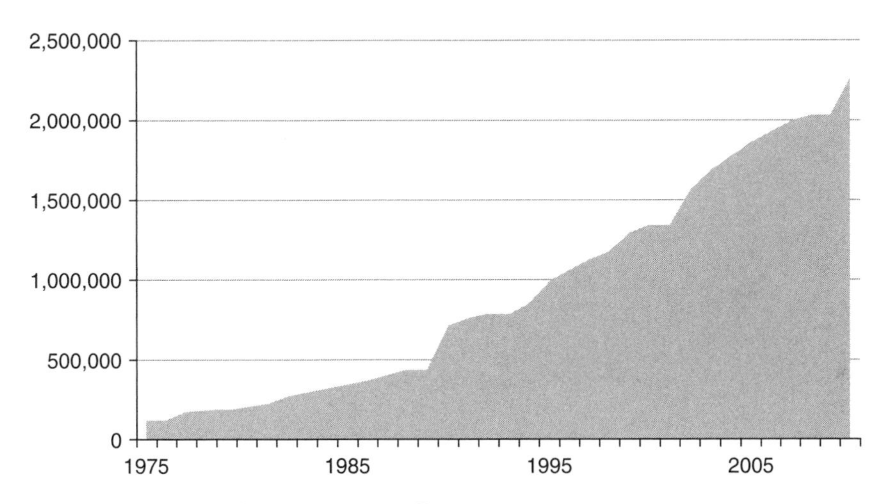

Figure 6-12 Proven oil reserves (1,000 m^3).
SOURCE: Empresa de Pesquisa Energética (EPE) (2011), citing Agência Nacional do Petróleo, Gás Natural e Biocombustíveis (ANP).

and urban migration. In more recent years, jobs and economic development in rural areas were the priority behind the launch of the new Biodiesel Program (de Almeida, Bomtempo, and de Souza e Silva, 2007; Rodrigues and Accarini, 2009).

Environment

Environmental concerns have advanced biofuels substitution, with this driver increasing in significance over time. In the 1980s, environmental concerns in Brazil centered on urban air quality, which improved with ethanol use (de Almeida et al., 2007; BNDES and CGEE, 2008; Goldemberg and Lucon, 2010; Macedo 2005). The decline in ambient lead concentrations in the São Paulo metropolitan region, for example, from 1.4 μg/m^3 in 1978 to less than 0.10 μg/m^3 in 1991, is attributed to the use of ethanol (Coelho and Goldemberg, 2004; Goldemberg, 2008, citing São Paulo State Environment Agency, 2003). In the 1990s, awareness grew with respect to sustainability and climate change, manifesting in biofuels interests related to certification standards for sustainable practices and environmental zoning.

Pricing and Dual Use

Pricing and dual use matter for biofuel adoption in Brazil. Drivers use a working rule of thumb that finds ethanol more favorable than gasoline when ethanol prices are 70% or less that of gasoline. This convention allows for shifting between fuels, and is based on the energy efficiency calculation in which sugarcane-based biofuels yield about 30% less energy per unit than gasoline. Less ethanol also tends to be produced in years when global sugar prices are high as feedstock is diverted to sugar production. These dynamics are not the only ones which impact biofuels. Domestic gasoline price controls have, for example, undercut the biofuels industry in the past decade, as the business viability of partly state-owned Petrobras was also substantially diminished. Efforts are now under way at Petrobras and the Ministry of Energy to separate company fuel pricing from national economic policy by securing fuel price independence. Taxes can also work favorably or unfavorably for biofuels. Preferential tax regimes have generally assisted biofuels in the years when present, yet can also blur the comparative price advantage between ethanol and gasoline.[20]

Technology Performance and Competitiveness

In the initial years of *ProAlcool*, the performance of biofuels and neat vehicles was regularly seen as a barrier. Early versions of both technologies were put into circulation in a learning-by-doing fashion, so were not yet refined. Over time, standards were put in place, and auto manufacturers introduced modifications for technology issues like material corrosion and cold starts to account for such considerations.

20. In 2011, for example, taxes for gasoline were 35% and for ethanol 31%, producing a limited 4 percentage point difference between the two fuels (UNICA, 2011).

Uncertainty/Inertia

Uncertainty and inertia also played a role in various ways with the biofuels transition. The reservations on the part of finance institutions in the 1970s to engage in the roll-out of biofuels delayed implementers and interested users at a critical juncture of adoption (Barzelay, 1986; Interviews, 2010–11). Car manufacturers also did not robustly engage until around 1979 (Barzelay, 1986; Interviews, 2010–11; Demetrius, 1990), possibly holding back to gauge whether the overall program would gain full traction. The second oil shock appears to have clarified the playing field by producing conditions that were unfavorable for imported oil in the near-term. In the 1990s, uncertainty prevailed in policy terms as nearly all support for biofuels was removed, leaving open questions on the viability of the market and industry.

DEVELOPMENTS IN COST, SOCIETAL ACCEPTANCE, AND INDUSTRY

Cost

Production costs for sugarcane-based ethanol in Brazil decreased roughly 70% in the first three decades following the launch of *ProAlcool* (Meyer et al., 2014; Van den Wall Bake, Junginger, Faaij, Poot, and Walter, 2009). These cost reductions are largely attributed to increasing agricultural yields, decreasing industrial costs, and benefits of co-generation (Dal-Bianco et al., 2012; Meyer et al., 2014; Van den Wall Bake et al., 2009).

Prices paid to ethanol producers between 1980 and 2005 also decreased 1.9% at a cumulative annual rate (BNDES and CGEE, 2008, Figure 6-13). If one compares the prices paid to producers of Brazilian anhydrous ethanol relative to U.S. gasoline prices, more favorable fluctuations can be seen (Figure 6-14). Specifically, for the period of January 2000 to January 2008, prices for U.S. gasoline ranged from roughly \$0.15 to \$0.62/liter, whereas prices for Brazilian anhydrous ethanol overlapped in a range between \$0.17 and \$0.58/liter. Ethanol was not subsidized then, with prices at times lower than those for gasoline.[21]

Considered somewhat differently, Figure 6-15 shows cumulative ethanol production in Brazil, with ethanol prices decrease to a point below that of gasoline. Key for this topic is a price decline in the period between 1997 and 1999, when subsidies were no longer provided (Cuelho and Guardabassi, 2014).

21. Discussion continues over the choice of reference year and exchange rate for fuel prices and related costs in a period of economic turbulence. See, for instance, Meyer et al. (2014) and Van den Wall Bake et al. (2009).

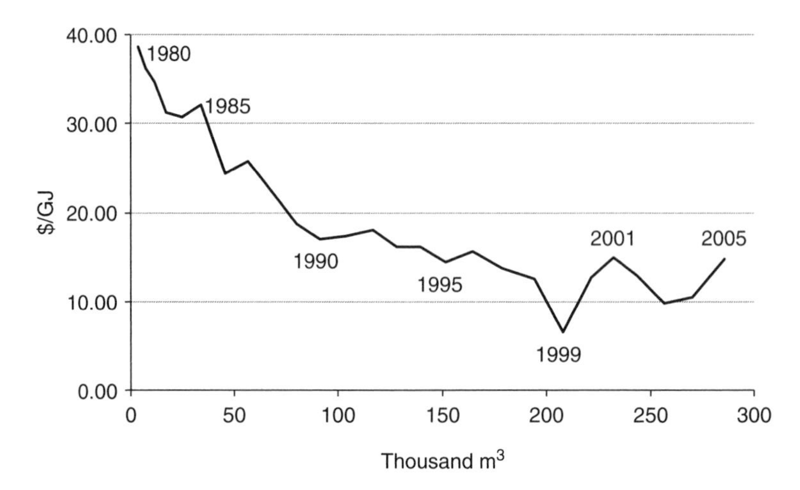

Figure 6-13 Prices paid to ethanol producers in Brazil.
SOURCE: CGEE and BNDES (2008), adapted from Goldemberg et al. (2005).

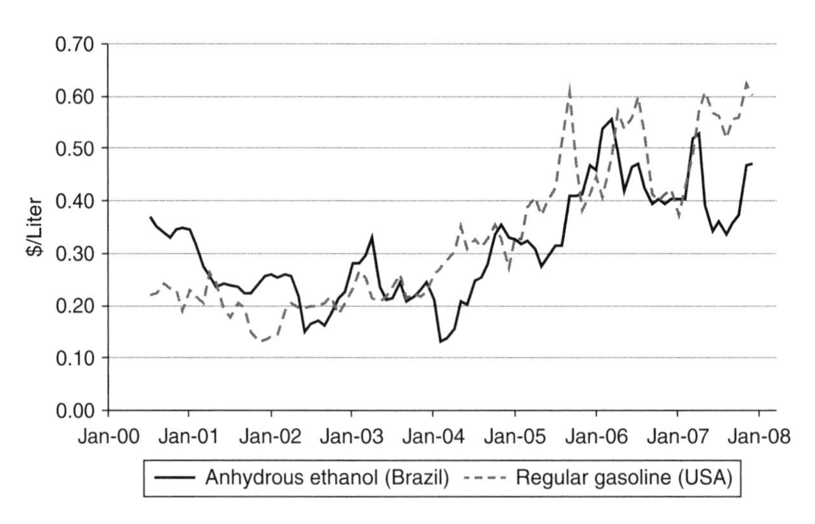

Figure 6-14 Prices paid to producers (ex. taxes) for U.S. gasoline and Brazilian anhydrous ethanol.
SOURCE: CGEE and BNDES (2008), using data from CEPEA (2008) and EIA (2008).

Compared to other forms of ethanol, Brazil produces the least expensive form (Arruda, 2011). In 2005, for example, Brazilian sugarcane ethanol reached a break-even point in which revenue equaled costs at roughly $33/barrel, whereas other feedstock reached this balance in the higher $50–80 range (Food and Agriculture Organization [FAO], 2008, citing FAO, 2006). During that time, oil sold for roughly $50/barrel.

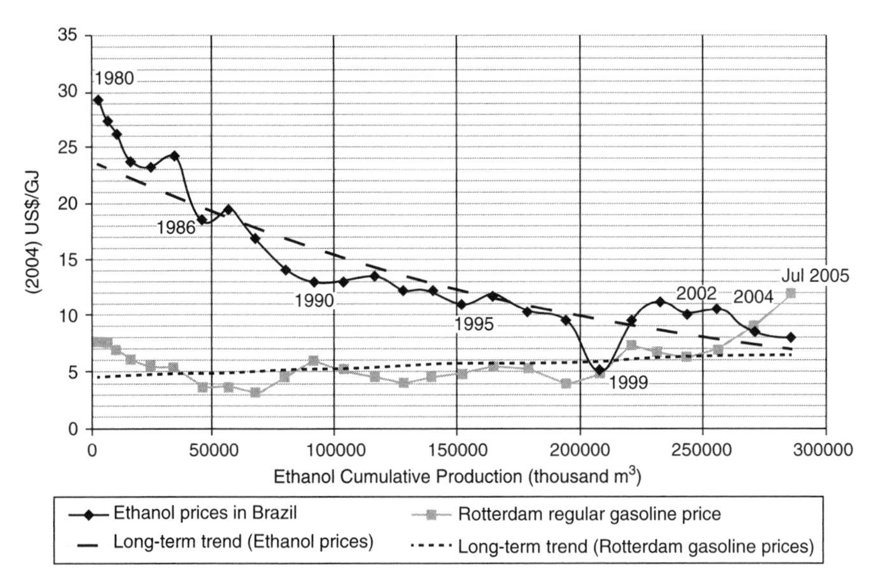

Figure 6-15 Ethanol versus Rotterdam gasoline prices.
CREDIT: Coelho and Guardabassi (2014).

Viewed in terms of foreign exchange savings associated with the displacement of oil imports by ethanol, Ricardo Dornelles, former Director of Renewable Energy in the Brazilian Ministry of Mines and Energy, estimated that the Brazilian government saved $42 billion between 1970 and 2009 (Dornelles, 2010). Particularly high savings were evident in 2008 and 2009.

Turning to a question of who paid, a simplified answer would point to automobile drivers who used gasoline particularly in the period from 1975 through to the late 1990s, because taxes on gasoline were used to cross-subsidize ethanol development at the time (Goldemberg, 2011; Interview, 2010). A more complex answer would point to the industries, actors, and fuel types that did not benefit, when support for sugarcane ethanol was promoted in lieu of other areas. Another component to the answer lies in the costs absorbed by the BNDES National Development Bank, the fiscal budget, and distributors, like Petrobras, which covered initiatives such as infrastructure conversion. When better access to relevant data becomes available, this area merits further study.

Societal Acceptance

The essence of the Brazilian energy choice in the 1970s was not a difficult decision: Brazilian society was compelled to support ethanol and domestic jobs or take an alternative path that included gasoline rationing and higher unemployment rates.

Despite the compelling aspects of biofuels adoption, acceptance was not always assured among automobile manufacturers and government ministers, financial institutions, Petrobras, and consumers. Automobile manufacturers, for instance, were regularly mentioned as being compelled by the government to produce neat vehicles in the late 1970s (Interviews, 2010–11). However, others, in interviews and published reports, describe the dynamic as one in which consensus was formed among auto manufacturers to mobilize their industrial capacity around a more stable fuel (Chen, 2011; Interview 2012). Without question, auto manufacturers had justification for maintaining conventional production lines with ICE-based vehicles, since ICEs were well-accepted worldwide and neat vehicles were a less-known, disruptive technology. Yet the persistent side effects of the global oil shocks were exceptionally high gasoline prices, a rising national debt, and a vulnerable balance of payment, as well as uncertainty about prospects for both the automobile industry and the economy. Here, neat vehicles that had been tested in the CTA lab provided an alternative path to sustain the automobile market. If the auto manufacturers were resistant to the change, they nonetheless worked with the negotiated arrangement, producing neat vehicles during the early part of *ProAlcool* and helping to co-create the new market. Auto manufacturers, rather than the government, later spurred the introduction of flex fuel vehicles that occurred in 2003.

Various government ministers were also, by some accounts, less than enthusiastic of Brazilian biofuels development (Barzelay, 1986; Interviews, 2010). This should not be surprising because the *ProAlcool* program in its early years required extensive resource mobilization and political commitment amid volatile economic times. Large-scale public expenditure, some with project financing at rates below inflation, can be risky, especially when high inflation and rising debt are prevailing conditions. One account of the senior level decision-making described a meeting before the launch of *ProAlcool* in which President Geisel told ministers they needed to be "on board" with the Program or should not plan to remain in their roles (Interview, 2010). Any remaining resistance by ministers appears to have been addressed through internal mechanisms of the top leadership.

Financial institutions, namely the Banco do Brasil and the BNDES, were also identified as early resistors of *ProAlcool* (Barzelay, 1986; Demetrius, 1990).[22] Once the Program was launched, these institutions were known for extensive delays in financing approved projects (Barzelay, 1986; Demetrius, 1990). This impedance appears to have changed to cooperation after 1979, when deepening political and industry commitment to *ProAlcool* appeared to solidify.

22. Other financial institutions charged with supporting *ProAlcool* were Banco do Nordeste do Brasil and Banco da Amazonia (Decree 76.593). There is no evidence showing they resisted implementation. They were based in regions that had strong need for economic development, so it is not likely that they (as regional development banks) would impede the process.

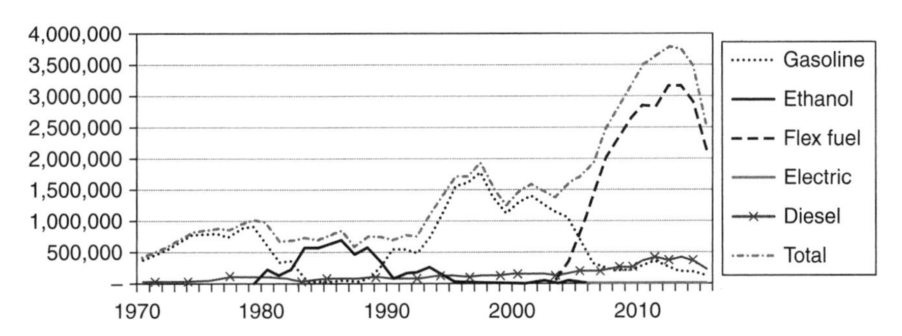

Figure 6-16 Registration of new automotive vehicles by fuel type in Brazil.
SOURCE: ANFAVEA data (2017). Note: This does not account for vehicles converted from internal combustion to neat technology.

Petrobras was also mentioned as resistant to the biofuels development plan (Barzelay, 1986; Demetrius, 1990, Interviews, 2010). Given that Petrobras was legally charged to manage liquid fuels for the domestic market and its primary focus was on oil, the prospect of an emergent biofuels market meant that the company's institutional strength could be diluted by the growing influence of sugar industry actors. Moreover, Petrobras was created in the 1950s to locate and exploit petroleum reserves, and, at the time of *ProAlcool*'s launch, the company had not yet proven itself (Barzelay, 1986). With *ProAlcool*, Petrobras served as the guaranteed buyer of set annual amounts of ethanol and had a crucial, albeit somewhat debated, role in adapting the transport fuel infrastructure. Up through recent times, Petrobras remains engaged in supplying ethanol and, now biodiesel as well.

The acceptance and resistance of automobile drivers to the biofuels transition can, in some respects, be gauged by the types of automobile that were purchased. Figure 6-16 shows a sharp rise in new, neat vehicle registrations in the 1980s and flex fuel ones in the 2000s. Notably, flex fuel technology placed choice in the hands of the automobile driver, and drivers responded overwhelmingly in favor of such technology.

Over the period of study, social acceptance varied and appeared to be moderate to significant. No notable protests occurred, and the most obvious resistance was inherent in consumer purchase choices, which were allowed to evolve. Less obvious resistance among state actors and institutions may have been managed behind the scenes.

Industrial Development

Fundamentally, the sugar and automotive industries that existed in 1970 were repurposed, enabling scaled production and the wider use of ethanol.

Ethanol production rose by a factor of roughly 130 between 1970 and 2014, with Brazil becoming a top, global ethanol producer and exporter during the period. In production terms, the Brazilian biofuel industry is currently the second largest in the world (F.O. Licht, 2016). In connection with this, foreign participation has grown in the Brazilian ethanol-sugar market. In 2009–10, 22% of the mills were owned by foreign capital, relative to 7% in 2007–08 (UNICA, 2010). Brazil has also served as a leader in technology transfer related to biofuels by offering training both domestically and abroad (Interviews, 2010–11). In addition, Brazil is a founding member of the Global Bioenergy Partnership (Hira, 2011).

Specific to the automobile industry, the introduction of flex fuel technology that was tailored to the Brazilian market in 2003 has proved to be a commercial success (ANFAVEA, 2017; Kamimura and Sauer, 2008). Flex fuel is a dominant design in an industry that produced 2.1 million lightweight commercial vehicles in 2016, with 85% of newly registered cars and light commercial vehicles being flex fuel (ANFAVEA, 2017). Brazil also leads globally with the majority of the flex fuel engine market and the highest penetration rate for flex fuel vehicles (Technavio, 2016). As of 2013, 13 manufacturing brands and 163 models of flex fuel vehicles were available (Joseph, Jr., 2013).

CONCLUSION

Brazil's biofuels transition highlights how an energy market and industry can be rapidly mobilized and integrated by repurposing existing ones. In doing so, a simple product of the sugar industry became a mainstream transport fuel for use in a new class of vehicles. This case underscores the importance of industry taking the lead at times (and of government not impeding this).

In the last four and a half decades of the modern, Brazilian biofuels transition, incremental and disruptive innovations occurred in agriculture, energy, and transport, with dual-use capabilities representing an unusual feature of this system. Today, Brazilians have flexibility in the dual-use substitutability of sugar versus ethanol in agro-industry production, various fuel options in transport, and new opportunities for bagasse waste to be used in commercial power generation as well as cellulosic ethanol. This dynamic can enable or undercut the enduring nature of Brazil's biofuels system. It is here where continuing innovation and (at least) neutral policy will matter.

7

Danish Wind Power

Alternating Currents

The days of wind being costly and technologically immature are over.
—Fatih Birol (*Global Wind Energy Council [GWEC], 2016)*

INTRODUCTION

According to Michael Zarin, Director of Government Relations with Vestas Wind Systems, there is nothing "alternative" about wind power anymore (Biello, 2010). After all, wind generation is the most cost-effective option for new grid-connected power in markets like Mexico, South Africa, New Zealand, China, Turkey, Canada, and the United States (Renewable Energy Policy Network [REN21], 2016). At 433 GW of cumulatively installed capacity in 2015 worldwide, more than half was added in the past 5 years (REN21, 2016). This technology may be used by individuals, communities, and utilities. It can be grid-connected or off-grid, and be used onshore or offshore.

This chapter examines the influences and evolution of the Danish wind transition, highlighting how ingenuity and often less-obvious incremental advances produced a world-class industry. It reveals how citizens can be important catalysts of energy system change. The case also indicates that innovations can emerge in practices and policy, not just technology, science or industry.

PROFILE FOR DANISH WIND POWER

Denmark is a cultural and traditional technology leader for modern wind power.[1] This country of roughly 5.6 million people and GDP of approximately $65 billion in 2016 (ppp) (Central Intelligence Agency [CIA], n.d.) is where today's dominant, wind turbine design was established and where state-of-the art wind technology testing centers are based. It is also the site of the first, commercial-scale offshore wind farm, built in 1991.

Denmark has a world-class hub for wind energy technology (Megavind, 2013; State of Green, 2015; Renewable Energy World, 2016). Top-ranked companies like Vestas, LM Wind Power, Siemens Wind Power, A2SEA, and MHI Vestas Offshore Wind are among those that base core parts of their global operations in Denmark. A close network of wind engineers and their professional affiliates drives the industry, which includes ancillary services and subcomponent supplies. Wind energy technology also represents one of Denmark's top-ranked exports (United Nations Comtrade, n.d.).

Currently, Denmark has more wind power capacity per person than does any other country in the world (REN21, 2017). This Northern European nation is on track to derive 50% of its electricity from wind power by 2020. With more than 5,000 wind turbines in Denmark, the turbines generate more electricity on windy days than is demanded domestically (Danish Energy Agency [DEA], 2016; Lilleholt, 2015), allowing for export or storage.

BACKGROUND

The use of wind energy in electricity has deep roots in Danish history. Poul la Cour, Denmark's counterpart to Thomas Edison, established a wind turbine test station at the Askov Folk High school in 1891. It was there that he installed some of the world's first electricity-producing wind turbines and trained early wind electricians (Ackermann and Soder, 2000; Gipe, 1995; Redlinger et al., 2002).[2] La Cour was also behind the founding of the world's first *Journal of Wind Electricity*, the Danish Wind Energy Company, and the Danish Wind Electrical Society in

1. Wind power has been used for centuries in navigational travel and agriculture. Wind energy was employed for propulsion on the Nile River as early as 5,000 BC (Department of Energy [DOE], n.d). By 200 BC, wind power was utilized for pumping water in China and grinding grain in Persia and the Middle East (DOE, n.d.). This chapter focuses on the modern use of wind energy in electricity.

2. Some students of la Cour's wind electrician course later built turbines for the F. L. Smidth company during World War II (Krohn, 2002*a*).

1903 (Jorgensen and Karnøe, 1991; Pederson, 2010). A la Cour-based wind turbine design was commercialized around the time of World War I and, by the end of that war, 120 Danish rural power stations utilized wind turbines with a rated power of 20–35 kW, producing roughly 3% of Danish electricity (Andersen, 2007; Meyer, 2011; Redlinger et al., 2002).[3] The use of wind for electricity complemented wind utilization for mechanical energy. Around the same time, 30,000 windmill units were producing mechanical energy with a wind power equivalent of 150– 200 MW (Meyer, 1995). This early use of wind energy would not sustain over the next few decades, but would provide vital cultural reference points for future adopters.

During the early period of wind development, wind research occurred in regions like Denmark, the United States, the United Kingdom, and Germany (Jorgensen and Karnøe, 1991; Meyer, 1995). Throughout the interwar period and World War II, Danish companies F. L. Smidth and Lykkegaard Ltd. leveraged emergent research in aerodynamics to produce 30–70 kW windmills (Andersen, 2007; Gipe, 1995; Jorgensen and Karnøe, 1991; Pedersen, 2010). Danish energy consumption continued principally with fossil fuels. Nonetheless, experimentation continued, such as with Danish power utility engineer and former la Cour student Johannes Juul who initiated a research and development program on wind utilization. Key to the research was a 200 kW turbine with a 24 meter rotor that he installed in Gedser, 90 miles south of Copenhagen (Andersen, 2007; Jorgensen and Karnøe, 1991; Pedersen, 2010). This turbine, known as the Gedser model, operated from 1959 to 1967.[4] When Danish entrepreneurs later mobilized in the 1970s, they revisited elements of the Gedser model,[5] which would become the basis for the dominant design that still exists today.

MODERN TRANSITION: 1970–THE PRESENT

Anyone considering Denmark in 1970 would not likely have foreseen the wind energy transition that was yet to unfold (Figure 7-1). After the Gedser demonstration project was concluded in the 1960s, wind energy development

3. By 1918, roughly 3 MW of installed wind capacity was in place relative to the total Danish electricity capacity of 80 MW (Meyer, 1995).

4. The program supporting Juul's turbine was discontinued in 1962 when the Wind Commission concluded (based on a cost-effectiveness calculation of saved fuel rather than customer price) that the intermittency of produced power would reduce fuel consumption, but not the necessary capacity of conventional power plants (Jorgensen and Karnøe, 1991).

5. This three-bladed, stall-regulated, upwind turbine was designed with a horizontal axis, automatic yaw system, fixed pitch, asynchronous motor, and pitchable blade tips to modulate overspeed (Jorgensen and Karnøe, 1991; Pedersen, 2010; Pedersen and Xinxin, 2012).

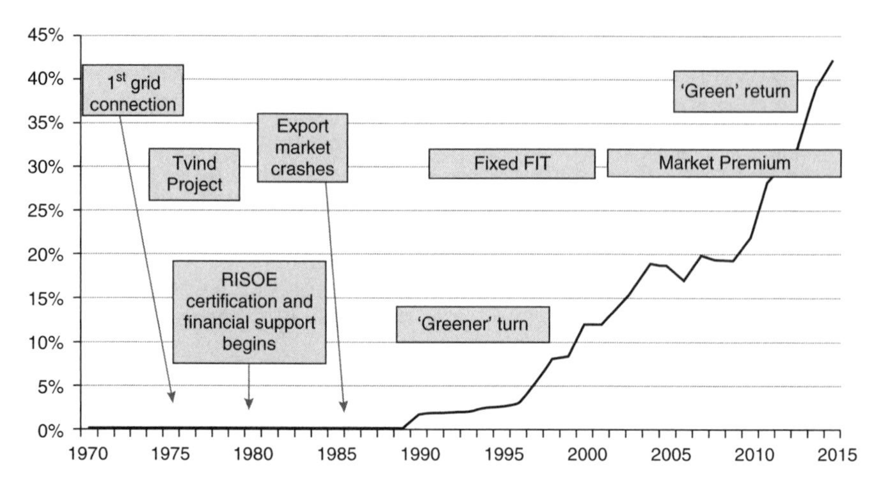

Figure 7-1 Danish wind power development (wind power as a share of total Danish electricity).
SOURCE: Compiled from Energinet (2016); Danish Energy Agency (DEA, n.d.); and author's estimates.

had receded to negligible levels until the first oil shock. At the time, Denmark was heavily reliant on oil imports, with 92% of the country's primary energy mix fueled by oil, obtained largely from the Middle East (Maegaard, 2009; Sawin, 2001).[6] New industries and sources of income, including construction, the chemical industry, electronics, and pharmaceuticals in addition to oil refineries and power stations, now began to overtake traditional sources of income—namely agriculture, small to medium-sized industrial craft-based enterprises, and shipbuilding (Jamison, Eyerman, and Cramer, 1990; Van Est, 1999). These conditions set the stage for the energy system shift that would commence.

Entrepreneurial Mobilization and New Market Formation: The 1970s–Early 1990s

Briefly sketched, the oil crisis of 1973–74 was a wakeup call. Petroleum imports for Denmark more than tripled in nominal currency terms from 1972 to 1974

6. Denmark did not make its first discovery in North Sea oil until 1966, in what became known as the Kraka field. Danish production later began with the Dan field in 1972 (IEA, 1980). Contract renegotiations by the Danish government and oil and gas industry in 1980–81 led to increased activity. Development followed in the Gorm, Skjold, Tyra, and Rolf fields, with gas production becoming operational in 1984 (IEA, 2011*c*).

and increased by a factor of 7 between 1972 and 1980 (UN Comtrade, n.d.). Denmark had not yet matured as an oil or gas producer in the North Sea, so in response, Danish energy consumers turned down their heat, insulated their homes, and began car-free Sundays (Energinet.dk, 2009*a*). The government assigned responsibilities for energy policy matters to the Minister of Trade and Industry and announced that Denmark would accelerate nuclear development (OECD/NEA, 2007; Vasi, 2011). Fuel for electricity shifted from oil to coal, and the electric utility ELSAM released a list of potential locations for siting a nuclear power plant (Vasi, 2011). The palpable sense of urgency surrounding energy spurred Danish inventors, scientists, environmentalists, and communities to develop alternatives to energy imports.

Entrepreneurs and activists were some of the first to mobilize around the idea of scaling wind energy. Inventive citizens, including farmers, blacksmiths, machinery manufacturers, and environmentalists, drew upon technical skills, historical familiarity, and available components to begin tinkering with wind technology. One such entrepreneur was carpenter Christian Risiinger. He used interchangeable parts from other equipment, such as bicycles, to build a wind turbine with an asynchronous generator that he then connected to the power grid in 1976 (Andersen and Drejer, 2008; citing Jensen, 2003; Grove-Nielsen, n.d.; Karnøe, 1990). Accounts vary on whether his model was connected with or without permission (Andersen and Drejer, 2008; citing Jensen, 2003; Energinet.dk, 2009*a*; Grove-Nielsen, n.d.; Karnøe, 1990). In any case, the turbine drew widespread attention, with Risiinger later pioneering serial production of wind turbines.

Environmental groups also stepped into action. The *Organization for Information about Nuclear Power* (OOA) focused on stopping the adoption of nuclear power through public debate, guidance about alternatives, and advocacy of decision-making that should occur in parliament rather than within central authorities (Vasi, 2011).[7] The *Organization for Renewable Energy* (OVE) grew as an offshoot of the OOA, focusing on energy-related policy and demonstrating the feasibility of renewable energy technologies (RETs). Its members participated in government committees and hearings on energy, published policy papers, and organized early knowledge exchanges in the tradition of old Danish folk schools (Vasi, 2011, citing various).

Among the mobilizers, dedicated amateurs at the Tvind school of Western Jutland also set out to build the world's largest wind turbine (Tvind Internationale Skolcenter, n.d.; see also Karnøe, 1990). Partnering with scientists and others, the

7. The Swedish Barseback nuclear power plant, based 20 kilometers outside of Copenhagen, provided an early point of reference for opposition by the OOA. The Barseback plant, a boiling water nuclear plant, went into commercial operation in 1975.

volunteer Tvind crew constructed a community turbine between 1975 and 1978. As the project neared completion, an estimated 77,000 people visited the site over the following months (Tvind Internatioinale Scolcenter, n.d.; Vasi, 2011). The resulting Tvind turbine still operates today (as of the writing of the book) at around 1 MW, providing evidence that collective ingenuity can overcome imported energy dependence (Grove-Nielsen, n.d.; Interviews, 2010–12; Vasi, 2011).[8]

Meanwhile, in 1976, the government released its first national energy policy, *Danske Energipolitik*. To insulate Denmark against energy supply crises, the plan prioritized fuel switching to nuclear and coal, in tandem with conservation (IEA, 2002*a*; Meyer, 2004; Nielsen, 2005, citing Handelsministeriet, 1976).[9] Taxes were imposed on electricity to support R&D for renewables. The Danish Energy Agency (DEA) was also formed to assist public authorities in overseeing the development, supply, and use of energy (OECD/NEA, 2007; Sawin, 2001).

Around the same time, scientists from a number of Danish universities put forward an alternative plan, *Skitse til alternative energiplan for Danmark*,[10] outlining a path to energy self-sufficiency that would leverage renewable energy and efficiency. The Danish Academy of Technical Sciences (ATV) also released two wind energy reports in 1975–76. The first report proposed a wind energy program for Denmark, concluding that 10% of Danish electricity could be derived from wind power without altering the existing energy system (Gipe, 1995, citing Peterson, 1990). The second report outlined a 5-year action plan for adopting wind energy (Blegaa, Josephsen, Meyer, and Sorenson, 1977; Meyer, 1995, citing Danish Academy 1976).

Public research also prioritized energy. The National Energy Program supported, among other aims, research on large-scale turbines by the Technical University of Denmark (DTU), electric utilities, and the government

8. Another network of people in Western Jutland teamed up in the founding of NIVE to develop alternative forms of energy for power and heating. NIVE or the North-Western Jutland Institute for Renewable Energy, led by Preben Maegard, was a precursor to the Nordic Folk Centre for Renewable Energy. For more in depth discussion of actor networks during this time, see Dykes (2013).

9. In conjunction with the Heat Plan, which was also released in 1976, the Energy Plan highlighted: conservation targets to reduce the average energy growth of total primary energy; district heating to cover 20% of heating by 1985; natural gas to contribute an increased amount of the energy supply; imported coal to contribute 65% of electricity in 1995, compared with 35% in 1975; and, finally, nuclear energy for power production after 1990 (Nielsen, 2005, citing Handelsministeriet, 1976; IEA, 1980 and 1998*b*). No decision on nuclear energy was expected before 1981, hinging on disposal solutions for waste management.

10. The *Sketch for Energy Plan in Denmark* (1976) is cited in GWEC and IRENA (2012).

(Nielsen, 2005).[11] The Development Program was also formed to provide grants for projects, such as the establishment of the Nordic Folk Centre for Renewable Energy in Jutland that began in the early 1980s (Nielsen, 2005; see also the later section "Drivers and Barriers"). In 1978, the Danish government initiated research on small-scale wind turbines with a test station at the Risoe Laboratory (Andersen, 1998).[12] The initial wind power objectives for the station were to assess existing designs and assist in the design and development of evolving models. In conjunction with wind power initiatives, a knowledge network developed with meetings held for wind turbine producers, the DEA, and turbine owners (Anderson, 1998, citing Handelministeriet, 1978).

In the private sector, two industry organizations formed: the *Danish Wind Power Owners Association* (later *Danish Wind Turbine Owners Association* [DWTO] or *Danske Vindkraftvaereker*) and the *Danish Wind Mill Manufacturers Association* (DWMA) (later the *Danish Wind Industry Association* [DWIA] or *Vindmolleindustrien*). Both would support information-gathering, advocacy, and negotiation efforts to safeguard member interests. One outcome of the association activity was the production of the Danish industry paper *Naturalig Energi,* which reported on turbine model performance. It would provide early insight for users and producers as designs were adapted (Karnøe, 1990; Garud and Karnøe, 2003). Another offshoot of association activity was partnering by the DWTO with Risoe Laboratory to pressure wind turbine manufacturers to improve the technical reliability of early turbines (Karnøe, 1990). The more secure brake system enhancement which was developed from these efforts would be a critical distinguishing characteristic of Danish wind technology in the ensuing years.

By the late 1970s, many of today's Danish wind experts already demonstrated early progress in the emergent industry, with a dozen small Danish companies, including Vestas, Danregn (later Bonus, then Siemens), and Nordtank producing and selling wind turbines (Grove-Nielsen, n.d.; Interviews, 2010–12; Karnøe, 1990).

11. In conjunction with the National Energy Program, the original Gedser turbine of the 1950s and 1960s was refurbished. Aided by computer design and aerodynamic blade theory, findings from the Gedser tests served as a basis for two 630 kW test turbines: Nibe A (with stall regulation) and Nibe B (with pitch regulation) were constructed for further testing (Nielsen, 2005, citing Handelsministreriets og elvaerkernes vindkraftprogram, 1981, and Karnøe, 1990).

12. Ironically, Risoe Laboratory was established in the 1950s to conduct research on nuclear energy, in connection with the work of physicist Niels Bohr (Andersen and Drujer, 2008).

When the second oil crisis struck in 1979, Denmark elevated energy oversight to the ministerial level. Heavy taxes were imposed on petroleum and electricity, and a policy substituting coal for oil continued (IEA, 1980; IEA and IRENA, n.d.).[13] As economic pressures heightened, the Danish parliament passed legislation to stimulate employment and new energy technologies with the *Energipakken*. A key tenet of this legislation was an investment tax credit, paying up to 30% of capital costs for energy technologies, such as wind power, solar power, biogas, and heat pumps (IEA and IRENA, n.d.; Interviews, 2011; Sawin, 2001). In order for wind turbine owners to claim the credit, they were required to use the Risoe-certified turbines. This approach paired technology certification with financial support, creating regular interface for industry, turbine owners, and Risoe Laboratory to work together (Madsen, n.d).[14]

The government mobilized along additional lines. Parliament postponed a decision on whether to adopt nuclear power pending findings on nuclear waste management studies (IEA, 1980, 1981; Vasi, 2011). The center left, Social Democrat-led government engaged in a renegotiation of an oil and gas concession contract for North Sea exploration and development (Interviews, 2011–12; Sawin, 2002). The Minister of the Environment also wrote to local authorities indicating that the national government would support wind turbine installations by private individuals and encouraged local authorities to manage their permitting processes without undue delays (Madsen, n.d.). The Ministry also ordered utilities to provide grid access for small wind turbines and to arrange a fair deal on power generation payments (Madsen, n.d.).[15]

13. During this period, North Sea natural gas and oil constituted some of the largest indigenous sources of conventional energy that could be scaled into the energy mix for Denmark. However, use of North Sea gas required heavy investment and renegotiation of contracts (IEA, 1980, 1981; Interviews, 2011–12). In planning terms, one estimate of the time suggested that Denmark would be able to meet one-third of its total primary energy needs by 1990 with stepped increases of oil and gas output in the North Sea (IEA, 1981 and n.d.).

14. During the lifetime of the credit, approximately 2,567 turbines were supported (Nielsen, 2005, citing Energimiljoradet, 1998). The credit was reduced over time and phased out in 1989 (Meyer, 1995). An estimate of this policy's total outlay in 1995 was roughly $58 million (Sawin, 2001; see also Appendix "Country Timelines").

15. At the time, grid connections were arranged on an *ad hoc* basis. (Interviews, 2011–12; Tranaes, 2000). Grid connection rules were later established with a voluntary agreement in 1984 between utilities and wind power producers in which utilities agreed to pay 70–85% of the net customer price, excluding taxes and charges for wind power (Ibid; Meyer, 1995; Nielsen, 2005). The net price roughly equaled the utility production plus distribution costs, which varied somewhat in relation to contemporary coal prices (Meyer, 1995).

To facilitate the new wind turbine approval scheme, Risoe scientists developed codes of practice and design standards based on their increased understanding of manufacturer and turbine owner needs (Nielsen, 2005, citing Rasmussen and Jensen, 1999). Drawing upon certification assessments, Risoe scientists shared early technical insights with manufacturers and focused their research on problems that were discovered in the certification reviews (Interviews, 2011). Risoe scientists also developed a pioneering wind atlas in 1981, using computational measures to estimate wind resources in complex terrain (Meyer, 1995, citing Petersen, Troen, Frandsen, and Hedegaard, 1981; Meyer, 2004; Nielsen, 2005). In turn, manufacturers applied the knowledge gained from the wind resource map to their production of small turbines (Nielsen, 2005).

Building on the work of the Risoe Laboratory, Danish environmental authorities, in conjunction with the Danish Energy Agency, developed a national assessment that integrated a wind resource appraisal with specifics of the environmental landscape and planning conditions for onshore and offshore wind sites. This information was made available to local planners and formed the basis for subsequent work by the Wind Siting Committee (Nielsen, 2005, citing Planstyrelsen, 1981–86).

In policy terms, governmental support of wind development became more concrete in 1981, when the government introduced a subsidy for wind power on a per kWh basis (Sawin, 2001, citing Madsen, 2001). This subsidy was linked to energy taxes and would remain until 1992 when it was replaced by a package of subsidies and taxes. A fuller energy plan, *Energiplan 81,* included socioeconomic aims such as the decoupling of energy from economic growth, the securing of low-cost community energy, together with environmental considerations (Danish Ministry of Energy and the Environmental Ministry, 1997; Nielsen, 2005; Sawin, 2001; van Est, 1999). The new plan incorporated a wind power production target of roughly 1.3 TWh to be attained by 1995 (Nielsen, 2005, citing Danish Ministry of Energy, 1981) and emphasized continued conservation, extension of natural gas utilization, production of combined heat and power (CHP), and the acceleration of oil and gas exploration in the North Sea. Similar to actions following the release of the 1976 Danish energy policy, the announcement of the 1981 energy plan was followed by the release of alternative energy plan *Energi for fremtiden* that was produced by independent energy experts from Danish universities (GWEC and IRENA, 2012, citing Hvelplund et al., 1983; Meyer, 2005).[16] Based on conservation potential and new scenario methods, the alternative plan laid out a strategy without nuclear energy that

16. Wind power was a central component for both the 1976 *Sketch for Energy Plan in Denmark* and the 1983 *Energy for the Future: Alternative Energy Plan* as a key alternative to nuclear power (GWEC and IRENA, 2012).

entailed sourcing CHP, conservation, and RETs, such as wind power (Meyer, 2004, citing Hveplund et al., 1983).

Between 1982 and 1991, the government-sponsored Committee for Promoting Renewable Energy Systems became a central supporter of projects focused on wind power, solar power, and biomass (Meyer, 2004; Nielsen, 2005).[17] Under the leadership of solid-state physicist Niels Meyer, projects included early analysis for offshore wind farm programs (Meyer, 2004; Vasi, 2011).

As the wind industry gained traction in the new domestic market, opportunities for Danes in wind development also appeared internationally. Favorable policy and economic conditions in the United States provided an early and strong basis for exports (Madsen, n.d; Interviews, 2011).[18] The investment boom in California's wind power market motivated Danish wind turbine manufacturers to scale their businesses from batch production to mass production. By 1985, 900 MW of installed wind capacity had been added in California from Danish and other sources (Guey-Lee, 1998). Within this export market, Danish turbines grew from 5% to 50% of the market share between 1983 and 1985 (Karnøe, 1999, citing Stoddard, 1986). To address the end-to-end needs of international clients and expand market share, many Danish wind technology companies began to expand vertically, creating parallel business lines in financing, insurance, and post-sales services (Karnøe, 1990).

While wind turbine manufacturing was growing rapidly during years of the California wind boom (1982–1986), wind technology was also making significant progress. Advances in turbine size and performance occurred, with rated capacity increases from 55–65 kW to 75–99 kW (Karnøe, 1990). More specialized and less expensive components also came into use as suppliers accommodated the needs of the turbine producers (Karnøe, 1990). A notable technical advance was the development of a microprocessor control system in 1984 that

17. Once its activity began, the Committee worked with a budget of roughly €4.6 million per year or €30 million for the total period (Meyer, 2004).

18. The Public Utilities Regulatory Policies Act (PURPA), passed by the US Congress in 1978 as part of the National Energy Act, promoted greater use of RETs. Under PURPA provisions, electric utilities were required to purchase power from qualifying facilities if the cost was less than what the utility would otherwise pay by producing the power or purchasing it from an external source (i.e., avoided cost rate). Additional federal and state measures established investment tax credits applicable to wind generation. When federal and state tax credits were considered together, tax savings could equal 50–55% of the investment (Righter, 1996).

Furthermore, the California Public Utilities Commission set high avoided-cost rates for RET energy sources (to encourage RET utilization) in Standard Offer 4 contracts that were guaranteed for 10 years (Guey-Lee, 1998). In tandem with PURPA, this policy was an early precursor to feed-in tariffs. Together with these policies, favorable exchange rates also made it profitable for Americans to import Danish turbines (Karnøe, 1990).

enhanced turbine performance and enabled centralized control of several hundred wind turbines (Karnøe, 1990).

According to reports from the time, Danish turbines had the best technical performance record in the market (Karnøe, 1990, citing Stoddard, 1986; and Karnøe, 1990, citing California Energy Commission, 1988; Meyer, 2004). One leading wind engineer from the United States summarized the wind technology experience in 1986 by pointing to the technology uncertainty of aerodynamic loads and dynamic motions (Karnøe, 1990, citing Stoddard, 1986). Danish models reduced such design risk by limiting aerodynamic load exposure, letting inertial forces balance the loads, and preventing dynamic motion (Karnøe, 1990, citing Stoddard, 1986).

The wind technology boom in California was short-lived. As energy policy and economic conditions evolved, the California market collapsed around 1986[19] with Danish wind turbine manufacturers suffering more than 20 bankruptcies, and the industry consolidating (DEA, 2011*b*; Interviews, 2011; Madsen, n.d.). Reorienting production for the domestic market, Danish wind turbine manufacturers returned to product development (Karnøe, 1990; Madsen, n.d.). By 1987, seven of the top ten wind turbine producers worldwide were Danish (Karnøe, 1990, citing *Wind Power Monthly*, 1988).

Up to this point, owners of wind turbines in Denmark consisted primarily of farmers, cooperatives, and environmentalists (DWTO, n.d.; Nielsen, 2005). This style of local development was facilitated by rules that limited turbine ownership to community residents living nearby (Hvelplund, 2011; Interviews, 2011–12). The second half of the 1980s, however, brought another demographic group into the ranks of wind turbine owners—utilities. The government made two deals with electric utilities Elkraft, ELSAM, and the Association of Electric Utilities—one deal in 1985 and a second in 1990—which obligated Elkraft and ELSAM to collectively install at least 100 MW without subsidies (DEA, 2011*b*; Interviews, 2011; Karnøe, 1990; Madsen, n.d.; Tranaes, 2000).[20] These deals would catalyze the nascent market for the struggling wind industry as it recovered from the disappearance of the California market.

19. American federal investment tax credits expired in 1985, with California support expiring shortly thereafter (Karnøe, 1990). The 1986 collapse in oil prices and improved conditions for natural gas meant that avoided costs were less favorable for RETs after the 10-year Standard Offer contracts expired (Guey-Lee, 1998).

20. Accounts of these deals vary. One view holds that the arrangement in 1985 was precipitated by utilities wanting to reclaim control of power generation inputs. The government agreed to set limits for independent power production but also required utilities to cover part of increased wind production without the price guarantee available to independent power producers (Karnøe, 1990). Somewhat differently, a number of interviewees indicated that the utilities were reluctant participants at the time (Interviews, 2011–12).

Outlooks for energy pathways in Denmark in the 1970s and early 1980s seemed to diverge between decentralized configurations of renewables with efficiency characterizing one path versus centralized nuclear energy, and coal characterizing the second path. The former attracted a wide range of community supporters and entrepreneurs, whereas the latter resonated with power companies and industrial actors (Interviews, 2011–12; Vasi, 2009) because nuclear energy and coal-based generation aligned more closely with the centralized nature of the existing infrastructure. Forces favoring nuclear energy, however, were overcome by persistent pressure from groups like Danish nongovernmental organizations (NGOs) OOA, OVE, and Friends of the Earth; general opposition by Danes to a Swedish nuclear plant sited nearby; and the nuclear accident at Three Mile Island in the United States. In 1995, Parliament voted, under a conservative government, to remove nuclear power as an option in Denmark's energy planning (IEA, 1988; Meyer, 2004). This decision was reinforced a year later by the nuclear accident at Chernobyl.

Throughout the rest of the 1980s, utilities pursued plans with wind generation, but encountered delays due to siting issues (Karnøe, 1990). These would be addressed with new approaches to planning in the 1990s. Consideration also began to look toward offshore wind (IEA, 1988), with a government committee publishing a report on potential siting locations for offshore wind turbines (IEA, 1989). The report recommended that two offshore wind projects be built to test the economic, technical, and environmental aspects of development (Sawin, 2001). Implementation of such a project would occur in the not-too-distant future.

Overall, the 1970s and 1980s reflected the establishment of a modern Danish wind industry and market. Early experimentation and technology development occurred alongside standard-setting and policy development. While there was little, overtly discernable progress domestically in installed capacity, wind generation, and shares of the energy mix within Denmark (Figures 7-1, 7-2 and 7-3), this would change in the ensuing period.

Rapid Growth and Adaptation: 1990–2001

The 1990s and early 2000s reflected a very different character for Danish wind development. Two robust energy plans framed the period by focusing on sustainable development and carbon dioxide (CO_2) mitigation. An active Minister of Energy and the Environment, significant adaptation and learning with planning processes, and the early stages of liberalization in the electricity sector also shaped the focus of the times.

In 1990, a new, energy plan, dubbed *Energy 2000*, reinforced priorities in wind energy, by calling for the installation of 1,300–1,500 MW of wind power

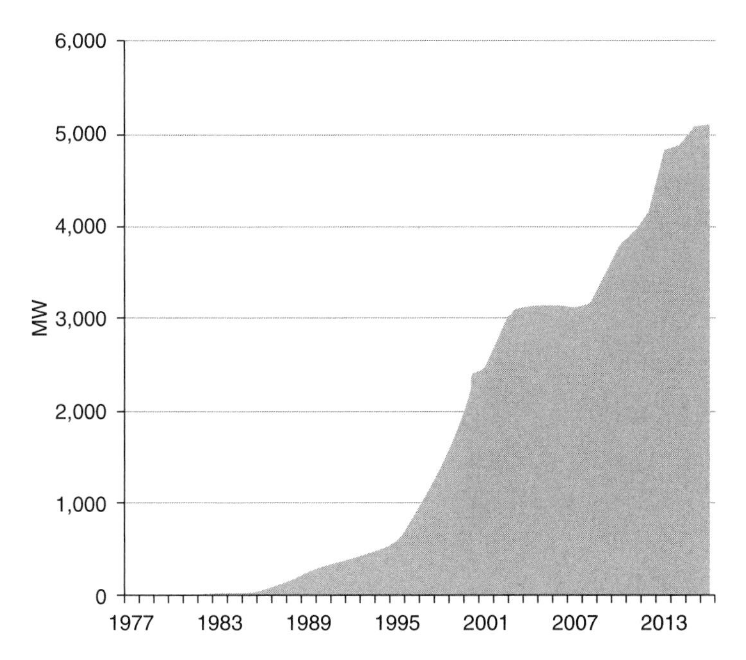

Figure 7-2 Installed wind capacity (MW).
SOURCE: DEA data (n.d.).

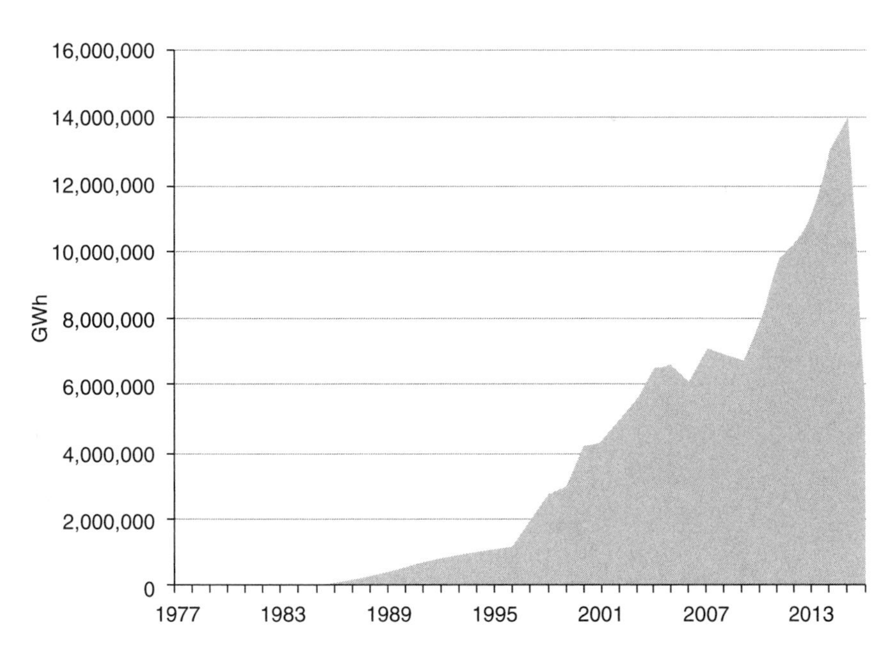

Figure 7-3 Wind power production (GWh).
SOURCE: DEA data (n.d.).

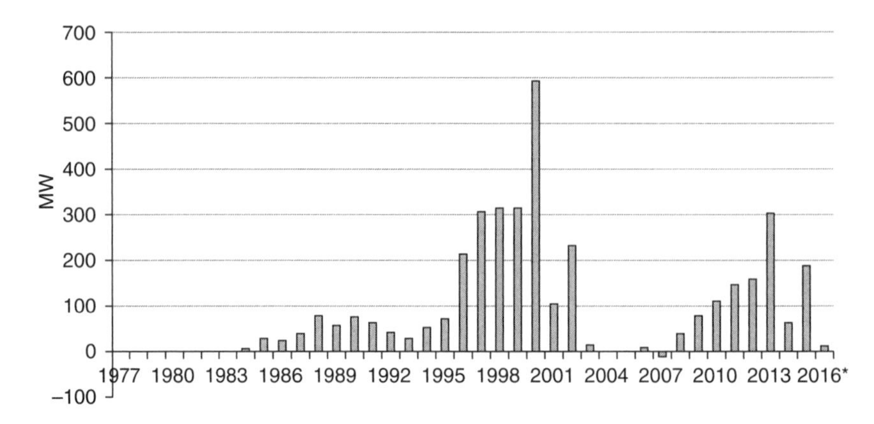

Figure 7-4 Year-over-year installed capacity change (MW).
SOURCE: DEA (n.d.). Note: 2016 estimate.

by 2005 (Sawin, 2001).[21] Despite the supportive plan, domestic installations of wind turbines slowed between 1991 and 1993 (Figure 7-4). Much of this is attributed to unresolved issues with planning, a 10% drop in electricity prices affecting the price of power purchases, and a breakdown in negotiations between the wind associations and utilities on power purchases and grid costs (Interviews, 2011–12; Madsen, n.d.).

Responding to concerns over the increased size and quantity of turbines, the national government established a siting committee that recommended greater coordination of planning and siting (Madsen, n.d.; Sawin, 2001). The national government also released a wind map with Danish wind resource estimates to assist local authorities in planning (Krohn, 1998).[22] The map would later be extended and updated in 1997 and 1999 (Sawin, 2001).

A key hurdle for Danish wind development was an impasse between the utilities and wind associations in renegotiating voluntary power purchase prices and grid connection costs. The Ministry of Energy and parliament took up the issue in 1992, setting a formal feed-in tariff, which maintained

21. The Energy 2000 Plan broadly focused on emissions reductions while maintaining a reliable, efficient, and economic supply of energy. Restructuring of the power sector was key to meeting reductions in CO_2, sulfur dioxide (SO_2), and nitrous oxide (NO_x) (IEA, 1993). The plan was to do so, by increasing CHP to replace district heating and for use in industrial applications, shifting toward lower carbon fuels, using demand side management, and developing wind power (IEA, 1994, 1995).

22. The committee also identified an additional 1,000–2,800 MW of wind potential (Interviews 2011–12).

the previous, voluntarily agreed price at 85% of the residential electricity rate.[23]

Specific to financing and economic support, a Conservative-led government in the early 1990s joined with the Danish turbine manufacturers and two, Danish finance companies to establish the Danish Wind Turbine Guarantee (Sawin, 2001). Through public underwriting of this initiative, large projects using Danish wind turbines could take advantage of government loan guarantees at favorable rates (a 2.5% premium was added to the interest of the debt) (Sawin, 2001). The government also replaced the performance-based RET subsidy of the 1980s with a set of policies, including a revised RET subsidy (0.17 Dkk per kWh), a CO_2 tax, and an environmental subsidy linked to the CO_2 tax (Interviews, 2011; Odgaard, 2000; Sawin, 2001).[24]

In what might be described as a "greener" turn, the national government shifted from the Conservative leadership of the 1980s to a left-leaning, Social Democrat-led coalition in 1993. The Ministries of Energy and the Environment were merged under one person, Svend Auken, who for 9 years robustly advanced Denmark's deeper move toward green growth in the context of sustainable development. The national government also instituted new measures requiring that localities develop plans for wind turbine siting (Krohn, 1998, citing ENS, 1994)[25] and identify suitable sites for zoning by 1995. In addition, a study was initiated to evaluate the potential for upgrading badly sited or inefficient turbines in conjunction with a 3-year repowering program to replace such turbines (IEA, 1995; Sawin, 2001). Planning for wind turbines generally factored for 20-year lifespans and, with the above developments, planning became more integrated with energy policymaking.

Another key area for Danish wind development emerged in offshore wind power projects. The government asked utilities ELSAM and Elkraft to build two

23. The cost of grid connection was also fixed so that turbine owners paid to connect to the low-voltage distribution grid, whereas utilities were required to pay for any strengthening of the high-voltage transmission lines (Nielsen, 2005; Vasi, 2011). This new arrangement was maintained until 2000, when (as a result of liberalization) a fixed nominal price was put in place (IEA Wind, 2007; Vasi; 2011).

24. At this time, the European Commission was more actively engaged in the environmental and energy policies of its member states, and approved this set of wind support policies that would accompany a new rule on RET power purchases (Sawin, 2001, citing EC 1992).

25. The passage of the 1994 Electricity Supply Act also stipulated that the management practices of electricity utilities incorporate integrated resource planning approaches, including least-cost planning that accounts for conservation and development, plus information-sharing with the government and other utilities (IEA, 1999; Sawin, 2001).

offshore wind projects (Sawin, 2001) resulting in the launch of Vindeby and Tuno Knob wind farms in 1991 and 1995, respectively. Vindeby represented an important breakthrough as the first commercial-scale, offshore wind farm worldwide.

In social and business terms, trends in private turbine ownership began to shift by the mid-1990s from cooperatives to farmers and other groups as laws opened new opportunities for individuals, such as farmers (DWTO, n.d.; Interviews, 2011).[26] Ownership and residency rules of the mid-1980s were relaxed with liberalization to then be partly revived around 2008 to assure opportunities for communities (Interviews, 2011).

In 1996, another energy plan, *Energi 21*, was released. It focused on efficiency and RETs while maintaining a coal moratorium and tackling the transport sector for persistent growth in CO_2 emissions (Miljø & Energi Ministeriet, 1996; Maegaard, 2009). The plan confirmed wind targets of 1,500 MW (12% of electricity consumption) for 2005 and set a new wind target of 5,500 MW (40–50% of electricity) by 2030 (IEA, 1998*b;* IEA, 2001). The national government also forged another deal with utilities to add 200 MW of wind power by 2000 (IEA, 1998*b*).

By 1997, an important energy milestone was reached when Denmark became self-sufficient in energy on a net basis (IEA, 2000*c*). New offshore planning rules were also put in place establishing a "one-stop" process for regulatory approvals (DEA, 2012). Utilities worked with the Minister of Energy and Environment's Energy Authority and Environmental Protection Agency to map potential offshore wind sites equal to 4,000 MW of technical potential (Interview, 2011). Utilities also initiated preliminary proposals for projects. Not long after instituting this strategy, a requirement was placed on power companies to install an additional 750 MW of offshore wind before 2008 (DEA, 2011*b;* Krohn, 1998).

During this time, the Danish wind industry re-engaged in the international market. Countries like Germany, Spain, India, and China instituted favorable policies to adopt wind power, and, within these new markets, Danish wind manufacturers exported turbines at levels that often exceeded domestic sales.[27]

The influence of the European Community's supra-national authority was also evident with the European-wide move to open electricity markets through liberalization. The Danish electricity market was deregulated by parliamentary

26. See the sections on "Innovation" and "Drivers and Barriers."

27. See the section on "Industrial Development."

decision.[28] As a test case for tighter regulations of state aid for RET energy, the Danish government and the European Commission negotiated Danish rates for funding wind generation.

In conjunction with the above measures, the Electricity Supply Act (Law No. 375, June, 2, 1999) established a basis for competition in the Danish electricity market. The Act reduced subsidies for wind power, abolished restrictions on turbine ownership, decreased public investment in RETs, established a CO_2 emission ceiling, and introduced renewable energy credits (RECs) to facilitate a transition to a RET market (IEA, 2001; Maegaard, 2009). As the 'old' policy window came to a close in 1999–2000, wind turbine installations increased (Figure 7-4).

Alongside the above, policy and instillation developments, the system and its players continued to evolve in technology and innovation terms.[29] One sign of network learning was evident in the strategy of Danish wind manufacturer Bonus (operating as Siemens Wind since 2005). Bonus attempted to leapfrog ahead of its peers with a larger turbine model and significantly improved MW yield (Andersen and Drejer, 2008, citing Andersen and Drejer, 2006). Bonus's bold move, however, failed because the novel demands placed by a single firm on collaborating suppliers and subcontractors in the Danish wind energy sector exceeded local technical capacity and risk tolerance. None of the collaborators wanted to assume the risk of supporting only one company, so Bonus postponed its generation-skipping design to wait for its competitors to catch up. Learning, risk, and industry progress were clearly intertwined in the *reverse salient* sense that Thomas Hughes would describe.

By 2000, Denmark's shift away from its energy import dependence was quite evident. The once heavily reliant country was 142% self-sufficient on a net energy basis (IEA, n.d; 2004*a*). At the close of 2001, Denmark had the highest share of electricity generation from CHP in the world and one of the largest district heating systems (International CHP/DHC Collaborative, 2005; IEA, 2004*a*). Due to its use of CHP, Danish energy intensity was below the IEA European average and under that of neighboring countries (Sweden and Finland) as well as its peers (Netherlands) (IEA, n.d).

More specific to policy and wind development, the EU approved Danish electricity reform except for regulations guaranteeing minimum prices for new

28. The Electricity Supply Act of 1999 was designed to complement EU guidance and Directive 96/92/EC of the European Parliament and of the Council of 19 December 1996. Through the Danish Act, full competition was in place in 2003, enabling consumers to purchase electricity from suppliers of choice (Energinet, n.d.; IEA, 2000*c*, 2006*a*; IEA and IRENA, n.d.).

29. See the section on "Innovations and Adaptations" for discussion.

wind turbine installations during the liberalization transition. Contingent on turbine age and accumulated production, price reductions from 0.60 Dkk/kWh to 0.43 Dkk/kWh were to occur in the near future. Regulated limits on private ownership were withdrawn (IEA Wind, 2001), and uncertainty existed about buy-back rates. Tax on wind income was largely the same as that on other income, and most of the 205 municipalities had wind turbine plans prepared (IEA Wind, 2001). The first offshore wind project at scale (40 MW) was commissioned at Middlegrunden. Denmark also released *Climate 21* in March 2000, assessing Danish climate policy in preparation for the ratification of the Kyoto Protocol (IEA Wind, 2001).

For the period from 1990 to 2001, significant growth in wind technology, infrastructure, and power generation was evident. From 1990 to 1999 alone, wind technology exports rose by roughly a factor of 18 and represented 52–95% of Danish wind turbine manufacturers' sales (Sawin, 2001). Wind power as a share of total electricity rose from 2% to 12%. In addition, wind generation and installed capacity also grew roughly eightfold to 4,312 GWh and 2,497 MW, respectively.

Domestic Stasis, Continued Export, and an Eventual Resumption of a Green Development Strategy: 2002–The Present

The next years in Danish wind development reflected major flux tied to politics, renewable energy policy, and the electricity market. The period began with policy uncertainty associated with a postponed shift to green certificates in 2001–02 and a government that was less politically supportive of renewables.[30] The Venstre government came to power in 2001 and, with a right-wing coalition, eliminated many of the RET support policies.

Transitional implementation issues of the green certificate policy also resulted in the introduction of a market premium (IEA Wind, 2005; Interviews, 2011–12; Meyer, 2004). Through the premium-based approach, renewable generation was to be sold at market prices with additional support to be paid by electricity consumers. This premium, in conjunction with repowering measures from 2000 and 2003,[31] became the cornerstone of

30. The green certificate plan was in anticipation of an EU-wide shift from feed-in tariff-styled policy mechanisms, which did not materialize (Interviews, 2011–12; Mendonca et al., 2009).

31. Repowering, as noted earlier, replaces old turbines with more efficient ones.

economic support in the near-term. Support hinged on the original year of grid connection, the onshore-offshore status of the project, and the turbine size.[32]

In 2002, the new conservative Danish government rescinded a previous executive directive to build three additional offshore wind farms (DEA, n.d).[33] With the exception of some repowering[34] and offshore development, the domestic wind market stalled between 2004 and 2007.

In market terms, the Risoe-based certification process was opened to international competition in order to comply with liberalization objectives (IEA Wind, 2004).[35] Additional market changes, such as the appearance of balance responsible players, would allow wind power producers to collectively manage their power flows.

Specific to research and development, Risoe Laboratory formed a consortium with DTU, Aalborg University, and the Danish Hydraulic Institute in 2002. A test station was also established for large, multi-MW turbines at Hovsore. In 2006, the Megavind partnership was established to maintain Denmark as a globally leading hub in wind power. Another major area of R&D was the EU-backed Project, Upwind, which was the largest public–private partnership for wind technology and which focuses on studies related to very large MW turbines.[36]

32. In recent years, the premium for new generation was approximately $0.05/KWh for the first 22,000 hours of full load for wind turbines connected to the grid as of February 19, 2009 (DEA, n.d; DWTO, n.d.). An additional $.01/kWh is paid during the production lifetime of a grid-approved turbine to compensate for the cost of balancing. Private wind turbines under 25 kW that are connected from a residence receive a fixed feed-in tariff of $.13/kWh (DEA, n.d).

Repowering includes additional coverage. Development and demonstration of new energy technologies are also supported by a government fund that distributed approximately $126 million (750 million Dkk) in 2009 and $168 million (1 billion Dkk) in 2010 and each year thereafter (DEA, 2012; see also the Appendix).

33. This was partly reversed in 2004 by a political agreement with six coalition parties to install two offshore wind farms of 200 MW each (DEA, 2011; Maegaard, 2009).

34. The first repowering scheme ended with the replacement of 1,200 older, small turbines by 300 larger ones (IEA Wind, 2004 and 2005).

35. Private enterprises could be authorized to perform approval services, certify, test, and measure (IEA Wind, 2004). The approval scheme also underwent revision to an internationally accepted scheme with the International Electrotechnical Commission (IEC) and European Committee for Electro-technical Standardization (CENELEC) standards (IEA Wind, 2004).

36. Risoe Laboratory led and coordinated this 5-year project (IEA, 2007; Upwind, n.d.; Webb, 2012). Based on modeling to compare theoretical designs with existing turbine technology, the group reported that a 20 MW turbine is feasible (Upwind, n.d.).

By roughly 2005, the Danish wind industry accounted for 70% of energy exports and roughly 40% of the global wind power market (DEA, 2005; Interviews, 2011–12). Denmark also met its 2005 wind target in advance, launching a second repowering scheme. Danes paid roughly 0.2% of gross domestic product (GDP) for RET support at the time (IEA, 2006*a*). The DEA also estimated that the all-in costs of onshore wind fell from about 10 cents per kWh in the 1980s to 6 cents per kWh in 2004.[37] Global sales of Danish wind turbine manufacturers had also increased from roughly 200 MW/year in the prior decade to 3,000+ MW/year (IEA, 2006*a*).

In terms of institutional change, the Danish government transferred the energy ministerial functions to the Ministry of Transport and Energy. Within the electricity market, Danish transmission system operator Energinet.dk, was formed with liberalization and the merger of Eltra, Elkraft System, Elkraft Transmission, and Gastra. This independent public enterprise is owned by the state and represented by the Ministry of Climate, Energy, and Building. At the local level, Danish municipalities were also consolidated in 2007, reducing the number of municipalities from 271 to 98 (Miljominsteriet Naturstyrelsen, 2012).

By 2007–08, major turning points could be observed with the wind industry. Denmark experienced an unusual low for installed capacity as more wind turbines were decommissioned than installed in 2007 (Maegaard, 2009). The following year, a significant policy pivot began, with the Government once again recognizing RETs as a priority. A new feed-in tariff-like, market premium was introduced in 2008. With this, wind power adoption was spurred (Vitina, 2015). From 2008 to 2012, four offshore wind farms were commissioned: Horns Rev II, Rødsand II, Sprogø, and Avedøre Holme (Vitina, 2015). Parliament also passed an energy agreement in March 2012 for the period 2012–20 with 95% approval. The agreement centered on an ambitious green transition of energy savings, jobs, and the harnessing of more wind, biogas, biomass, and the like to convert all of Denmark's energy supply (power, transport and heating) to renewables by 2050.[38] The agreement also set a target to derive 50% of electricity consumption from wind energy

37. Estimates of 10 Euro cents (1980s) and 4.9 Euro cents (2004) for support (IEA, 2006*a*) are converted with average exchange rates of 1.02552 $/€ for the 1980s and 0.804828 $/€ for 2004; see http://fxtop.com.

38. Some of the highlights for 2020 include the goal of more than 35% of total energy from renewables, nearly 50% of electricity from wind, and CO_2 emissions of less than 34% of 1990 levels. Wind power aims for 2020 entail 600 MW offshore wind turbines at Krieger's Flak and 400 MW at Horns Rev with an additional 500 MW of near-shore wind generation by 2020, as well as strengthened planning for new wind generation with a total cap of 1,800 by 2020 (including decommissioning); see http://www.kemin.dk/Documents/Presse/2012/Energiaftale/Faktaark%201%20-%20energiaftalen%20kort%20fortalt%20final.pdf).

by 2020, with specifics for onshore and offshore wind and local ownership rules for offshore wind projects (Global Wind Energy Council [GWEC], 2015).

In accordance with Growth Plan 2014, new government policies revised the wind power incentive, extended the timeframe for offshore wind farms, and reduced capacity for a near-term auction (Vitinia, 2015). Onshore wind projects had a ceiling of $.09 per kWh for the combination of market price and premium. Caveats applied to turbine type, swept area, and full load hours, with a reduction in the premium once the market price exceeds $.05 per kWh. Offshore wind development continued to be driven by auctions that award the lowest bid with a tariff for 50,000 full-load hours (GWEC, 2015).

In 2014, all seven test stands for turbines were rented at the Osterlind R&D facility. These included R&D on 6–8 MW turbines (GWEC, 2015). The following year, wind generation in Denmark broke a new global record, equaling 140% of demand on a given day (State of Green, 2015).

EXPLAINING CHANGE

Denmark's wind energy transition can be described as a back-loaded shift for the period under consideration. The types of sectoral intervention and readiness are described next.

Mode of Intervention

The early stage of Danish wind development commenced with a combination of contributions, led by civil society engaged in knowledge-sharing and community/cooperative projects, working alongside a nascent industry's mobilization. Governmental intervention began with Risoe oversight and economic support, rendering the overall early stage a *bona fide* mix of sectoral inputs. Government, industry, and civil society all continued to contribute in the 1990s through to the early 2000s. Government input, in particular, was through standards, information, and continued economic support, enabling the overall shift to be characterized as emergent-encouraged-mandated change.

The following period reflected an unusual shift, with the government pulling out of almost all wind energy development until around 2008. During this time, more or less no growth in wind turbine installations occurred other than that from previously committed utility projects and repowering. Industry actors, specifically utilities, the new quasi-public TSO Energinet.dk, and wind turbine manufacturers, however, sustained some momentum. For manufacturers, activity focused on sales in international markets and R&D. For the utilities and grid operator, progress was evident in the management of continually increasing shares of total electricity derived from wind, as well as partnering on grid integration and harmonization with regional counterparts. In more recent years, a greener turn by government once again reflected hybridized change through multisectoral contributions. Overall, the Danish wind energy trajectory represents a hybrid mode of intervention with significant bottom-up orientation.

Model of Readiness

Turning to the models of readiness, the Danish transition since 1970 displays strong attributes of a developmental model in which a new market and industry were created through the harnessing of indigenous resources.

Specific to the infrastructure, expertise, and existing energy pathway, some qualification is warranted. Denmark historically had all three of these constituent parts of the readiness model that are mostly associated with the expansionary model. However, after a period of inactivity, these elements, particularly the wind energy pathway, largely disappeared. Although the nation possessed an extensive track record in research on wind technology, Denmark lacked the skilled wind power engineers and entrepreneurs that it had in earlier eras. Moreover, the level of entrepreneurial activity and knowledge-sharing in the 1970s suggests that significant new expertise was developed during the period. Hints of the reconstitutive model can also be seen with agricultural/crane machinery, blacksmiths, and ship-building craftsmen/companies moving into the new wind energy market.

INNOVATIONS AND ADAPTATIONS

Primary areas for innovation or critical adaptations in the Danish transition included: *wind turbines, components, offshore technology, analytical tools, cooperative ownership, advanced policy and planning, and decentralization.*

Wind Turbines: Size and Output

Danish turbines are well-known for their incremental improvements over the period of the study (DEA, 2009; Hansen and Andersen, 1999; Meyer, 2004). In terms of size, early Danish turbines often generated less than 10 kW on average and were 40–70 meters in height; now, they range from roughly 2 to 8+ MW and stand at a height of up to 200 meters (Hvelplund, 2011; IEA, 2005, IEA Wind, 2005, 2007; Madsen, n.d.; see Table 7-1). By 2004, turbines were producing about 100 times as much electricity as the first ones in 1980 on a per turbine basis (IEA Wind, 2005).

Table 7-1. Existing Turbines by Output and Installation Year

Timeframe	0–225 kW	226–499 kW	500–999 kW	1,000+ kW	Total
1978–84	91	1	0	0	92
1985–89	425	43	6	0	474
1990–94	616	169	65	0	850
1995–99	218	91	1687	73	2,069
2000–04	44	2	812	526	1,384
2005–09	33	0	26	150	209
Total	1,427	306	2,596	749	5,078

SOURCE: DEA (2009).

Components and Specialization, Design, and Materials

Changes in the performance of Danish wind turbines can be attributed not only to shifts in size, but also to improvements in wind turbine components, materials, and design. Specific to components, early Danish turbine models used off-the-shelf parts that were interchangeable with other industries, such as the automobile industry (Heymann, 1998; Maegaard, 2009; Pedersen, 2010). Substitutability allowed for significant early testing without the added cost of producing dedicated components. Over time, however, components became more specialized. This specialization occurred not only with turbines, but also with the associated industries and services for logistics, installation, and transport (European Wind Energy Technology Platform [EWETP], 2008; Heymann, 1998; Pedersen and Xinxin, 2012). Previously, installation might have entailed a couple of trucks and a crane. Today's turbines, however, are the largest rotating structures in a built environment and at sea (EWETP, 2008). Crane designers now ask wind turbine producers about their expected size needs 1–2 years in advance. Moreover, today's cranes can now lay down narrow belt tracks for travel on small roads and then widen them for the installation of a turbine designed specifically for the wind industry.

Design and material enhancements played a role in the move toward stiffer and lighter carbon fiber blades and away from more conventional glass fiber and other materials. This advance, embraced by companies like Vestas of Denmark (and Gamesa of Spain), means that turbines can have larger blades (i.e., wider swept space) with less robust turbine and tower components, thereby translating to higher energy output, efficiency, and overall savings.[39] The Vestas V112

39. Carbon fiber material is more expensive, but this is offset by cascading costs savings, such as that from adding 16 feet in blade length without weight gain.

3 MW model incorporated the stiffer and lighter, carbon-fiber, blade feature for low to medium wind strength regions. This model is characterized by 179-foot blades at a blade width similar to its 144-foot blades, but the new model sweeps an area that is 55% larger with substantially greater energy output (Wood, 2012*a*).[40] Not all companies view carbon as the way forward, given the supply concerns and higher prices currently related to the material. LM Wind Power (formerly LM Glasfiber), a top-ranked global blade manufacturer based in Kolding, Denmark, is finding new ways to work with glass fiber. It installed 240-foot blades made of a composite of glass fiber and polyester on a 6 MW turbine offshore in France.[41] The 240-foot blade weighs 20 metric tonnes but employs 35–40% more attachment bolts into the same 11-foot blade root diameter as the competition's 202-foot blade model (Wood, 2012*a*). This allows the blade root to support blades that are up to 20% longer without changing the root diameter.

One additional development of note was the introduction of the more secure brake system that enhanced the reliability of the Danish design. It likely played a role in the appeal of the Danish model for export markets.

Offshore Technology

Denmark launched the world's first, commercial-scale offshore wind farm in 1991 with its Vindeby project, and has since completed the installation of 12 more projects (Table 7-2). These "marinized" versions of the onshore models must, among other technology challenges, handle the robust conditions of the sea and the corrosive effects of seawater. In the most basic configuration, they require the connection of the turbine to a foundation and use of a subsea cable to transmit the power supply to shore. The distinct character of the resource assessment, and the installation, support structure, water depths, irregular waves, aerodynamics, hydrodynamics, mooring, control system, access, and maintenance are a number of areas that require procedural and technological

40. Currently, carbon is typically used in the spar, a structural part of large blades (≥148 ft). The increased stiffness and reduced density of carbon fiber allows a thinner blade profile with stiffer and lighter blades. Greater use of carbon, however, entails significant challenges, since blades made of carbon require perfect fiber alignment and curing (Wood, 2012*a*, citing Schoeflinger).

41. LM Wind Power's approach, like that of Siemens Wind Power, includes an enhanced manufacturing process with automation, studies of aerodynamic and load design, and more integrated designs of turbines and rotors (Wood, 2012*b*).

Table 7-2. Offshore Wind Farms

Project	Description
Vindeby (1991)	11 turbines, 5 MW
Tuno Knob (1995)	10 turbines, 5 MW
Middlegrunden (2000)	20 turbines, 40 MW
Horns Rev 1 (2002)	80 turbines, 160 MW
Samsø (2003)	10 turbines, 23 MW
Rønland (2003)	8 turbines, 17 MW
Fredrikshavn (2003)	3 turbines, 8 MW
Nysted (2003)	72 turbines, 165 MW
Horns Rev 2 (2009)	91 turbines, 209 MW
Avedore Holme (2009/2010)	3 turbines, 11 MW
Sprogø (2009)	7 turbines, 21 MW
Rødsand 2 (2010)	90 turbines, 207 MW
Anholt (2012)	111 turbines, 400 MW
Horns Rev 3 (under development)	400 MW
Vesterhav Syd (under development)	180 MW
Vesterhav Nord (under development)	170 MW
Kriegers Flak (under development)	600 MW

SOURCE: Buch and Kjaer (2015) and project websites.

adaptations for projects to advance (European Wind Energy Association [EWEA], 2009; Robinson and Musial, 2006).

As other regions adopt offshore wind power, data for cross-case comparisons indicate that Danish projects, including Middelgrunden, Horns Rev, Samoe, and Nysted, have some of the lowest costs (EWEA, 2009, citing DTU Risoe). The new, Danish offshore wind project Horns Rev III, which is projected to become operational in 2018, is estimated to deliver power at the lowest offshore costs to date with a feed-in tariff of $.12/kWh (Shankleman, 2016).

Analytical Tools

Wind analysis and modeling tools for resource assessments and forecasting have advanced considerably over the past 30 years (EWEA, 2009; Interviews, 2011–12). In the 1980s and 1990s, Risoe Laboratory created a new atlas method that has been refined over time to produce more accurate analysis of wind resources. Risoe Laboratory also uses numerical and mathematical models based on aerodynamics, structural dynamics, and control to develop design improvements that enhance the maximum energy output from the wind under optimized load conditions (Interviews, 2011–12).

With analytical tool development, grid operator Energinet has also played a role. It has done so, particularly in the innovation of advanced forecasting tools that can be used in system operation and day-ahead/cross-border congestion management, coordinating the commitment and economic dispatch of controllable power plants, performing contingency analysis, as well as assessing grid transfer capacity and the need for regulating power (Interviews, 2011–2012). Energinet developed an operational planning system tool for integrating forecasts of wind and CHP into system planning up to 2 hours ahead of the time of operation (Chandler, 2012). This tool, *Drift Planlaegnings*, shows power flows among the Danish, German, and Swedish systems.

Cooperative Ownership

The application of cooperative ownership to wind turbine adoption is an important social and business innovation in Denmark. With it, people voluntarily partnered to meet common social or economic goals through a jointly owned and democratically-controlled enterprise. Specific to wind power cooperatives, they often included a few hundred small and local investors owning a cluster of turbines (Nielsen, F., 2002). When projects were held by partnerships, risk was reduced because bylaws required that such turbines be insured.

In modern Danish wind development, cooperative ownership was a critical factor for early learning and social buy-in. For the period from 1985 to 1993 (Figure 7-5), new cooperative acquisitions of turbines provided important traction for the nascent wind market, especially following the collapse of the export market in California and industry bankruptcies. Political pressure from cooperatives also factored in the maintaining of rules and support for wind turbines in the face of utility opposition. According to the *Danish Wind Turbine Owners Association*, roughly 15% of Danish turbines have been owned by cooperatives in recent years, including the well-known Middlegrunden offshore wind farm and Samsø Island development that is discussed below (DWTO, n.d.) (see Box 7-1).

Advanced Policy and Planning

Innovative approaches to policy and planning have also been brought to bear with Danish wind development. A new planning "architecture" was established, designating distinct, institutional oversight for onshore and offshore wind.

For offshore wind projects, a *one-stop permitting process* was implemented with the Danish Energy Agency (DEA, n.d.). Project developers now work with the DEA to secure permits, and the DEA in turn interfaces with other governmental

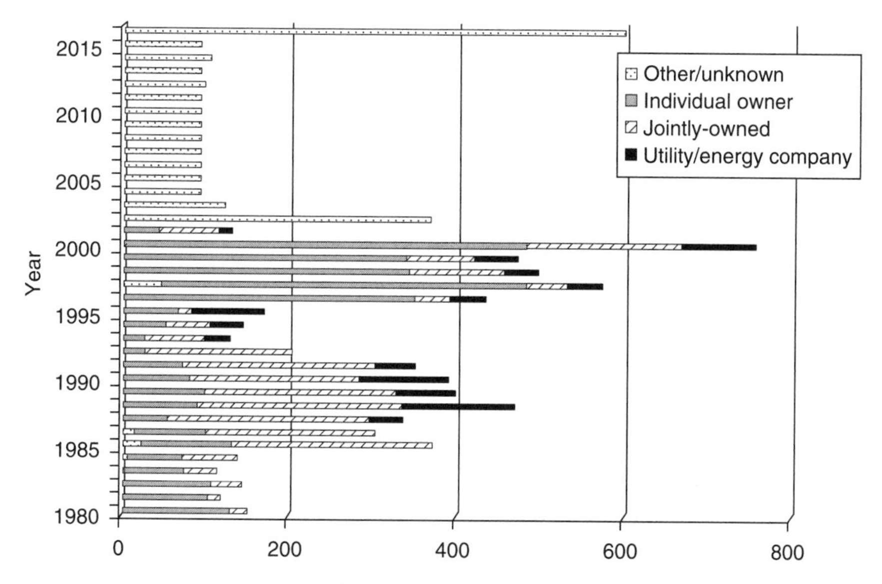

Figure 7-5 Numbers of Danish turbines (additions).
SOURCE: DWTO data (n.d.) and Communications with DWTO (July 5, 2016).
NOTE: Specific data on ownership by type are available for 1980–2001.

Box 7-1

KEY DANISH ENERGY PROJECTS BASED ON COOPERATIVES

Middlegrudden and Samsoe Island are important examples of cooperative energy projects.

Middlegrunden is an offshore wind farm that was built 3.5 kilometers outside of Copenhagen in 2000. This "largest offshore wind farm of its time" is 50% owned by 10,000 owners of the Middelgrunden Wind Turbine Cooperative and 50% owned by DONG Energy (Middelgrundens Vindmollelaug, n.d.). The 40 MW project has become one of the most well-known wind farms worldwide with its arc of turbines built along Denmark's historic maritime protection zone.

Samsø Island reflects a second, Danish energy project that is based on cooperative engagement. In 1997, the island won a competition held by the Danish Minister of Energy and the Environment to become Denmark's 100% renewable energy island. Since then, Samsø Island has implemented a 10-year plan to transform its energy balance. Through onshore wind (cooperatively owned), sustainable district heating, and private distribution systems, among other measures, the Island has become a net energy exporter (Samsø Energy Academy, n.d.). One project within the Island energy system roll-out includes three 1 MW wind turbines near the village of Permilille, owned by local farmers and a wind turbine owners association with roughly 450 members. Together with eight other onshore turbines, a normal year's production is 25,300 MWh, roughly equal to the consumption of 6,500 households (DEA, n.d.).

---

bodies to address competing interests involved in implementation. The scope of the DEA authority includes strategic environmental appraisals of the offshore regions and a monitoring of environmental impacts, as well as assessments for future locations for offshore wind projects. While the character of offshore wind projects shifted from directly negotiated/imposed governmental-utility deals to competitive auctions, an "open door" policy in which a developer can propose an unsolicited project is also maintained.[42]

In terms of onshore wind oversight, the local municipalities hold principal responsibility for projects (DEA, 2012). Guiding principles are put forward at the national level, and local authorities interpret or supplement these at their level. Rules are now in place for height, distance from residences, sound, habitat, and nature considerations, as well as for environmental assessment and cultural heritage site preservation (Miljoministeriet Naturstyrelsen, 2012).

To assist with local planning, the national government created a *Wind Secretariat* to advise local authorities and gather information. A working group was also established to address the future of wind power planning with representatives from the national government including the Ministry of Environment, Climate and Energy, TSO Energinet, DWIA, DWTO, and others. These approaches to planning are now being promoted and adapted in other areas of the world, including the UK and Australia (Northern Territory of Australia, 2012; United Kingdom Renewables, 2012).

Another key innovation in policy and planning has been the use of *community-level competition to mobilize regional demonstration* of clean energy. The 1997 competition that Samsø Island won (Box 7-1) required that the winning region become energy self-sufficient, employ readily available technology produced in Denmark, and provide local matching funds to complement national investment. Over the next decade, the Danish government invested nearly $90 million in the Island (Biello, 2010). This region became a net energy producer in 2005 and now earns revenue from its surplus energy as well as from tourists who visit the renewable energy island.

Decentralization with Distributed Generation

The makeup and management of the Danish power system has evolved considerably since the 1970s in conjunction with changes in the sector, technology, and market. Structurally, the power sector shifted from a highly centralized one based on fossil fuels, to a decentralized model harnessing significant distributed generation (Figures 7-6 and 7-7).

42. Executive Order No. 815 of August 28, 2000; EU EIA Directive 97/11/1997.

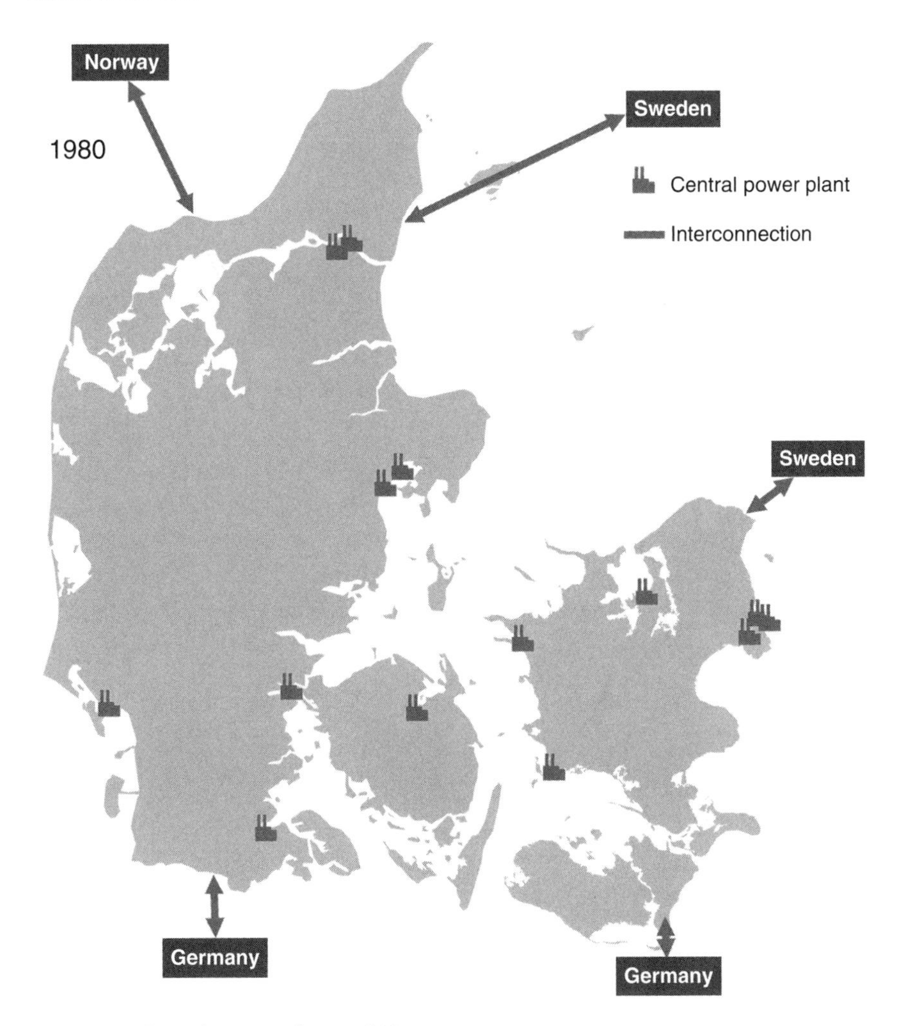

Figure 7-6 Danish power plants, 1980.
CREDIT: Energinet.

This shift entailed countless adaptations in infrastructure, institutions, and practices—two adaptations of which are highlighted here.

Danish grid operators have become expert in managing system integration and the deployment of large-scale intermittent power in the electricity mix with some of the highest security of supply in Europe.[43] Utilities use

43. For years, these same Danish grid experts have also provided technical guidance to decision-makers and technical experts from other regions, such as Africa, China, or Spain, seeking to scale distributed generation, develop new systems, coordinate monitoring, or develop market systems that utilize interconnections (Interviews, 2011–12).

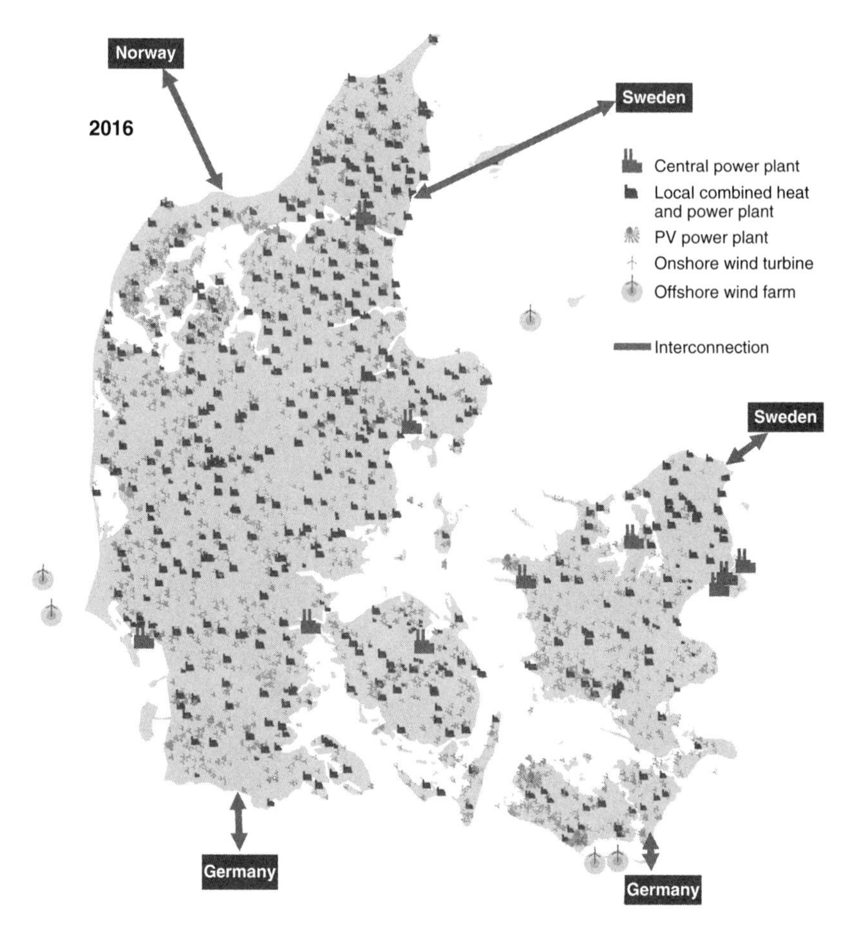

Figure 7-7 Danish power plants, 2016.
CREDIT: Energinet.

advanced system operation practices, regularly employing dynamic applications of large interconnection capacity with neighboring countries and the flexible use of thermal generation (IEA, 2016*a*).[44] In September 2015, the Danish power system functioned absent central generation units for the first time, with grid stability maintained in part through synchronized condensers (IEA, 2016*a*). Other technical achievements in the grid system include greater wind power yields associated with benchmarking, performance monitoring, and increased focus on maintenance during low wind periods (Interview, 2012). Such changes in practices contributed to the rising share of

44. The presence of large hydropower in Norway is also optimized through interconnections. The large-scale adoption of electricity boilers in CHP similarly plays a role in the economic attractiveness of large-scale intermittent power (IEA, 2016*a*).

wind power in Denmark's energy mix over the past decade even as installed capacity remained largely flat.

Alongside the above changes, the shift to competitive markets allowed power trading to become more varied. What was previously carried out as bilateral and long-term contracts now includes a more diverse set of transactions, thus enabling physical power and capacity assurances to be arranged in real time, intraday, day-ahead, and over other time spans.[45] The current, Nordic power pool is regional in scope, involving Sweden, Finland, and Norway (Nord Pool, n.d.). This power pool allows for the disposal of power transfers through implicit auctions, typically including spot and hourly trading; financial market activity, such as derivatives and long-term contracts; and other forms of over-the-counter and bilateral trading (Nord Pool, n.d.). With this pool, the scope of coverage and range of options extends well beyond that which had been readily available in the pre-liberalized period.

KEY DRIVERS AND BARRIERS

Major determinants to Denmark's energy transformation are outlined here. These reflect drivers and barriers of significance that were identified through the analysis of cases, historical records, interviews and data trends.

Reduction of Oil Import Dependence: Improved Balance of Payments, Foreign Exchange Savings, National Energy Security, Self-Sufficiency

The intertwined goals of energy self-sufficiency and a strengthened trade balance (through reductions in petroleum imports) combined to be a force for change in the Danish wind transition. At the time of the first oil shock, Denmark depended on petroleum imports for 87% of its total final energy consumption (IEA, n.d.). The import costs rose in nominal terms from $0.41 billion to an early peak of $3.71 billion in 1980 (Figure 7-8). Related to this, the

45. In Denmark, the shift from a vertically integrated regime to a competitive market began with the 1999 Danish Electricity Act, which required the unbundling of generation, transmission ownership, grid operation, distribution, and supply as separate legal entities (IEA, 2003). In addition to the transmission system operation, local distribution remained as a monopoly (Energinet, 2009). Organized wholesale market activity occurs within a mandatory pool model based primarily upon (1) Denmark's entry into the regional power exchange; (2) the Danish Competition Act, which produced the Danish Energy Regulatory Authority; and (3) TSO Energinet's market framework rules for electricity suppliers and balance-responsible market players (Energinet, n.d).

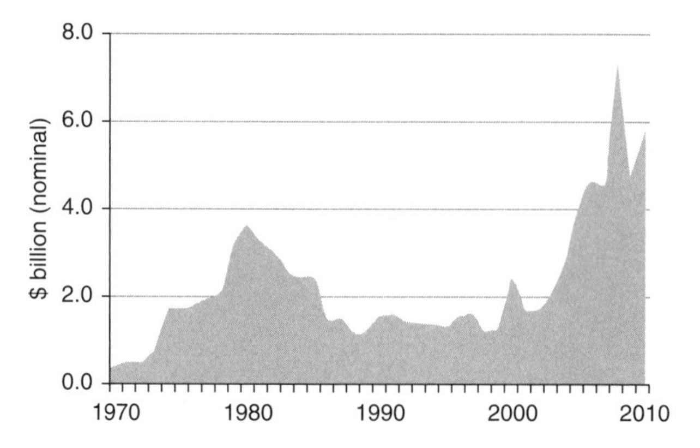

Figure 7-8 Import costs of petroleum for Denmark (billion, nominal $).
SOURCE: UN Comtrade data, SITC REV 1, Petroleum and Petroleum Products (n.d.).

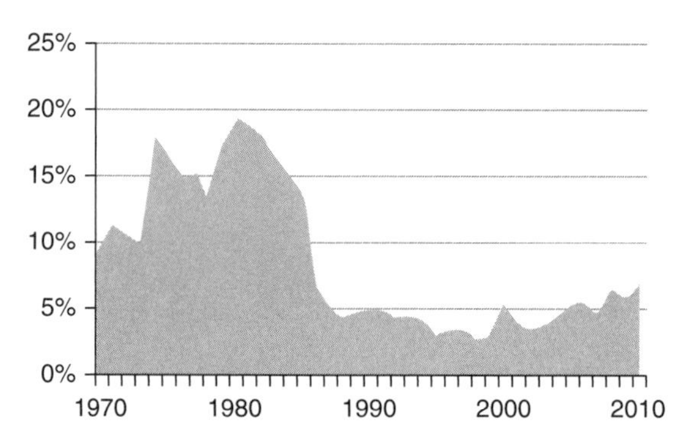

Figure 7-9 Petroleum as a share of total import costs.
SOURCE: UN Comtrade data, SITC REV 1, Petroleum and Petroleum Products (n.d.).

---

cost of petroleum as a share of total import costs increased from 10% in 1973 to an early peak of 19% in 1980 and 1981 (Figure 7-9).

Local Engagement

Danish wind development would not be what it is today without the local engagement of individuals like Danish farmers and of cooperatives. As Figure 7-5 demonstrated, such groups led the way in the early adoption of wind in Denmark. Importantly, utilities did not participate as owners until around 1987.

Among early owners of wind turbines, what generally appeared to attract individuals and cooperatives were environmental concerns, philosophical interests in decentralization, a pragmatic penchant for providing one's own energy, and the opportunity to profit from the sale of wind power to the grid (Interviews, 2011–12; Vasi, 2011). In the period from the mid-1980s through liberalization, residency requirements protected local ownership opportunities and limited absentee ownership of turbines by external developers. Rules also existed on the amount of ownership any one investor could have in a cooperative. Up to roughly 2002, wind energy coops or individual farmers owned more than 80% of the 6,300 wind turbines operating at the time (Krohn 2002*a*). This translated to approximately 150,000 families participating in wind energy production (Maegaard, 2009). With liberalization and the repowering and/or decommissioning of smaller turbines, alongside the increasing size of turbines and projects (requiring larger financing more suited to developers and large-scale energy companies), there has been growth in other models of ownership, including public–private partnerships and utility-owned models. Rules still exist obligating wind project developers to offer shares to local owners (GWEC, 2015) so that the trend in coop ownership may continue. Ownership approaches appear to also be evolving in other respects, with Danish pension funds now investing in the wind turbine industry and project ownership.

Long-Term Political Support for Green Development

Danish political parties and their general position on energy and the environment have provided a strong base of support for Denmark's wind power transition for most of the time since the mid-1970s. Denmark had 11 state ministers in the period between 1970 and 2016 (Statsministeriet, n.d.).[46] During this time, wind power, CHP/efficiencies, conservation, and renewables benefited.[47] This was based in part on a political balance in parliament that was open to wind development. Even during the 1980s, when Conservatives grew in strength, support was provided to wind power on the basis of its progress and potential associated with exports, jobs, and industry, among factors (Interviews, 2011–12). The disruptive shift in policy in 2001–09 is tied largely to changes in ruling political coalitions and liberalization.

46. The Danish political spectrum can be distilled into four categories—Left, Center Left, Center Right, and Right—with the left-leaning groups typically favoring RETs and right-leaning groups favoring free markets (Interviews, 2011–12; Karnøe and Buchorn, 2008). For the period of this study, the Center Left Social Democrats regularly partnered with the Center Left Social Liberals to support RETs. The more leftist end of the spectrum consisted of even stronger proponents of RETs. This cluster of all left-leaning actors provided stability to Danish energy policy from 1977 to 1991, despite the absence of a formal Green Party (Karnøe and Buchorn, 2008). By contrast, right-leaning Liberal, Conservative, and the Danish People's Party's (Folke Party) positions ranged from those who were moderately to strongly opposed to skeptical of RETs and government support within the energy sector (Karnøe and Buchorn, 2008; Interviews, 2011–12).

47. In Denmark, politics are based on consensus. Within the multiparty parliamentary framework, no political party has held absolute power since the early 20th century and only four governments have held majorities since 1945. With such conditions, passage of legislation is typically predicated upon negotiation and compromise among parties.

Activist Scientists, Experts, and Select Specialized Organizations

Activist experts and organizations were prominent agents of change in the Danish wind transition. In terms of NGOs, the OVE and OOA were instrumental in elevating the public discourse on renewable energy pathways as an alternative to nuclear energy. Their specialized engagement, analysis, and advocacy of energy options effectively brought pressure to bear on the political process, influencing the renouncement of nuclear energy by Parliament in 1985 (Vasi, 2011).

Often working within or in conjunction with the NGOs, activist experts from Danish labs, schools, and industry centers provided early technical guidance, which likely affected societal acceptance. The Tvind school turbine and its many contributors remain a symbol of Danish activism and ingenuity. Experts also guided wind studies in the mid-1970s with the ATV committee (Vasi, 2011; Karnøe and Buchorn, 2008) and worked in the 1980s and early 1990s through the Steering Committee for the Promotion of Renewables to fund small to medium-sized enterprises interested in RETs and demonstration projects (Vasil, 2011; Meyer, 2013). Such groups, which included leaders like DTU's Niels Meyer, were instrumental in producing expert reporting on alternative options for national energy plans and for shaping some early-stage demonstrations of renewable energy projects.

Groups like those working with the Wind Turbine Owners and Manufacturers, Nordic Folke Center for Renewable Energy (NFCRE), the former BTM Consultancy, and the Poul la Cour Museum were critical cultural and technical guides in the energy transition. The NFCRE, for instance, is known for its demonstration of renewable energy projects, training, and information dissemination in partnership with NGOs, companies, and governmental authorities from Europe, Asia, North and South America, and Africa (Nordic Folkcenter for Renewable Energy [NFCRE], 2012; Vasi, 2011). The Poul la Cour Museum has also served as a key cultural foundation for the Danish wind industry. Similarly, BTM Consultancy was a hub for global wind technology expertise, and the Danish wind associations have served in knowledge-sharing and advocacy for nearly 40 years.

Local Entrepreneurship and Open Learning

Local entrepreneurship and open knowledge-sharing are additional drivers that characterized the Danish wind transition. Bottom-up, trial-and-error style learning was fundamental to the transition, in which local entrepreneurs (often practical and tech savvy blacksmiths, mechanics, and other individuals) tinkered with small-scale turbine design and implementation (Karnøe and Garud, 2003; Kamp, Smits, and Andriesse, 2004; Vasi, 2011). This bricolage-styled approach to technology development, as Garud and Karnøe describe it, occurred in an environment of collective learning and with a mode of accessible communication in meetings, journals, and demonstration projects (Heymann, 1998; Nielsen, 2005).

Periods of Uncertainty Related to Planning or Policy Change

The most prominent barrier to Danish wind energy adoption noted in this study was uncertainty related to planning or policy pivots at two, distinct junctures: the early 1990s and the period from 2001 to 2008. The issue in the earlier period was associated with increases in turbine size and proliferation, necessitating revised planning that would more fully involve both the local and national authorities. This was corrected once a siting study and local zoning assessment were completed with rules and processes put in place to clarify suitable steps and locations. The second period of uncertainty was associated with liberalization and the entry of a market-oriented government in 2001 that did not favor renewable energy. During this time, wind installations came to a standstill, with the exception of repowering and completion of utility deals.

Grid Connection

Limits associated with grid connection were also a consideration for the wind energy transition, particularly in the early years. This challenge drew the attention of groups like the DWTO to advocate for independent power producers.

Incremental or Stepwise Learning

Incremental or stepwise learning could be considered both a barrier and a driver of the Danish wind power transition. Unlike the breakthrough style of technology development adopted by the United States and Germany that initially centered on large-scale turbines, Denmark generally adopted a more incremental approach of scaling from smaller models with the trial-and-error learning noted earlier in a close network of users, developers, and certifiers. This approach provided strong feedback loops and, by many accounts, enhanced industry and technology development as well as societal acceptance (Interviews, 2011–12). However, it also served as a brake on radical departures. As discussed in detail earlier in the chapter, manufacturer Bonus disruptively attempted to advance in the area of turbine model size in the 1990s, however management with Bonus found that its network of suppliers and subcontractors within the Danish wind industry hub were not willing to assume the added risks associated with the support of a unique design venture (Andersen and Drejer, 2008, citing Andersen and Drejer, 2006). Bonus decision-makers then opted to wait for the industry to catch up.

DEVELOPMENTS IN COST, SOCIETAL ACCEPTANCE, AND INDUSTRY

Costs

Broadly speaking, the economics of the Danish wind industry have improved markedly since the 1970s. The DWIA reports that production cost per kWh has been reduced by more than 80% in the past two decades (Danish Wind Industry Association [DWIA], n.d.). Onshore wind power is now quite competitive on an LCOE basis in Denmark (IEA, 2015*c*, Table 7-3 shows recent LCOE estimates for onshore wind at $85.36, compared to large, combined heat and power (coal) at $131.84, and large, solar photovoltaics (ground-mounted) at $166.35 (IEA, 2015*c*). Danish wind investment costs per kW also declined dramatically, from $2,700 around 1982 to less than $1,500 in 2009, a 44% decrease (Nielsen et al., 2010; Wiser et al., 2011; see Figure 7-10).

More specific to the project life of the wind energy plants, Danish levelized costs are some of the lowest for Europe (IEA, 2015*c*) and below that of the United States for 2008 and 2012 (Hand, 2015). From 2008–2012, costs declined 18% for typical onshore projects in Denmark from $71/MWh to $58/MWh (Vitina, 2015). In roughly the same period, Danish electricity prices decreased 40–50% (Bondergaard, 2015).

Considering the learning curve costs for Danish wind technology for the period from 1981 to 2000, another study evaluated the price of wind turbines (Euro/kW) versus cumulative capacity produced (MW), specific production costs (Euro/kW) versus cumulative capacity produced, and levelized production costs versus cumulative capacity produced (MW) (Neij, Andersen, and Durstewitz, 2004). For each measurement, an improvement was evident. Progress ratios showed that, after the first doubling of production, (1) the prices

Table 7-3. Levelized Costs in Denmark (10% Discount Rate)

Fuel	Onshore Wind	Offshore Wind	CHP Medium Natural Gas	CHP Large Natural Gas	CHP Large Coal	Solar PV, Large Ground-mounted
$/MWh	85.36	170.33	107.79	108.32	131.84	166.35

SOURCE: IEA (2015*c*).

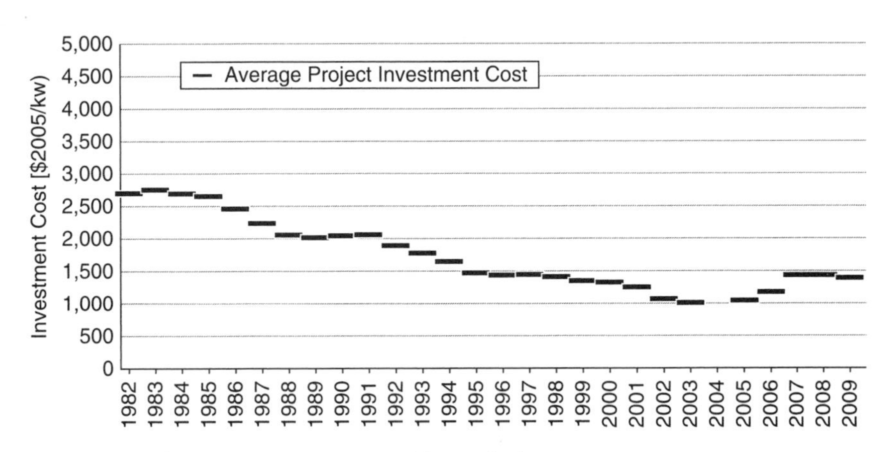

Figure 7-10 Change in investment costs ($2005/kw).
SOURCE: Wiser et al. (2011), referencing, Nielsen et al. (2010).

of turbines, (2) production costs, and (3) levelized costs declined to 92%, 86%, and 83%, respectively, relative to their original numbers (DWIA, 2011/2012).

Looking at economic support during the core stages of scale-up, estimates for the period 1970–2000 indicate that $709 million ($1995) was spent for subsidies and R&D in Denmark versus $1,924 in Germany and more than $3,000 in the United States for the same period (Figures 7–11 and 7–12; Sawin, 2001).[48] More recently, changes associated with the shift in economic support toward a public service obligation (PSO)/market premium imply that costs are substantially reduced per kWh and spread across electricity users rather than tax payers. Figure 7-13 shows the total PSO/premium amounts essentially for clean energy through 2004, with individual technologies broken out separately thereafter (DEA, 2016, referencing Statistics Denmark). Focusing strictly on 2005–14, where wind technology is clearly indicated, one can see that wind received increased support on an aggregate basis beginning around 2012–14. This aligns with greater wind power utilization.

If one then looks at Figure 7-14 for revenue from energy taxes, the $2.8 billion earned through CO_2, sulfur, and electricity taxes in 2014 far exceeds the $0.7 billion in support costs for wind power.[49]

Another way to compare wind economics is to examine energy R&D spending on wind technology by Denmark, Germany, and the United States from 1975 to 2014. These three countries represent early, wind technology leaders

48. Data should be viewed as incomplete estimates, based on the often complicated and non-transparent nature of funding processes. In the case of Denmark, for instance, the estimate does not account for revenue losses associated with tax depreciation; long term loan-repayment guarantees or export assistance (Sawin, 2001).

49. The select taxes for 2014 equaled 15.8 billion Dkk. Wind power support equaled 4.1 billion Dkk. (Exchange rate: $1:5.6125 Dkk; CIA, 2016.)

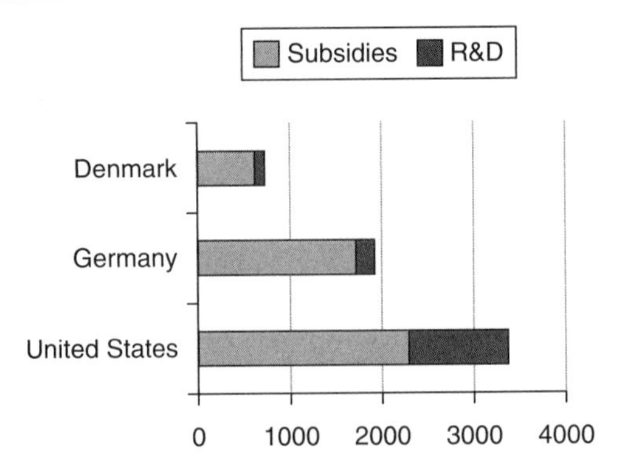

Figure 7-11 Economic support ($1995).
SOURCE: Adapted from Sawin (2001). Note that estimates do not necessarily cover all policy elements.

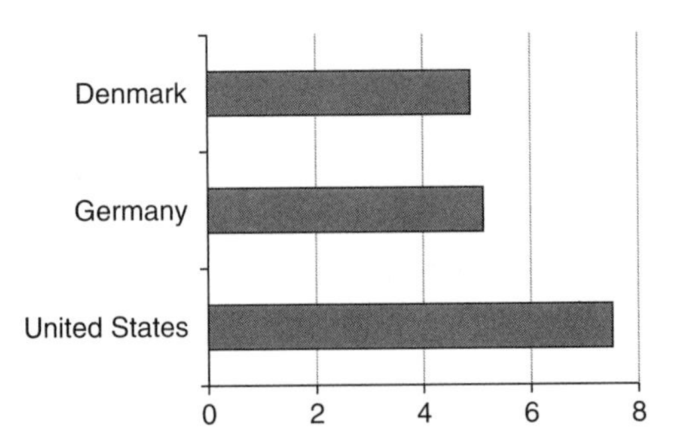

Figure 7-12 Economic support per unit of power generation (cents $1995/kWh).
SOURCE: Adapted from Sawin (2001). Note that estimates do not necessarily account for all policy elements.

in the period of study. In Figure 7-15, one can see that, Denmark spent substantially less in absolute terms. During this period, the Danish Gedser-based model became the dominant design and the Danish wind industry emerged as an international leader in wind turbine manufacturing (Nielsen, 2005, citing Heymann; Righter, 1996; Sawin, 2001).

If one asks who paid, the answer entails utilities, electricity consumers, and taxpayers for economic support plus R&D. How the distribution differs pre- and post-liberalization is a subject that warrants further study.

Societal Acceptance

Turning next to societal acceptance and wind power supporters, engaged groups included a diverse constituency of individuals: environmentalists, farmers, inventors, those favoring decentralized living and/or Danish energy independence,

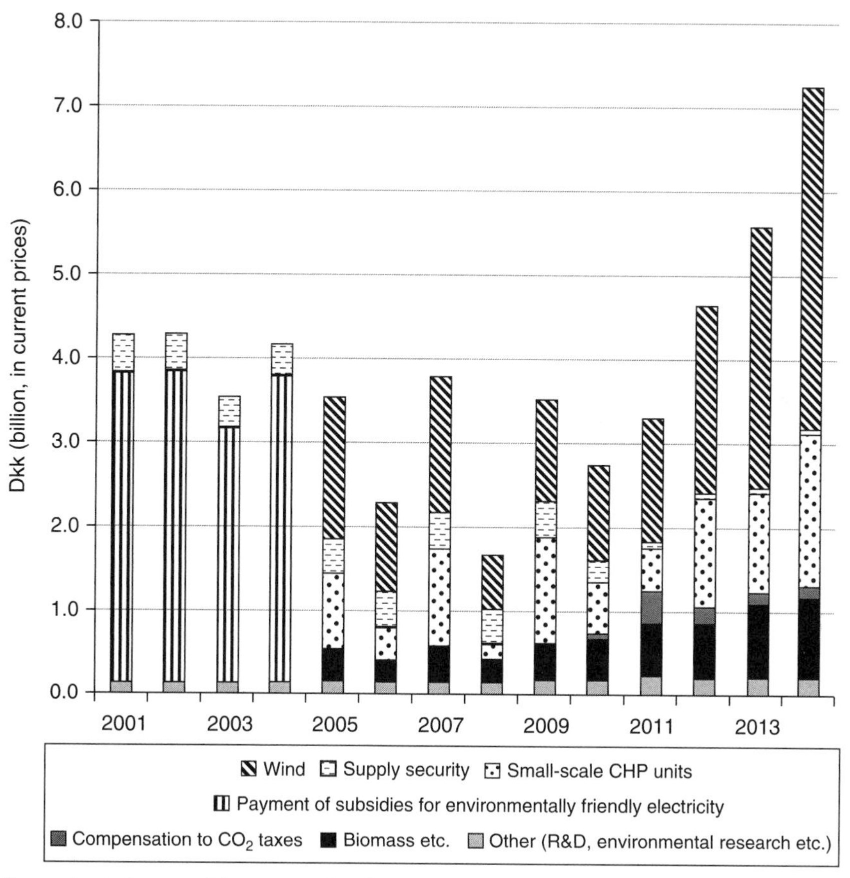

Figure 7-13 Service obligations related to electricity (billion Dkk current prices).
SOURCE: DEA (n.d.), referencing Statistics Denmark.

scientists and academics, pro-green politicians, schools, and communities. Those who resisted wind power development encompassed utilities (early on), nuclear energy supporters, some residents opposed to specific projects, taxpayers who questioned the profits of wind power producers, and the Danske Folke party in the past decade. Some highlights are discussed next.

Position of the Utilities

The utilities have a complicated and mixed history in Danish wind development. This group resisted wind power adoption and were in favor of nuclear development, particularly in the early years, yet they were also engaged in finance and research from the beginning with wind R&D (Interviews, 2011–12; Krohn, 1998; Nielsen, 2005; Sawin, 2011; Traneas, 2000). After rules for grid connection and Risoe standards were put in place by the government in 1979, indications showed that not all utilities were supportive of

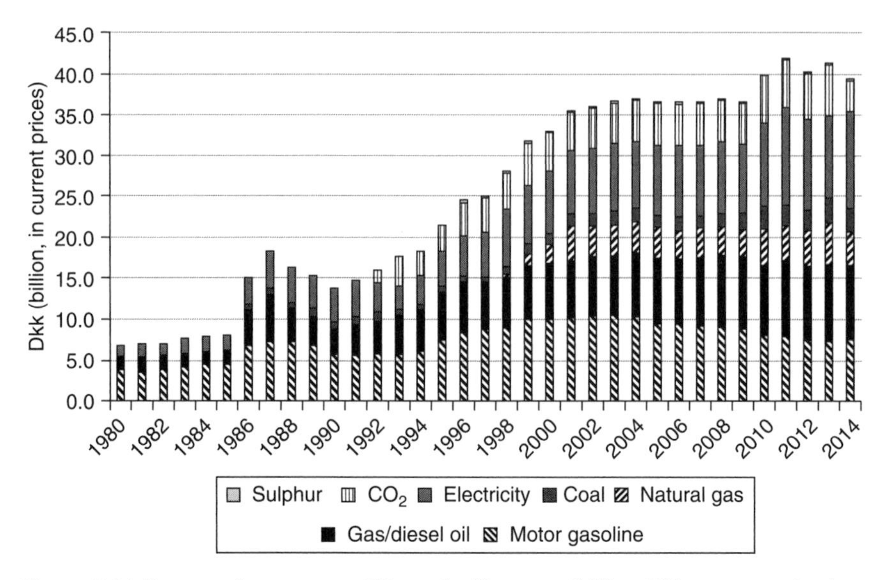

Figure 7-14 Revenue from energy, CO_2, and sulfur taxes (billion Dkk, current prices).
SOURCE: DEA (n.d.), referencing Statistics Denmark.

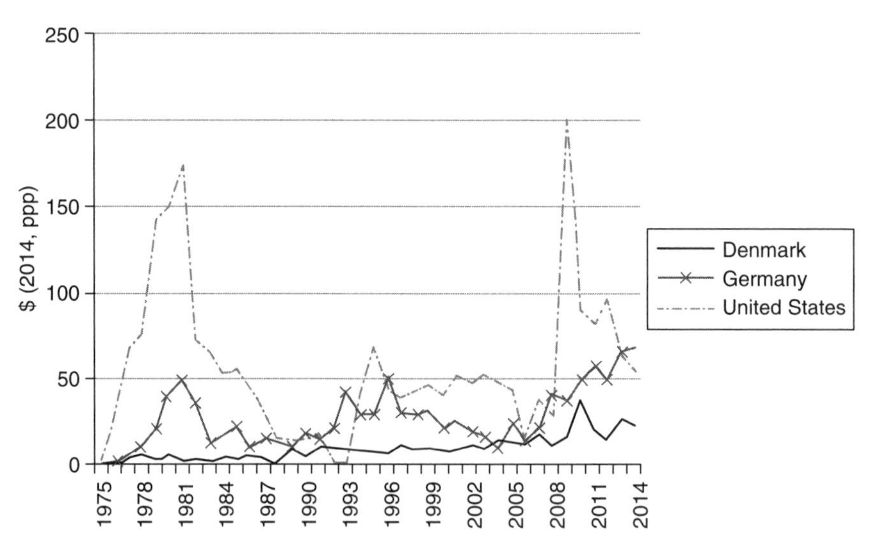

Figure 7-15 Total RD&D expenditure on wind technology ($2014 prices, ppp).
SOURCE: IEA (n.d.).

connecting wind turbines to the grid (Interviews, 2010–11; Tranaes, 1997). Nonetheless, an agreement between the utilities and wind associations was reached in 1984 for power purchase and grid connections. In the 1990s, by contrast, the government stepped in, formally setting rules when negotiations for a new agreement reached an impasse.

With regard to offshore wind, utilities now frequently lead the development in this area, albeit with strong early pressure from the government (i.e., required deals) and to some extent from scientists/academics (reports showing the feasibility of offshore wind). Fundamentally, utilities learned how to manage large-scale wind on the grid and with offshore dynamics, now leading in this technology.

Some may say that the Danish utilities eventually "came around" on wind power once they realized that this form of energy was a fairly cost-competitive way to meet environmental and regulatory aims. Importantly, utility learning by the late 1990s enabled this group to install wind capacity at lower costs than elsewhere in Europe (Interviews, 2011). In today's more liberalized environment, these same actors are now applying expertise on large-scale wind management to other markets—either in international development (foreign aid) efforts or in business activities (Interviews, 2011–12). Partly privatized DONG Energy, for instance, has made green growth and clean energy priorities the principal dimensions of its business strategy, working internationally from this position (DONG, n.d; Interviews, 2011).

It is also worth highlighting that utilities prior to liberalization were generally not-for-profit. They also typically had a consumer- or municipality-based form of ownership. This feature may have lessened opposition to wind power, particularly if local residents wanted wind power. If these companies were to lose market share to independent power producers like local wind cooperatives, the situation might have been palatable since the utilities were generally answerable to local residents, rather than to market analysts and remote shareholders.

Pro-nuclear Supporters

Before Parliament voted to eliminate nuclear power from Danish energy planning, pro-nuclear supporters were often at odds with pro-wind supporters and others supporting "soft energy" options of efficiency, conservation, and RETs.[50] The regular production (by RET advocates) of alternative energy plans that excluded nuclear energy kept the RET versus nuclear energy divide in sharp focus. This issue was largely put to rest in 1985.

Additional Points on Resistance

It bears noting that there were points in the period after 1970 when some resistance was directed at turbines owned by nonresident developers, poorly sited turbines, and larger turbines. The national government responded by setting residency and ownership requirements, putting repowering incentives in place

50. By some accounts, nuclear energy proponents used professional and personal attacks against individuals who advocated for RETs (Interview, 2012).

to decommission older units, and strengthening the planning process with local authorities. Public participation and transparency, often in the form of public comment periods on project environmental impact assessments, have been fundamental to project planning (DEA, n.d.). Denmark is now robustly embarking on a plan to scale wind power to 50% of total electricity by 2020 (DEA, n.d.). This plan is expected to rely on larger turbines and significant off-shore development.

In the 1990s, some opposition to Danish wind development was voiced about the level of profits that wind power producers were making. Since 1992, wind generators would earn a guaranteed buyback rate of 85% from utilities, amounting to around 0.33 Dkk or 5.4 cents/kWh (contingent on coal prices) in addition to a CO_2 tax-related subsidy of roughly 0.10 Dkk/kWh.[51] In conjunction with these, private producers of RET generation would also receive a RET generation subsidy of 0.17 Dkk per kWh. In total, this equaled 0.60 Dkk, or roughly 9.8 cents per kWh. Earnings from these three forms of support can appear significant, especially if charged to tax payers. However, earlier analysis of comparative prices paid to wind power producers indicates that Denmark ranked in the middle range of four countries evaluated for such support (Söderholm and Sundqvist, 2007).

One group, in particular, had a high profile in the past decade for catalyzing a policy shift away from wind power (Interviews, 2011–12). The Danske Folke party, a far-right conservative party, provided votes for a coalition government and held a minority position in the government from 2001 to 2007. During this period, government support for wind development dwindled.

Industrial Development

The Danish wind industry has been a world leader in wind turbine production and knowledge development for decades. In the 1980s and much of the 1990s, when the globally installed capacity grew slowly, it was not uncommon to see four to six Danish manufacturers ranked among the top ten largest manufacturers of wind turbines worldwide (Sawin, 2001). The emergence of leadership in installed wind capacity and manufacturing by other countries, including that of Spain, Germany, India, China, and the United States, has many direct links to Danish technology, knowledge-sharing, and/or financing (Interviews, 2011–12; Lewis and Wiser, 2007; Maegaard, 2009; Sawin, 2001). Gamesa, for instance,

51. A European Commission press release at the time indicated approval for the combined support, noting it equaled roughly 55% of the building and operating costs of the turbine (Sawin, 2001, citing EC, 1992).

is a Spanish wind turbine producer and wind power project developer that is an outgrowth of a joint venture with Denmark's Vestas.

In 2016–17, the global market remains highly competitive, and is showing the pull of new markets with rising actors in India, China, Mexico, Brazil, and Africa (GWEC, 2016; Renewable Energy Policy Network [REN21], 2016, 2017). Worldwide, Danish manufacturer Vestas and (historically Danish) Siemens Wind Power remain among the top-ranked wind turbine manufacturers (Levring, 2017; REN21, 2017).

CONCLUSION

The Danish wind transition is emblematic of an evolving innovation system. Scientists and communities, industry and nongovernmental organizations, as well as public actors and other interested citizens all influenced wind technology adoption, amidst changing conditions. Danish society drew upon its historical familiarity and inner resourcefulness to establish a new industry, market, and energy pathway. In doing so, the Danish government provided fairly consistent support with strategic policy adaptations to address uneven economics, technical change, and planning. Denmark has now embarked on a multi-decade green growth plan with wind power poised to be central to the strategy. As wind technology continues to evolve with more advanced opportunities, like offshore wind at scale, Denmark's international leadership should be a source of knowledge for years to come.

8

Comparative Analysis, Tools, and Questions

> We are like tenant farmers chopping down the fence around our house for fuel when we should be using nature's inexhaustible sources of energy. . . . I hope we don't have to wait until oil and coal run out before we tackle that.
>
> —THOMAS EDISON SPEAKING WITH HENRY FORD AND HARVEY FIRESTONE IN 1931 (*Newton, 1987*)

This chapter returns to the overarching questions of this book, namely, how can national energy transitions be explained, to what extent do patterns of change align and differ in the transitions of this study, and how does policy play a role, particularly with innovations that emerged amid the transitions. To broadly answer, the four cases are comparatively examined here. The conceptual tools from Chapter 3 are also elaborated based on the findings. Implications of the results are discussed, and will serve as a basis for further discussion in Chapter 9 on how to think about energy transitions as a planner, decision-maker, and researcher.

Among the more significant findings are the following. Greater energy substitution (in relative terms) occurred initially within the countries that extended or repurposed existing energy systems versus the country (i.e., Denmark) that developed a new energy system from a nearly non-existent one. Cost improvements were evident in all cases; however, a number of caveats are worth noting. Among the energy technologies and their services that were studied, only Icelandic geothermal-based heating was competitive in its home market in the 1970s; nonetheless, the remaining energy technologies that were studied later became cost competitive. As the national industries of this book became globally recognized, increases in the quality of living within the given countries also occurred, as gauged by the Human Development Index (HDI). With respect to timescales, substantial energy transitions were evident in all cases within a period of 15 years or less. In terms of technology complexity, this attribute was not a confounding barrier to change. Finally, government was instrumental to change, but not always the driver.

EXPLAINING CHANGE: TENDENCIES, TOOL REFINEMENTS, AND INDICATORS

There are countless ways to compare national energy transitions. This section illustrates ways of doing so, first by describing broadly observed, socio-technical patterns with the tool typologies outlined in Chapter 3. A discussion of tool refinement follows. The section then turns to more systematically assess key, qualitative and quantitative dimensions of the four transition cases.

To begin, Table 8-1 outlines the types of intervention, policy mixes, and the nature of technological and systemic shifts that were observed in the four cases. Initial explanations of how national energy system change occurred are summarized as follows:

- *Iceland*: Iceland's shift to geothermal energy represented a "hybrid" form of sectoral intervention, consisting of key contributions from the national government, civil society, and industry as noted in Chapter 4. The expansionary model of readiness was evident in the period between 1970 and the present with the build-out of a continuous transition that was underway since the early 20th century. Co-evolution of heating and power occurred with different timescales and, at times, with distinct influences, yet the overall, national energy system was enlarged by extending existing infrastructure and industries. Market players and infrastructure were added via municipal energy companies, and Landsvirkjun leveraging indigenous capabilities and expertise, such as that of the National Energy Authority. The market changed noticeably with privatization. Policy largely occurred through direct implementation by public actors and deals with heavy industry. In terms of the technology and systems, radical change was evident with spillover applications into fish farming, de-icing, health and tourism.

Table 8-1. Summary of Intervention and Technology-System Change Attributes

Country	Model of Readiness/ Structural Change	Sectoral Intervention *(Top-down, bottom-up, or hybrid)*	Key Policy Mix	Technology and System Change
Iceland	Expansionary	Hybrid for space heating Bottom-up for CHP Hybrid for power	**Deployment:** Implemented primarily by state and municipality actors **Encouragement:** Deals for heavy industry; loans **Information:** Resource assessments; promotion	**Technology:** Incremental and radical change with carbon mineralization, exploration and drilling, site-specific process engineering and plant (re)design **System:** Radical change evident with industry spillovers in fish farming, de-icing, tourism, etc.
France	Expansionary-Reconstitutive	Top-down for the Messmer Plan	**Deployment:** Implemented by state actors	**Technology:** Radical change in some PWRs for load-following and MOX, 3- to 4-loop design, size; Incremental change in safety and other functions **System:** Radical conversion to 58 PWRs in 25 years, also with development of the industrial fuel cycle and load-following practices

Brazil	Reconstitutive	Top-down led (early) with *ProAlcool* for ethanol Bottom-up led (late) for ethanol/flex fuel) Top-down led (late) for biodiesel	**Deployment:** Implemented in part by state actors Petrobras, BNDES, deal with auto-manufacturers, etc. **Mandates:** Blending requirements; zoning **Encouragement:** Soft financing **Information:** Resource maps	**Technology**: Radical change with neat and flex fuel technology, new seed varieties; Incremental and radical changes in agricultural practices **System:** Radical change with fleet conversions, agricultural yields
Denmark	Developmental	Bottom-up led w/ top-down steering	**Deployment:** Grid management **Mandates:** Equipment and grid connection standards; government-utility deals; requirements **Encouragement:** Tax credits **Information:** Resource maps	**Technology:** Mostly incremental advances with wind turbines and wind analysis tools over time **System:** Radical change included offshore wind technology and the power system shift to distributed generation

NOTE: Criteria for distinguishing radical or disruptive change from incremental development can be based on the extent of observed change in physical hardware and direct use or be based on observed process outcomes. A composite is used here.

- *France*: In contrast to Iceland's hybrid form of sectoral intervention the French shift toward nuclear energy reflected a "top-down" approach, as detailed in Chapter 5. Here, the national government developed and largely deployed the entire transition. Specific to readiness, the French nuclear shift represented growth *and* repurposing of the existing system (i.e., expansionary and reconstitutive models), allowing for rapid substitution of nuclear power. Triggered by the Messmer Plan in 1974, the French nuclear developments leveraged an existing transition largely with incumbent institutions and expertise. The French government accelerated the adoption of nuclear energy through the mobilization of key state actors, namely the Atomic Energy Commission (CEA), Életricité de France (EDF) and Framatome. The expansion of the nuclear fleet and fuel cycle was accomplished by repurposing institutions and knowledge capabilities that were in place for the civilian and military-based nuclear programs. Policy that underpinned this transition was embodied in (at times non-codified) direct actions of public actors in the early days, but evolved with privatization to include more diverse and explicit public rules in later years. Radical change in the technology and systems was observable in the standardization of the nuclear fleet, load-following practices, and the development of a commercial- and industrial-scale fuel cycle.

- *Brazil*: Brazil's biofuels transition reflected two, unique growth stages, as described in Chapter 6. In the early stage, the Brazilian government catalyzed the shift with a mix of support from industry and civil society. In the second growth stage, by contrast, Brazilian industry led the way. Overall, Brazil's transition reflected a hybrid form of sectoral intervention, similar in some respects to Iceland's. When explaining readiness, the Brazilian biofuels transition represented a repurposing of industries that created a cross-cutting new industry and market. The transformation involved adaptations in infrastructure and practices that were associated with significant learning in automotive technology, horticulture, and farming. In the more than four decades that were studied, the relevant policy mix was extremely varied. The early years reflected rapid, market realignment with direct deployment by public actors, fuel blend mandates, and soft financing. The approach shifted in later years to include less of the direct implementation by public actors but a continuation of the fuel blend mandates and an introduction of resource zoning maps. Overall, radical change was evident at the technical level, with new sugarcane seed varieties and vehicle design advances that translated systemically into radical change in agricultural yields and the conversion of the automobile fleet.

- *Denmark*: Finally, in Denmark, a wind energy transition was launched by entrepreneurs, local communities, nongovernmental organizations (NGOs), and scientists in a "bottom-up" led approach, as was outlined in Chapter 7. This was strengthened with governmental steering and industry adaptations that overall can be characterized as a hybrid shift. In terms of structural readiness, the Danish transition illustrates a developmental approach in which a transition emerged from a negligible point of commercial use.[1] Mobilization of the transition included early development of novel wind turbine technology and adaptations from the discontinued Gedser design of an earlier era. Alongside these changes, a new industry and domestic market emerged. Novel types of actors also emerged, including wind power cooperatives, associations for wind turbine owners and manufacturers, and balance-responsible market players. In tandem with these changes, the policy mix also evolved with greater market orientation over time. The Danish policy mix was largely maintained for more than 30 years, consisting of standards, incentives (i.e., tax credits and a FIT/market premium),[2] resource mapping, government–utility contracts, and some forms of public deployment, particularly in grid management. In this transition, wind power substitution occurred primarily after 1990. Since that time, the development of offshore wind technology and the shift to a highly distributed power system configuration represent radical changes in the Danish power landscape.

Advancing the Conceptual Tools

Insights from the cases provide opportunity to emphasize nuances in how these conceptual tools may be applied. Specific to structural readiness, for instance, the Danish case brought to light a "disrupted" transition that was

1. Some qualities of repurposing were evident with the adaptation of agricultural machinery/crane manufacturing companies, like Vestas, and the application of ship-building expertise to wind turbine development. However, the predominant transition features point to the developmental model of readiness.

2. One could classify a FIT/market premium as a regulation rather than an incentive since a price is stipulated for the generation provided. For the purposes here, the policy tool is treated as an incentive since the covered action (provision of generation) is optional. As with other elaboration, here, the choice of scope and orientation frames the segment of the transition being explained.

initially launched in the late 1800s. If the targeted timespan for evaluating Danish wind power were extended from the early 1900s to 2015, for instance, then the energy transition would have reflected expansion, retraction, and late, "new" development, mirroring aspects of the biofuels Brazilian transition. Likewise, if the Brazilian transition were viewed for shorter periods, such as 1970 to 1990, and 1991 to the present, one would need to distinguish quite different models of change. This highlights a design challenge for transition scholars who seek to sharpen evaluative tools. Naturally, the points in which a transition is delineated have a bearing on the broader explanation of change.

Recognizing the *types* of public actors that intervene during a transition also provides opportunity to elaborate on nuances. With "deployed" change, for instance, the original description in Chapter 3 presumed that government, in the traditional sense of central agency actors, would be in charge. However, state-owned or partly state-owned entities and the use of government-contracted third parties reflect alternatives to this scoping. For the purposes of this study, all of these variants were classed as "deployed" by government. They, however, offer interesting opportunities for further scholarly refinement.

The definition of government can also be differentiated into tiers of oversight. Energy system change can, of course, occur at the local, regional/state, national, or supra-national levels. Municipalities, for example, were a key locus for the geothermal energy transition in Iceland, where geothermal-based district heating was sought. Given the national orientation of this study, municipalities were classed among bottom-up groups (i.e., civil society) however, interactions across tiers of jurisdictions present an area for theory-building and refinement.

The hybrid category of intervention raises still another point to highlight, as different kinds of hybrid combinations can have distinct, political and financial implications. A "pure mix" of government, industry, and civil society inputs that are evenly balanced throughout a transition will differ markedly from a hybrid example in which government drives the early stages of conversion and industry later leads, as was the case in Brazil. This matters for decision-makers who also may be thinking about phases, levers and overall engagement.

Select Indicators of Change

This section reviews key, qualitative and quantitative indicators of change.

Timescales and Robustness of Change

To begin, timescales and robustness of change are critical indicators for energy transitions. Table 8-2 shows peak inflection periods in each of the four cases. By evaluating substitutions in energy mixes, one sees that Brazil and France shifted 50–62 percentage points in their respective sectoral/fuel class balances in under 15

Table 8-2. Energy Displacement and Growth Periods

Relative Change in Specific Fuel Mixes for Various Periods	Iceland	France	Brazil	Denmark
Peak growth periods for sectoral, fuel class, or total primary energy** substitution (relevant years noted)	1997–2007 7% to 30% (Power) 1970–1984 43% to 83% (Space heating) 1973–1981 34% to 53% (TPES)	1976–1986 8% to 70% (Power)	1976–1988 1% to 51% (Automotive Fuel)	2009–2015 19% to 42% (Power)
Full period for sectoral, fuel class or total primary energy** substitution	1970–2015 Negligible to 29%$_{2015}$ (Power) 1970–2015 43% to 90%$_{2015}$ (Space heating) 1970–2015 34% to 66% (TPES)	1970–2015 4% to 76%$_{2015}$ (Power)	1970–2014 1% to 34%$_{2014}$ (Automotive Fuel)	1970–2015 Negligible to 42%$_{2015}$ (Power)

SOURCE: France and Iceland (IEA, n.d.); Brazil (MME, 2016); Denmark (DEA, n.d.).

NOTE: Space heating for Iceland refers to coverage of the population.

** Total primary energy supply (TPES) was included for Iceland to capture the multiple energy sectors and services that were met by geothermal energy.

years, while Denmark and Iceland shifted 23–40 percentage points, also in less than 15 years. These timescales for energy transformation show that sectoral/fuel class levels of energy sources within a country can be altered considerably (i.e., roughly 20–50% in relative fuel mix) in just 10–15 years.

Energy Self-Sufficiency

Disruption from the oil shocks of the 1970s was a clear trigger for the four transitions of this study. In the energy shifts that followed, each country in this study became much more self-sufficient, with Denmark demonstrating the greatest progress (Figure 8-1), aided also by the development of North Sea oil and gas, and deployment of district heating solutions, alongside its progress in

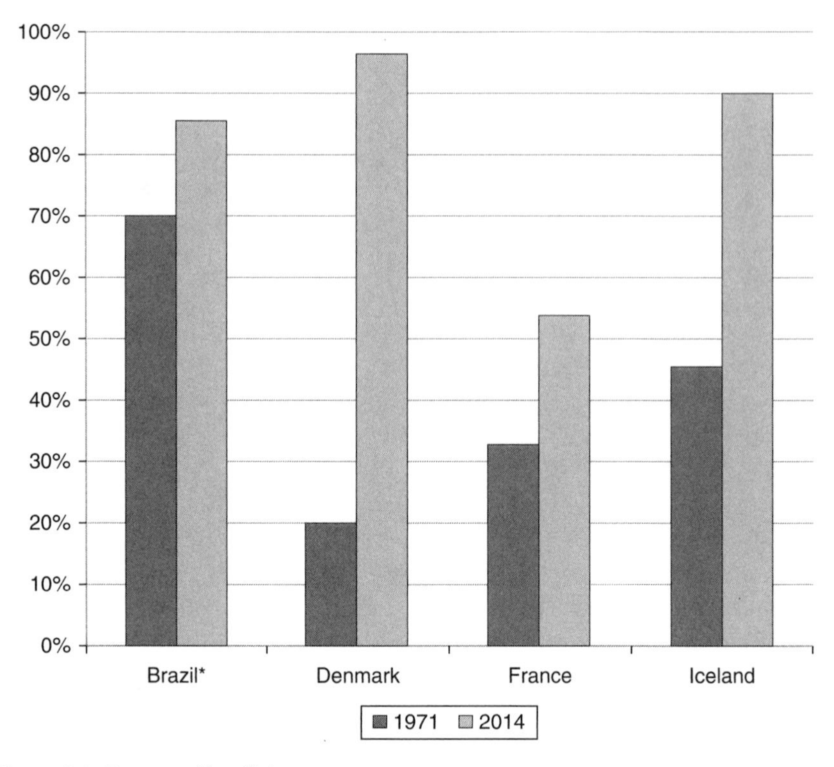

Figure 8-1 Energy self-sufficiency.
SOURCE: IEA (n.d.). Note: Brazil reflects 1972 and 2013.

harnessing on- and offshore wind power. In fact, Denmark now has a status of "negative net importer" for oil, meaning it exports more than it imports (Figure 8-2). In Iceland, the increase in both geothermal energy and hydropower contributed to the improvement of its overall, energy self-sufficiency status. For France, the rise of nuclear energy is principally responsible for the country's enhanced energy independence, although uranium feedstock imports must be factored in today's energy planning. In Brazil, energy self-sufficiency is now characterized by a more diverse range of domestically-sourced, energy, including oil, nuclear, and renewables. Recently, the types of renewables that are utilized have expanded to include forms, such as wind power, alongside the country's traditional hydropower strengths.

Trade Balance Exposure

As noted in Chapter 1, the share of a country's trade balance that is tied to a specific commodity offers insight into strengths and vulnerabilities. If, for example, a large portion of a country's total imports is derived from one commodity in monetary terms, the condition may translate as dependency that is susceptible

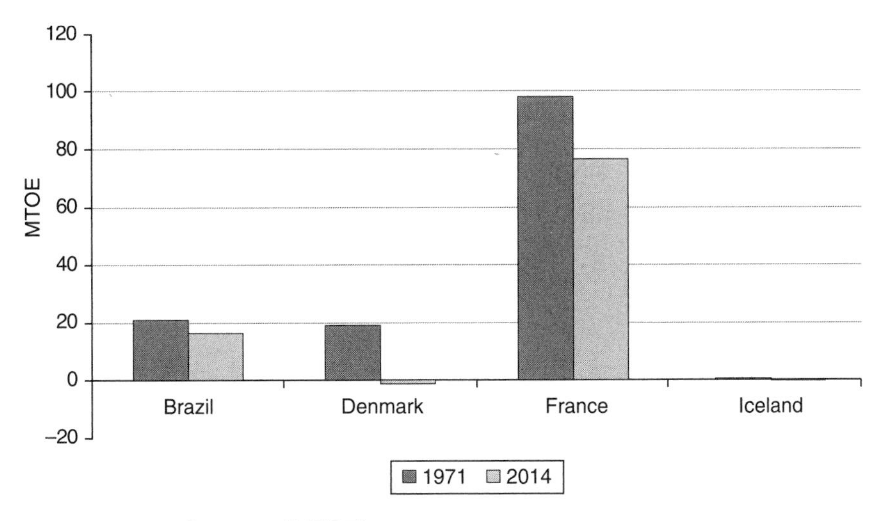

Figure 8-2 Net oil imports (MTOE).
SOURCE: IEA (n.d). Note: Brazil reflects 1972 and 2013.

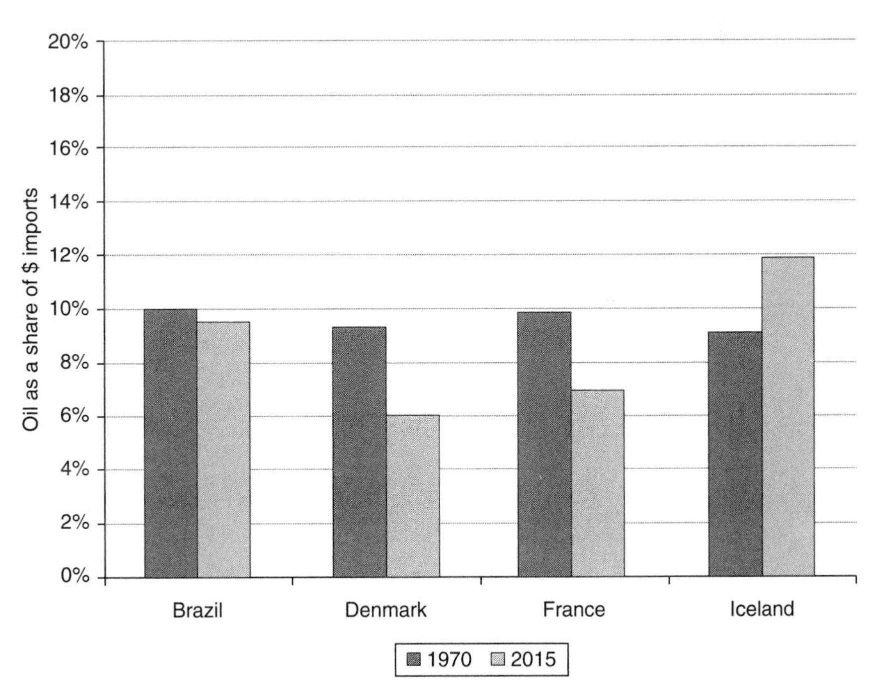

Figure 8-3 Oil as a share of $ imports.
SOURCE: UN Comtrade, Petroleum (n.d).

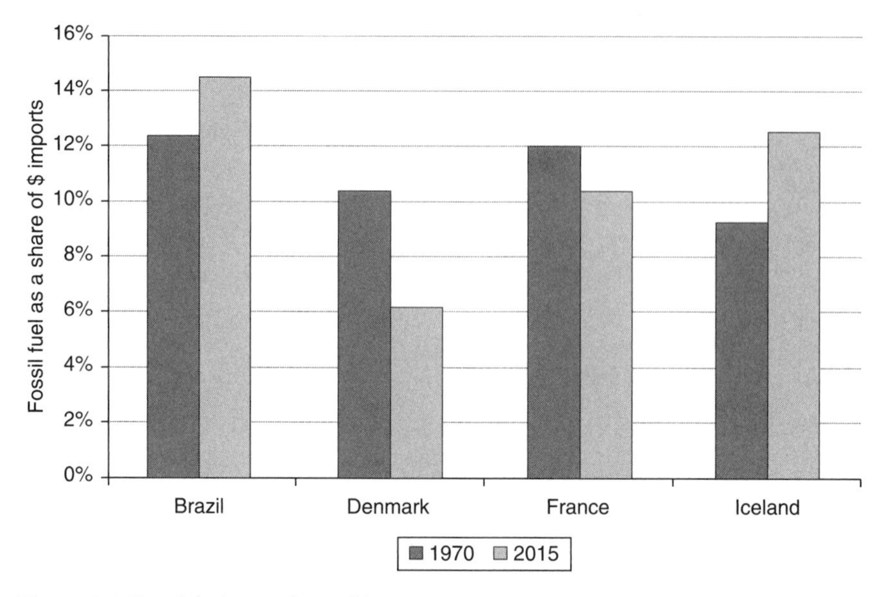

Figure 8-4 Fossil fuels as a share of $ imports.
SOURCE: UN Comtrade, Petroleum, gas and coal products (n.d.).

to price shocks or shortages, particularly if systemic agility is limited. Specific to fossil fuel imports, 1970 and 2015 provide "early and late snapshots" of trade balance exposure with fuel as as share of $ imports (Figures 8-3 and 8-4). Denmark and France show declines in their share of total imports from oil as well as all fossil fuels. Somewhat differently, Brazil showed a slight decline in its share of imported oil in the trade balance, but also an increase from overall fossil fuels, which merits further study. Iceland showed an increase of roughly 3–4% in shares of oil and all fossil fuels within the trade balance.

Carbon Intensity

Carbon intensities, as discussed in Chapter 3, can be measured in terms of units per energy, per capita or per gross domestic product (GDP). For the years 1971 and 2013, the carbon intensity of total primary energy decreased in Denmark, France, and Iceland, whereas for Brazil, carbon intensity grew by 23% (Figure 8-5). This set of trends is consistent with broad directional tendencies for member countries of the Organization for Economic Cooperation and Development (OECD) and non-OECD countries. Overall, Iceland reflected the greatest shift in carbon intensity of total primary energy with a reduction of 78%.

Looking next at carbon intensity per person (Figure 8-6), similar directional patterns are evident with Brazil aligning alongside non-OECD countries and the

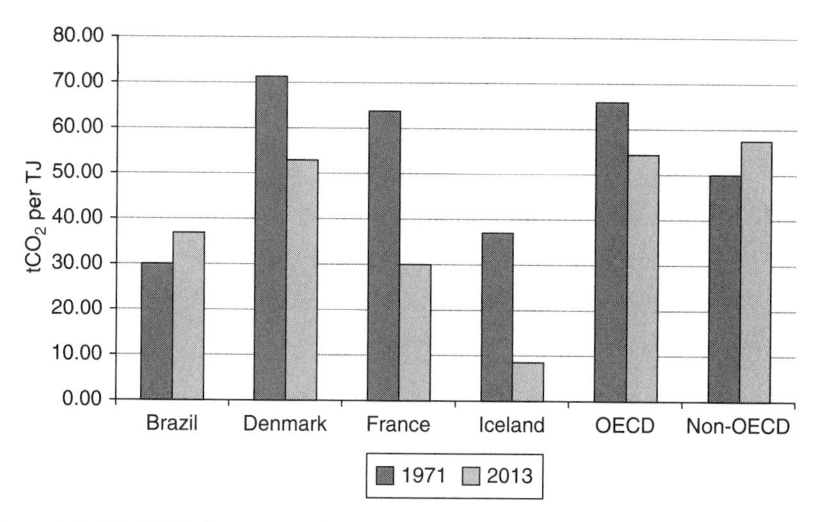

Figure 8-5 CO_2/TPES (tCO_2 per TJ).
SOURCE: OECD iLibrary (July 2016).

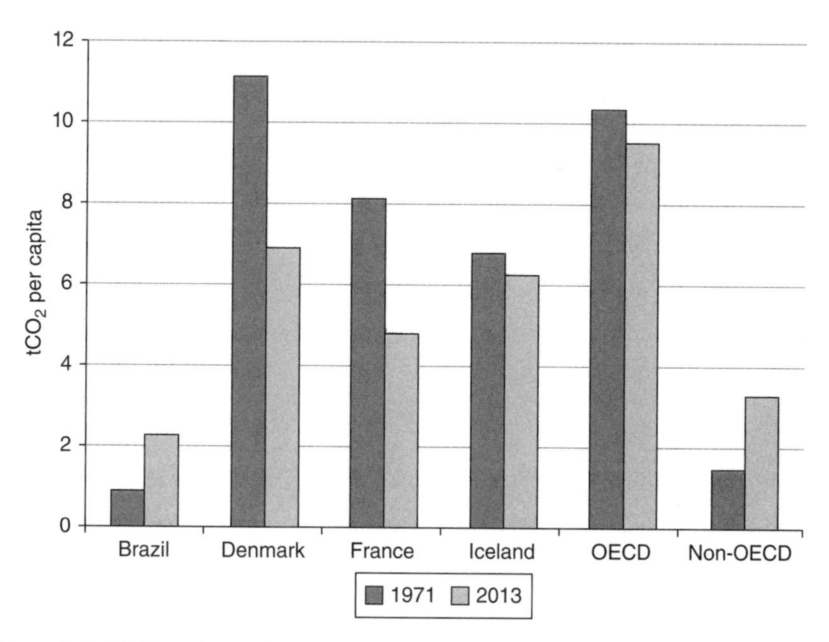

Figure 8-6 CO_2/Population (tCO_2 per capita).
SOURCE: OECD iLibrary (July 2016).

remaining three countries of the study tracking with the OECD countries. Most striking here is the 154% decrease in carbon dioxide (CO_2) emissions per capita by Denmark (OECD iLibrary, 2016).

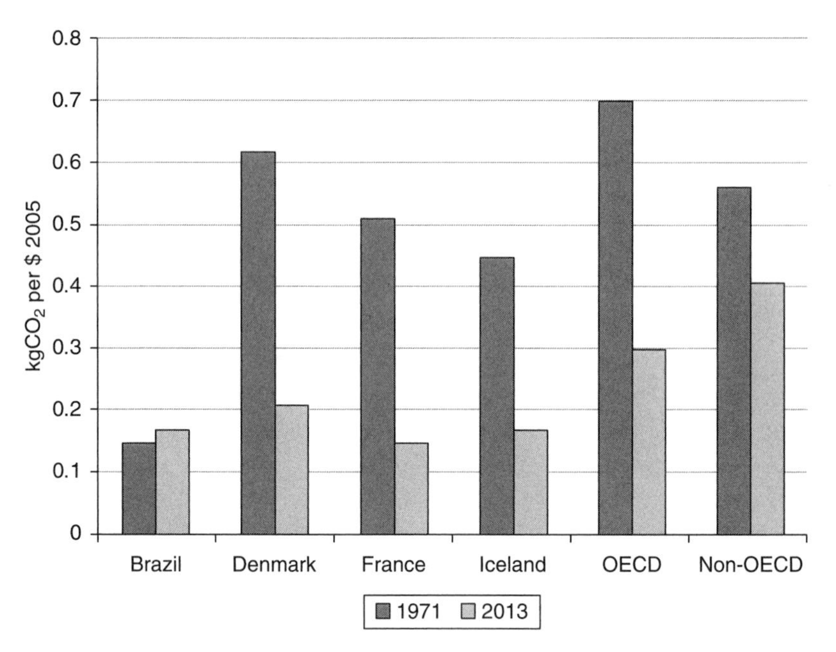

Figure 8-7 CO_2/GDP PPP (kgCO_2 per $ 2005).
SOURCE: OECD iLibrary (July 2016).

In terms of carbon intensity of GDP, levels have been steadily declining, (Figure 8-7), with one exception: Brazil shows a 13% increase. This highlights how a country can shift to greater shares of a low carbon energy, and yet not decarbonize its economy. The subtleties of the metrics and scoping once again matter.

Cost Competitiveness

Cost competitiveness is a dynamic that often captures the dynamic between incumbents and new entrants. New and/or less used technologies are often at a comparative disadvantage in relation to what are deemed as incumbent technologies (Bergek, 2002; Rosenberg, 1994). However, substantial declines in costs can be expected to occur after a significant amount of commercial trial and error with industrial development, standardization, and mass production (Grubler, 1998).

Recognizing the uneven nature of cost data for cases of this study, relative cost competitiveness was considered for specific markets. Across the cases, no energy technology of interest, with the exception of Iceland's geothermal energy in space heating, was competitive at the beginning of the 1970s. Since then, Brazilian ethanol became competitive when evaluated in relation to international gasoline prices.[3] For Danish wind power, costs recently became

3. Note that the more, newly emergent biodiesel is not yet competitive compared to diesel in terms of cost.

competitive in the domestic market (DWIA, 2011/2012; IEA, 2015*c*). Similarly, costs for Icelandic geothermal generation are now competitive domestically. For French nuclear power, the status is more nuanced. Costs associated with production from existing plants are competitive, yet the sharply higher costs associated with new plants merit fuller scrutiny. This tempering effect on nuclear costs is evident in other regions as well.

Industrial Progress

In conjunction with industrial progress, all the countries of this study demonstrated real growth in international leadership for their respective low carbon industries. The current outlooks, however, differ. The French nuclear industry and the Brazilian biofuels industry have recently been experiencing pressures from financial and market consolidation. How each realigns to be more competitive and resilient in the near term will influence the durability of the transitions. The agility of the Brazilian biofuels industry to work with multiple technology markets, for example, may prove to be a key, safety net. Somewhat differently, the French nuclear industry may gain its most important traction by leveraging its historical strengths in science, large NPP-fleet and fuel cycle management, and plant decommissioning. Looking to the Danish and Icelandic national industries of this study, challenges exist, yet stronger potential exists. The Danish wind industry is being tested internationally by newer entrants, some of which benefitted from Denmark's open model of knowledge-sharing. Nonetheless, the Danish industry is also poised to lead in knowledge-sharing and technology export with later adopting markets, having endured many market spurts in the past. The international leadership of Iceland's geothermal industry also continues subtly with its global consulting work, scientific advances, and principal role in training new geothermal experts. Potential, here, remains for Iceland to take a larger step forward with technology breakthroughs in deep drilling, carbon mineralization, and regional links to new markets.

Societal Acceptance and Human Development

Turning to more qualitative or quality-focused assessments of transitions, societal acceptance is an important yet complicated attribute to consider. If a transition is not accepted by a society, whatever shift that occurs may recede or move in an alternative direction eventually. The four cases of this study offer useful, albeit incomplete, insights on this count. Broad acceptance was evident in the national energy transitions of Brazil, Denmark, and Iceland, with France currently reflecting somewhat, mixed acceptance. In Brazil, acceptance of biofuels appears to be widespread. While use can be attributed in part to blending requirements, demand often exceeds the required level of blends, revealing a greater degree of interest. For Denmark,

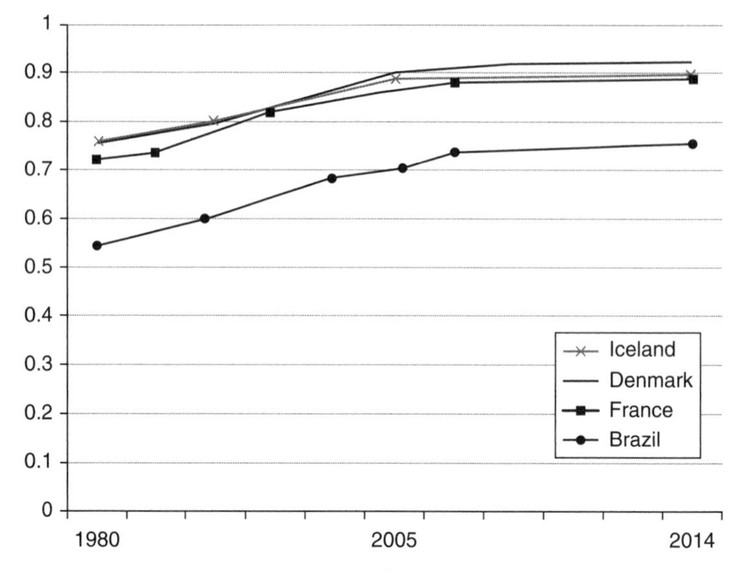

Figure 8-8 Human development trends indexed (Base year = 1980).
SOURCE: United Nations Development Program (UNDP, n.d.).

wind power also appears to now have fairly broad support as green development continues to break new records. In addition, the country is also on a path to attain new levels of green leadership. In Iceland, geothermal energy is viewed favorably, as demonstrated in the feedback from the Master Plan, although localized concerns about emissions and seismicity persist. In contrast to the other three country cases, French acceptance of nuclear energy is more complicated to explain, with a new energy transition toward renewables signaling a potentially diminished role for nuclear. There has been and continues to be some opposition to nuclear energy; nonetheless, discussion with French stakeholders and surveys point to a majority of the French public accepting some use of nuclear energy. An important takeaway here may be that a robust energy system transition can occur, despite a somewhat divided society.

Along with social acceptance, another critical, societal attribute of change is human development. In this study, each of the four countries demonstrated improvements in their performance within the Human Development Index, as reflected by standards of health, education, and life expectancy (Figure 8-8).

This underscores that energy transitions can occur without compromising the basic quality of life for a given society. It bears noting that the quality of life is highly subjective and often uneven—worthy of more research.

Looking across the above attributes (Table 8-3), one can see that the energy transitions began with energy technologies that were not yet cost

Table 8-3. Summary of Key Transition Trends

	Iceland (Expansionary)	**France (Expansionary-Re-constitutive)**	**Brazil (Re-constitutive)**	**Denmark (Developmental)**
Energy self-sufficiency	Improved	Improved	Improved	Improved, greatest relative increase of the cases
Carbon intensity	Decreased per TPES, capita, and GDP	Decreased significantly per TPES, capita, and GDP	Increased per TPES, capita and GDP in line with non-OECD countries	Decreased significantly per TPES, capita, and GDP
Cost competitiveness	Geothermal energy in space heating was and remains competitive; geothermal power became competitive	Nuclear generation is competitive for existing plants;	Ethanol became cost competitive; biodiesel does not appear to currently be competitive;	Wind power became competitive
Industrial progress	Expanded; knowledge leader	Expanded, yet undergoing restructuring; knowledge leader; top-ranked in % of power mix;	Expanded, some uncertainty tied to policy; knowledge leader; and one of the top-ranked exporters	Expanded; knowledge leader; top-ranked in % of power mix; one of the top-ranked manufacturers
Societal acceptance	Generally evident	Mixed: surveys indicate some level of acceptance / deference, however there are recent signs of change	Generally evident	Generally evident
Human development	Improved	Improved	Improved	Improved

competitive, but which became competitive. The transitions occurred in tandem with other gains, including advances in energy self-sufficiency, industrial growth relating to the technologies, and human development. While some of these priorities appear at times to be aspirational in today's discussions about energy transitions, the cases of this study indicate that these aims can, in fact, be attained.

Policy in the Context of Innovation

Innovation is often cited as a way to attain energy transitions. In the context of this study, one could ask whether policy played a predominant role in innovation. The following provides some answers, recognizing that policy and public actors were not always the driving forces.

Specific to Brazil, industry and related research institutes were the centers of innovation for biofuels. This was seen during times of heavy governmental involvement in the 1970s and 1980s, as well as in later years when government influence was largely absent. Favorable policies were readily apparent in the early commercialization of the neat vehicle and in some seed and agriculture developments. However, the flex fuel breakthrough and later-staged seed and agricultural innovations appeared to be driven largely by other sectors during a period when policy intervention was limited if not contradictory.

In Denmark, entrepreneurs mobilized with early innovations in wind power before government was largely focused on the wind technology or energy as a major priority. During the course of the roughly four and a half decades that were studied, innovations and adaptations in wind power emerged from across government, industry, and civil society in Denmark. Among the innovations and adaptations were continuous wind technology enhancements informed by feedback from turbine users, manufacturers, and the Risoe Laboratory. The application of local cooperative ownership models to community wind power projects, and early-stage grid integration by utilities reflect other wider-ranging technology and organizational developments that began in the early period. Within government, the innovative use of methods like one-stop permitting has increasingly been adopted elsewhere. Generally speaking, Danish policy mixes were favorable for wind development from the late 1970s to 2000, at which point support diminished considerably for a number of years. When this inflection in policy support occurred, wind technology advances still continued within industry and the Risoe Laboratory in areas, such as grid and power market improvements.

Specific to Iceland, innovations and adaptations in geothermal energy are traceable to government, industry, research labs, and civil society. During the early period of scale-up, the localities' pursuit of geothermal assessments and infrastructure combined with work by the state actor, Orkustofnun, to spur countless adaptations in exploration methods. In conjunction with these developments, the energy industry adapted drilling, site engineering, and process redesign, at times directly partnering with Orkustofnun. Resource parks and technology spillovers evolved and were led largely by industry anticipating or responding to civil society interest. Broadly speaking, over the course of the period that was studied, policies were mostly favorable (although not usually targeting innovation), with some early limitations in energy company charters that have since been relaxed.

In France, the government was fundamentally behind all relevant innovations and adaptations in nuclear energy since the actors driving most development were public employees, including scientists, utility managers, and reactor manufacturers. Key developments arose from industry and research centers, namely tied to state-owned EDF, Areva (and its predecessors), and the CEA. These developments included changes in reactor design, size, and early safety features; load-following and mixed oxide use; and the build-out of the fuel cycle. More recent safety modifications and design changes are linked to governmental safety directives, as well as partnering by state actors. Policy support, often in the form of outright action by state actors rather than codified measures, was generally constant throughout the period and favored nuclear energy.

Looking across the cases, government was not always the primary driver of innovation in the energy transitions. The early innovations developed by lay entrepreneurs and industry in Danish wind development or in Brazil's later flex fuel breakthroughs exemplify this point. In cases where public actors implemented energy transitions, the technology and institutional innovations that further propelled the transitions were often associated with cross-sectoral partnerships or entities that bridged multiple sectors. Importantly, innovations in civil society and in policy-making were seen as well, such as with Danish cooperatives and one-stop permitting.

It is reasonable to also remember that policy support for deployment and innovation can overlap, yet differ. Moreover, many of the advances that emerged from industry, academia, and non-public research centers were often indirectly supported by public funds. In such cases, governmental influence was significant to varying degrees—a subject that remains open for further exploration.

DRIVERS AND BARRIERS

The case chapters provided in depth discussion of drivers and barriers that were prominent influences. This section highlights commonly-shared drivers and barriers that were evident across the cases.

Among common and prominent drivers were: the pressure to reduce oil imports and to increase national energy security, the related aim to protect the balance of payments from oil price volatility, and the goal to improve environmental conditions or at least to reduce adverse environmental impacts. The interviewees for the cases consistently identified energy security and oil import reductions as principal policy aims. The dramatic drop in oil prices in 1986 temporarily reduced societal concern in this area, however the concern reappeared when oil prices rose. The focus on energy self-sufficiency also played a role within all four countries, yet the level of concern appeared to vary over time. Closely tied to energy security and oil import drivers was the corresponding aim of protecting the national balance of payments. This goal was seen as a key sensitivity in all cases.

Concerns about the environment were also evident across the cases, serving at times as a driver and at other times as a barrier to change. As a driver, environmental considerations were observed in terms of urban air pollution and the desire to reduce waste or to encourage conservation, along with worries about climate change. In Iceland, for instance, the early move toward geothermal energy for space heating was seen as reducing coal-fired pollution. In a somewhat different vein, early support for wind power in Denmark was propelled by public concern about nuclear waste, radiation, and broader safety issues. Interestingly, this same set of concerns about nuclear waste challenged French nuclear development, although the limited degree of early public engagement may have counterbalanced its effect.

Across the cases, uncertainty was a shared barrier. In Brazil, for example, uncertainty associated with policy, supply outlooks, and investment existed after the ethanol program was phased out in the early 1990s and when fuel supply shortages occurred. More recently, the use in Brazil of artificially low gasoline prices as an inflationary policy skewed energy market dynamics, creating uncertainty for investors in market-competitive ethanol. In Denmark, uncertainty was most evident in the period from 2000 to 2007, when a series of policy shifts and diminished support signaled a change of priorities. In France, some uncertainty lingers today in relation to the nuclear program, how the country will ultimately manage its long-term waste, and whether the industry will thrive in an ambivalent global market. Finally, in Iceland, uncertainty manifests in less overt ways in drilling and resource assessments, as well as in questions about projects like the subsea cable to Europe.

When considering generalizations across the four cases, the shared drivers and barriers do not sufficiently explain many of the key inflections in the different transitions. More idiosyncratic drivers and barriers provide some explanatory strength. Intervening variables or preconditions also carry weight in this regard.

PRECONDITIONS IN THE ENABLING ENVIRONMENT

In addition to the drivers of and barriers to energy transitions discussed above, this research explored a number of potential preconditions for such transitions, identifying seven in particular.[4] These included the instrumentality

4. This area can serve as a basis for more systematic testing in future studies.

of public actors, policy aptness, country-level readiness for change, institutional capacity, effects of strong networks, the historical familiarity of a technology for a society, and the conduciveness of a given energy technology for a specific society. The first four preconditions were found to have particular resonance and are discussed, here, by order of the strongest explanatory strength first.

The instrumentality of public actors was a precondition that was evident in all the cases that were studied. As noted earlier, cross-sectoral contributions played a role; however, public actors filled many important gaps and often engineered or led the change. In Brazil, for example, state-owned Petrobras facilitated the integration of ethanol into the supply chain, while development banks, such as Banco Nacional de Desenvolvimento Economico e Social (BNDES), provided early financing to modify distilleries and auto manufacturing lines. In the French energy transition, state-owned EDF, Framatome/Cogema/Areva, and CEA led the development of nuclear power plants and the nuclear fuel cycle in addition to the research and development underpinning technical redesign. In the case of Iceland, state-owned Landsvirkjun, as well as municipally owned Reykjavik Energy and Hitaveita Sudurnesja, were among the key public actors that developed and managed the necessary energy production and delivery systems. In addition, the Icelandic Orkustofnun was integral in identifying and advising on resource development. Specific to Denmark, municipally-owned utilities were early albeit (at times) reluctant enablers. Now, state-owned grid operator Energinet, the Danish Energy Agency, and the Wind Secretariat manage key lines of wind power adoption in a manner that allows continued scaling. In each case, public actors not only played critical roles in policy implementation, but, in some cases, they were the innovators or decision-makers behind the energy system change. What they share is a legislative mandate, but who defined how the mandate was implemented was case-specific.

The aptness of policy was another crucial precondition for explaining energy system change. This can be described as the appropriateness of policy in terms of aims, context, time, and societal orientation. It can imply that government diagnoses the right remedy and responds appropriately, when needed. Government might also be less involved if the original policy continues to address changing needs without added intervention. Aptness is often assumed within discussions of policy formulation and design. However, in practice, it can become secondary to the power politics and institutional capabilities that underpin the policy subsystems (Cahn, 1995; Galston, 2006; Immergut, 2006; Kingdon, 1995; Lowi, 1979). This concept of aptness intersects with ideas about policy learning (Borras, 2011; Sabatier and Jenkins-Smith, 1993); context (Rogge and Reichardt, 2016; Rist, 1998); and national characteristics[5] (Arentsen, 2005; Freeman, 1985; Linder and Peters, 1989), in addition to adaptiveness and multiple objectives (Howlett, Ramesh, and Perl, 2009).[6]

Importantly, policy aptness was not always constant in the cases of this study. In such periods, one might say the overall "momentum" of the energy system was tested. For Iceland and France, policy aptness manifested often in the direct and often rapid intervention of public actors as they built out or improved the existing energy systems during urgent times. When Icelandic energy-economic goals were challenged, for example, in the late 1990s and early 2000s with climate change concerns, the government secured special Kyoto Protocol allowances. This policy intervention could have occurred at various points in time (and still constituted policy adaptiveness); however, the Icelandic government acted within the treaty development process in a fairly timely and apt manner for near-term conditions on the ground. Similarly with France, when public concern over waste management reached a level of urgency in the late 1980s, the government instituted new measures of review and research in short order. For Denmark, the broader wind transition showed how changes in policy mix were typically well-attuned to prevailing needs. Course corrections with procurement deals and adjustments in permitting/grid rules during times of international market challenge or domestic discord enabled Danish wind development to sustain during many pivotal periods. With respect to Brazil, the goals of reducing both foreign oil dependence and balance of payment flux were met quite quickly as shocks from the oil crisis reverberated. The very robust formation of a national ethanol market with the *ProAlcool* program reflected a close syncing of policy approach with broadly based energy and economic needs. In line with the discussion of policy aptness, the dismantling of *ProAlcool* during a time of privatization and political recalibration demonstrated a period of policy "inaptness" for biofuels, yet perhaps policy aptness for other competing aims.

What may be more revealing in the Danish and Brazilian inflections is that new energy markets can sustain (at least for a while in "leaner form") when policy aptness is lacking if the markets have attained a certain threshold of viability or, in large technical systems terms, a momentum is gained. For both

5. The power of otherwise similar national institutions, for example, will often differ greatly in consensus versus majoritarian democracies (Adam and Kriesi, 2007, citing Lijphart, 1999). Interaction patterns are likely to be more cooperative in the consensus-based democracies and more competitive in the majoritarian ones (Adam and Kriesi, 2007). Denmark exemplifies a consensus-based model, whereas the French model is more majoritarian-based (Adam and Kriesi, 2007).

6. Other elements could include feasibility and actor capability.

cases, new growth was stymied for a period. However, continued problem-solving by industry and research labs enabled new innovation and efficiencies to be attained. These insights highlight areas for additional study of market momentum and industry longevity in the context of industrial leadership. Such lessons can provide insight on how an industry recalibrates, as is now occurring, particularly with the French nuclear industry.

Another often cited precondition for successful energy system change is institutional capacity, or the ability of organizations and their networks to serve societal needs. Institutions can be fairly well-defined, as with an electric power market, or less so, as with local cooperatives. Institutional roles can also vary, with some providing risk-sharing protection, connectivity, or incentive structures (Jacobsson and Johnson, 2000). In Brazil, for example, Instituto do Açúcar e do Álcool (IAA) buffered sugar producers from international market forces and was behind some technology modernization up to the 1990s. The collective efforts of IAA, Planalsucar, Copersucar, and the auto industry's ANFAVEA also ensured that industry issues were addressed at least on some level. In Denmark, cooperatives provided an early and very important means for guiding the modern uptake of wind power technology. These groups not only mitigated financial risk, but also served as a locus for learning and promoted societal acceptance. In France, widespread deference to the institutions of senior civil servants and groups like the Corps de Mines appears to have been vital for the robust and continued French nuclear trajectory. Finally, in Iceland, the deep involvement of the Orkustofnun, the Iceland Geosurvey, and their predecessors in geothermal energy development were fundamental for knowledge generation, policy development, and guidance. In each of these cases, institutions mattered, yet this point can be extended.

Linked to the condition of institutional capacity is the concept of readiness that was raised in Chapter 3. National decision-makers in all the countries that were studied were caught off guard by the speed and impact of the oil price shocks of the 1970s. Nonetheless, three of the four countries were able to rapidly mobilize with low carbon energy sources.[7] The shocks spurred action and a surge in capacity, in what Albert Hirschmann calls use of "slack" (1970).

Among the quick-to-mobilize countries were Iceland and France, which leveraged existing low carbon energy pathways and did not need to allocate significant start-up time or effort to establish new technology, expertise, or markets. This points to a somewhat obvious truism that having appropriate institutions and expertise in place to begin or amplify a transformation can increase the chances of its success.

7. Denmark mobilized more quickly in other areas.

The case of Brazil offers added perspective on the topic of readiness. Like France and Iceland, Brazil was able to rapidly mobilize its transition from niches and other markets, even though ethanol use was negligible in the period immediately prior to the first oil shock. What mattered was the presence of resources, institutions, systems, and a plan—even if they were in place for other purposes. Broadly speaking, when disruption occurs—whether economic, geopolitical, or even weather-related—readiness allows actors to quickly seize or create opportunity for change. This concept of system readiness resonates with an idea attributed to Benjamin Disraeli that the secret to success is being prepared when an opportunity arises.

NATURAL RESOURCE ABUNDANCE

When advancing explanations about energy transitions, there is a tendency to see national energy leaders merely as countries that are naturally endowed with a given energy resource; thus, we view the utilization of this resource as predetermined and obvious. Along these lines, it is accurate to say that Iceland has substantial geothermal energy potential and that the tropics of Brazil are well-suited for sugarcane production. Yet the mere existence of these conditions can in no way explain why or how the transitions to low carbon energy were so fully advanced or when they occurred.

In Brazil, the adoption of biofuels was far from predetermined. To begin, a proposal to expand ethanol production had been in place prior to the first oil shock, yet was not implemented. When the *ProAlcool* program was launched, countless individual initiatives in the sugarcane, transport fuel, and auto sectors had to be aggressively adapted and managed. Moreover, when policy attention shifted in the 1990s, significant retrenchment occurred, showing that the system dynamics relied on more than natural conditions.

The scale-up of geothermal energy in Iceland, like that in Brazil, was also neither predetermined nor certain. Nothing shows this more clearly than the fact that geothermal heating was competitive in the 1970s but was utilized by less than half the population. Greater knowledge of energy resources, improved financing, and the actions of public actors were all necessary for greater adoption.

France and Denmark did not even begin with a relative advantage in natural resources. France had some domestic reserves of uranium in an earlier period but not enough to support a significantly scaled program. Today, France imports uranium. Denmark has some good wind potential, but not to the extent of its neighbor, Scotland (Interviews, 2011–12; Troen and Peterson, 1989). However, the Danish wind energy mix and industry have progressed further than Scotland's (Scottish Renewables, 2017; DEA, n.d.). When the additional layers of decision-making, organizational learning, and system development are considered, it is clear that the national energy transitions of this study were far from preordained by natural resources.

Looking beyond these cases, other countries have potential and strengths related to a range of energy technologies. A core insight, then, is to make optimal use of the resources that are available—natural, human, and systemic.

IMPLICATIONS OF THE FINDINGS

The findings of this study allow certain inferences to be drawn in relation to prevailing ideas and literature. First, the timescales that are needed for a country to robustly shift a sectoral/fuel class mix and, in some cases, even total primary energy by 20–50% are 10–15 years. This runs against more widely held views that energy system change necessarily requires long periods. Previous efforts to predict energy substitution indicated that energy system shifts were typically multidecadal in nature, often requiring 50–100 years (Marchetti, 1977; Marchetti and Nakicenovic, 1979).[8] Since countries have institutions and cultures that can be transformed much more readily than counterparts at the global level, this insight should not be surprising; however, the debate wages on. Based on the insights gained, here, national decision-makers should focus on energy shifts using 5-year plans over a 15-year period.

Second, with respect to factors influencing rates of technology diffusion, complexity can present challenges without impeding the energy transition. In this study, nuclear technology was arguably the most complex technology of the four studied. However, French adoption was also the most rapid and robust. Caution with respect to the inertia of complex systems and network effects is perfectly reasonable (Grubler, Nakicenovic, and Victor, 1999*b*; Grubler, 2010, referencing Lovins, 1986; Rogers, 1995), yet this study revealed that there are other factors that overcome the potential lags of technology or system complexity. Moreover, complexity may be an enabler at times of energy system change when society is less engaged. If supported by subsequent studies, this insight could have considerable value for policy-makers and planners that are focused on energy transformation. Here, societal buy-in and the role of government will remain important for continuing study.

8. Within such work, examples of rapid transitions in oil, natural gas, and nuclear energy at the country and sectoral levels for earlier periods were also studied (Marchetti and Nakicenovic, 1979). Although the country-level examples may have received less attention in the literature, they resonate with the temporal findings of the current study.

Third, turning to factors that *accelerate* diffusion, this research confirmed the importance of pre-existing niche markets (Geels and Schot, 2007; Grubler et al., 1999*b;* Kemp, Schot, and Hoogma, 1998; Smith and Raven, 2012). The finding dovetails with the earlier discussion of readiness as a precondition, and the conceptual models that were introduced.

These findings offer new ways of thinking about a country's agility and the inertial effects of established infrastructure in the context of adaptive or absorptive capacities (Henderson and Newell, 2010/2011; Porter, 1990; Smil, 2010). In Brazil and France, for instance, traditional infrastructure was adapted to *facilitate* change, rather than encumber it. Given this, ideas on sunk costs and inertia of infrastructure should be qualified to recognize that existing infrastructure could be part of the solution. If suitable for adaptation, infrastructure might be leveraged for multiple purposes or to advance new directions. Naturally, a wider pool of countries with more varied systems will need to be examined to substantiate this. As for the notion that determinants of transition rates are likely to differ over time (Grubler, 1997; Rogers, 1995), this research fully bears this out.

Finally, the economics of energy system change also reflect some rather telling insights. Conversions to low carbon energy can begin with a focus on new energy technologies that are not currently cost competitive but that become competitive over time. Brazilian ethanol, French nuclear power, Icelandic geothermal-based electricity, and Danish wind power are now competitive. They weren't at the outset. This understanding is in marked contrast to the thinking underpinning least-cost economics, which is often the principal criterion in energy planning and a historical model for investment by public utilities (American Academy of Arts and Sciences [AAAS], 2011). Economic efficiency is, without a doubt, a powerful rationale, yet it can cause decision-makers to overlook wider objectives, systemic risks, and co-benefits, such as adaptability in otherwise irreversible infrastructure decisions, industrial strengths, environmental stewardship, enhanced public health, and the achievement of energy sustainability. It is worth emphasizing that each country of this study evidenced clear improvements in its human development trends as well as co-benefits, such as energy independence and industrial growth, as it was scaling low carbon energy.

LIMITS

As with any study, limits matter. Research of this nature, for instance, can benefit from lengthy periods of investigation, yet constraints on the memory of participants as well as with the availability of key information and access to people can complicate the endeavor. The choice of endpoints in the time span of the study can also empirically focus attention on differing maturity levels and respective attributes of transition pathways. The timeframe of this study was deliberately chosen to capture the shared global context of events, like the oil crises of the 1970s, as well as multidecadal learning. On other fronts, one must exercise care in inferring causality, when focusing on a small set of cases that are not randomly selected and that have key common traits. Moreover, the practical aspects of the research, including variations in the strength of data and use of translation, also impart natural limitations. Further, cultural and technology-specific differences can introduce idiosyncratic challenges for cross-case comparisons that are not generalizable. Here, I used different methods to enhance validity, when possible.

For those who seek answers largely in quantitative assessments, the more qualitative aspects of this study might also be seen as a limit. It is, however, in complex sociotechnical phenomena like energy transitions where quantitative and qualitative assessments can mutually inform. Applied history, strategic planning and management, as well as policy and other sociotechnical systems lenses, were among the approaches that informed this work. They will continue to be crucial for identifying as well as explaining interplays and influences that quantitative assessments may not capture.

9

New Paradigms

Lessons and Recommendations

There is an old saying that history is about revolution and evolution. This book considered history in the context of disruptive and incremental change within energy systems. In doing so, the research advances new tools and theory-building, while emphasizing broader lessons, particularly for policymakers, regarding the strategic management of energy transitions. This chapter discusses key insights from the study. It also identifies avenues for further research.

LESSONS

There is no set formula for a country to shift to low carbon energy (or, for that matter, to undertake any energy transition). Whether transitions emerge or are driven, there is room for strategic management.

- *Windows of opportunity:* Focusing events, like oil shocks, can provide an opportunity for the rapid mobilization of an energy transition, despite differences in views. Such windows of opportunity, however, have a limited shelf life. Common tensions between competing interests will re-emerge and can undermine progress. Here, cross-sectoral collaboration and learning can provide important traction amidst a transition for meeting longer term objectives.
- *Costs and benefits of change*: Least-cost economics can play a role in energy decision-making, yet policymakers should recognize that this approach does not adequately reflect all important objectives, costs or benefits. Co-benefits, including the flexibility to adapt in otherwise irreversible decisions, may matter for a society in an

energy transition. Such benefits can be difficult to value in planning and analysis, but warrant scrutiny. Here, analysts and decision-makers can be pivotal by ensuring that viable options which add important value are not crowded out.

- *The role of government*: It is clear from the preceding pages that governments have a role to play in the energy playing field, even if government is not the driving force. The fundamental importance of energy, the widely entrenched nature of such systems, and the intersecting aspects of energy-related challenges with other public priorities reinforce this point. Here, public actors and policy can be instrumental in bridging gaps at critical junctures in a way that no other individuals may adequately address.
- *Market mechanisms vs. direct deployment vs. additional means*: Societal views about ways to govern natural resources will factor in whether an energy transition depends on markets, government, or other means. Societies that are traditionally reliant on market-based approaches may focus on structuring incentive mechanisms in line with broader aims and, in doing so, signal where they see value. Market mechanisms are, of course, not the only or necessarily the optimal way to carry out change. Deployment with public actors can often attain objectives more quickly. The latter approach syncs with centrally-planned societies as well as with conditions warranting rapid and concerted response. The choice of approach, however, does not end there. Policymakers are wise to remember that additional tools of their trade include legitimization, leading by example, cooperation, and information sharing, among possibilities. Knowing how and when to align appropriate approaches with conditions is a sign of a savvy policymaker.
- *Modeled strategies*: The use of projections and other quantitatively-modeled strategies is a common practice in today's energy planning. Such approaches can simplify complexity, but also obscure critical value choices. It is incumbent upon decision-makers, analysts, and the broader base of citizens to ask the difficult questions that challenge the assumptions and scoping, and which recognize the less simply quantified.
- *Cultivating agility*: Effective readiness in a national energy system includes agility that is supported by expertise, capacity and institutions. Flexibility can be built into a system with measures, such as predetermined points of review. Strategic course-corrections can also be factored with upper and lower limits to performance,

and triggers for action. Here, integrated planning has an opportunity to foster important nimbleness through interlocking policy on resilience, the workforce, critical infrastructure, and science and technology.

In short, none of the early actors behind the energy transitions of this study could have anticipated the totality of the developments that would follow. This insight speaks to a larger take-away: that experimentation and learning are fundamental to energy system change. Importantly, such trial and error applies not only to technologies and systems but to policies and practices as well. If learning and uncertainty are understood to be an intrinsic part of fuller progress, then cross-sectoral problem solving might bring greater strengths to bear.

FUTURE RESEARCH

This book presented a comparative study of four prime mover countries, adopting different forms of low carbon energy over roughly four and a half decades. In terms of methodology, the work combined process mapping with interviews, case study, and historical record review to inductively identify patterns in energy system change. Specific to theory, it put forward conceptual models and logic to explain energy transitions in socio-technical terms. In a practical sense, this work shed historical light and provided an in depth look at the interplay of underlying developments and influences that are often missed in more traditional economic and political assessments.

Looking ahead, there are countless ways to advance a research agenda on energy transitions and systems change. Substantial space exists for research on the evolution in the built environment versus practices, dual purpose infrastructure, policy mixes and systemic levers, and market redesign. Parallel lines of questioning could consider the influence of agents of change, financing structures, or the role of prosumer buildings that produce energy. New cases could test and extend the analytical tools that were introduced or the explanatory strength for determinants of change. One could contrast energy transitions spurred by focusing events, like major energy-related accidents, with ones more closely linked to technological discoveries. Research could also consider the tradeoffs in sustainability, security and safety in energy system change. Other promising research includes the alignment of socio-technical attributes with culture in varying adoption patterns. Alternative units of analysis could be more fully explored at the local and global levels. Finally, the theory-building and tools put forward in this study could be applied and refined with different industries, such as biotechnology or communications.

FINAL TAKE-AWAY

Energy transitions, like other forms of change, can be proactive or reactive. While transformations can occur at times in response to challenging circumstances, windows of opportunity also exist to not only optimize an energy pathway, reduce environmental effects, and encourage industries, but to advance society itself. *Carpe diem.*

APPENDIX

Technology Primer

GEOTHERMAL ENERGY AND RELATED TECHNOLOGIES

Geothermal energy is derived from the Earth's inner heat. It is produced by the decay of naturally occurring radioactive isotopes (e.g. uranium, potassium and thorium) (Rogner, et al., 2012) and from the primordial heat associated with the Earth's early development. This form of thermal energy is contained in rock, as well as in the steam or fluid found in fractures and pores within the earth's crust (Goldstein, et al., 2012; Tester, Drake, Driscoll, Golay, and Peters, 2005).

Geothermal resources have varying potentials (Table A-1). One estimate of the earth's geothermal heat potential from the surface to a depth of 33,000 feet indicates an availability of 55,000 times more energy relative to what is available in oil and gas resources (Union of Concerned Scientists [UCS], n.d.).

The harnessing of this form of energy depends on the resource type, distance of a demand center from the resource, geological conditions, and costs, among factors (IEA, 2011*b*; Thorsteinsson and Tester, 2010).

Interestingly, the resource may be found practically anywhere, yet, the highest quality resource is linked to geographic areas with a significant amount of tectonic plate interaction and active or young volcanoes (UCS, n.d.; USGS, n.d.). One particular region known for this resource is the "Ring of Fire" or Circum-Pacific belt, a zone encircling the Pacific Ocean (USGS, n.d; Figure A-1). The area has both tectonic and volcanic activity, allowing for substantial amounts of heat to rise to the surface (USGS, n.d.).

To tap geothermal energy at scale, two primary means are employed: (1) the direct use of geothermal steam or heated water, and (2) the indirect use that converts geothermal energy into electricity (Kiruja, 2011; Tester et al., 2005). Applications of the resource depend largely on its temperature. For utilization as electricity, geothermal resources are generally 100–150°C or higher. By comparison, the geothermal energy used in heating can leverage a wider range of

Table A-1. Base Estimates of Worldwide Geothermal Resources by Type (Total thermal content in situ)

Resource Type	Total Thermal Energy(EJ)
Hydrothermal (vapor and liquid dominated)	**137,157**
Geopressured*	**569,730**
Magma**	**5,275,279**
Hot Dry Rock	**110,780,865**
Moderate to High Grade	**27,958,980**
Low Grade	**82,821,884**

SOURCE: Adapted from Tester et al. (2005), citing Mock et al. (1997), Armstead (1983), Armstead and Tester (1987), Duchane (1994), Rowley (1982). Converted to EJ, based on 1 quad = 1.055 EJ.

NOTES: *To depths of 10 km and initial rock temperatures of >85°C.
**To depths of 10 km and initial rock temperatures of >650° C.
This includes hydraulic and methane energy content. Estimates are based on scoping that differs from resource potential noted in Chapter 1.

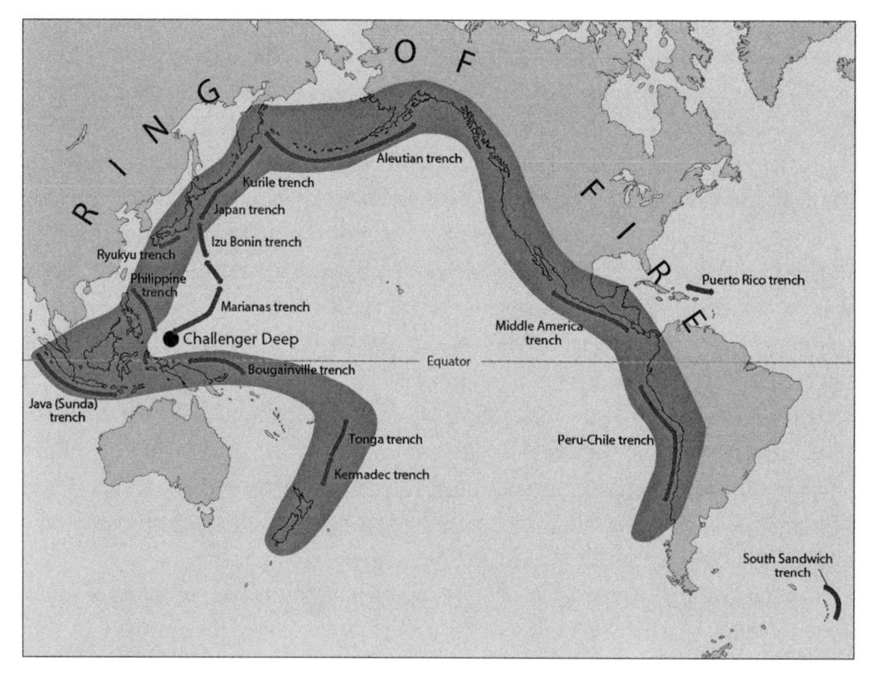

Figure A-1 Ring of Fire.
CREDIT: United States Geological Survey.

temperatures, making it employable for a number of purposes, including space heating, heated pools, greenhouse and soil-heated treatments, aquaculture, and the melting of snow (IEA, 2011*b*). In Iceland, low-temperature geothermal energy (<150°C) is used directly for heating. High-temperature geothermal energy (200–400°C) is utilized primarily for generating electricity and secondarily for heat in space heating, fish farming, and industry (Interview, 2012). Ground source heat pumps are another way to make use of geothermal energy. However, this mode is not typically employed in Iceland.[1]

Power Generation

Systems

Centralized as well as decentralized (i.e., distributed) systems can utilize geothermal energy for power generation. Unlike intermittent renewables, geothermal energy can be used as a base-load form of electricity to supply the minimum amount of power that is provided by a utility or distribution company (See Box A-1). Such an application is possible because geothermal energy does not generally have seasonal or weather-related flux and can be dispatched on demand (IEA, 2011*b*).[2] Geothermal energy could also be used to meet peak demand, but effective processes and methods for load following have yet to be developed. An important technical distinction between geothermal energy and its intermittent power sector counterparts is that high penetration of geothermal energy does not impose load balancing requirements on the system, as may be the case with wind and solar power.

Plant Types

Two primary types of geothermal power plants are in use: steam and binary cycle. Each plant extracts hot water and steam from a borehole, recycling warm water to allow the life cycle of the heat source to be extended (UCS, n.d.).

Steam cycle plants, of which the dry- and flash-steam design types are the most common, allow hydrothermal fluid (essentially high-temperature water) to boil. Steam is then separated from brine (i.e., solution of salt in water) and expands in a turbine (Valdimarsson, 2011). The brine can then be disposed

1. Since inexpensive geothermal space heating is available around most residential centers in Iceland, heat pumps are not widely in use. A notable exception may be in regions where geothermal resource water is unavailable at temperatures above 50°C. Communities in such locations are encouraged to replace electricity or oil with heat pumps (Orkustonun, n.d.).

2. Air-cooled binary geothermal plants are an exception. Their output is influenced by changes in air temperature (IEA, 2011*b*).

Box A-1

Base-Load in Power Systems

Base-load power plants operate at maximum output generally all of the time, except when taken offline or when reductions in power are required for maintenance and servicing. These plants typically produce the least-cost electricity in power markets. Hydropower, geothermal energy, biomass, fossil fuel plants, and nuclear power are fuel sources that often power base-load plants.

---

of by reinjection into the ground or vaporization (flashing) at a low temperature. Dry steam plants are used in about a quarter of today's global geothermal capacity and involve the piping of geothermal steam directly from a well to a turbine/generator (Valdimarsson, 2011; IEA, 2011*b*). In a flash steam model, currently used for about two-thirds of geothermal installed capacity, the hydrothermal fluid enters the well at high pressure and boils on its way up as the pressure is reduced (IEA, 2011*b*). The mixture of water and steam enters a flash tank where steam and water are separated. The steam then powers a turbine/generator equipment. The water is eliminated or reinjected into the ground together with condensed water from the turbine.

In contrast to the steam cycle design, the binary cycle design utilizes geothermal water/steam in such a way that it does not directly interact with the turbine/generator equipment, but rather warms a secondary fluid in a closed power generation cycle that employs a heat exchanger (UCS, n.d.; Valdimarsson, 2011). The heat exchanger conveys heat from the geothermal fluid to the secondary fluid, at which point cooled brine is eliminated or reinjected into the ground (Valdimarsson, 2011). This type of design represents the fastest growing form of geothermal power plant design because it draws upon low- to medium-temperature resources that are more widely available (IEA, 2011*b*).

Generally, the choice of plant type is a function of the site and resource characteristics along with cost considerations. A dry-steam model is most appropriate when steam is emitted directly from the well, while the flash-steam model generally works best when wells release high-temperature water. A binary cycle, however, often used in conjunction with a heat exchanger, is most suitable for lower-temperature sources (UCS, n.d.).

Enhanced Geothermal Systems

An emerging, complementary geothermal technology, *enhanced geothermal systems*, captures heat from hot, dry rock, tapping an energy source that the methods discussed above cannot. Hot dry rock reservoirs are generally located at depths below those of traditionally used geothermal sources and can be utilized by sending high-pressure water to break up rocks, similar to the hydraulic fracturing [fracking] process used with unconventional oil and gas extraction. Once the rock is fractured, additional water is injected that, when heated, can be used as steam. A key challenge for the commercial viability of EGS lies in the ability to efficiently and reliably stimulate multiple reservoirs (IEA, 2011*b*).

Direct Use for Heat

Beyond the use of geothermal energy in power generation, the resource can also be harnessed directly for heating. Hot springs, for example, may be used in aquaculture or to warm greenhouses. Direct utilization of geothermal energy includes applications like district heating systems, as is the case in Iceland.

Ground-Source Heat Pumps

Ground-source heat pumps offer another way to tap geothermal energy by moderating building temperatures. Such devices leverage the year-round temperature of 10°C (50°F) that is generally found several feet below ground. Employing subsurface pipes filled with a conveyant, like air or antifreeze, a pump recirculates the conveyant to push heat out in the summer and produce the reverse effect in winter (UCS, n.d.). Particularly in regions with temperature extremes, these pumps are the most efficient of heating/cooling systems.

Environmental Considerations

Sustainability

When thinking in sustainability terms, geothermal energy differs from many other forms of renewable energy. This is because the rate of geothermal energy use can outpace the rate of its replenishment. The key to its utilization is to balance surface-level releases of heat with the recharging of fluid and heat in the source reservoir (Krater and Rose, 2009, citing Rybach and Mongillo, 2006). Here, stepwise development can be used to ensure sustainability in a geothermal field while minimizing long-term production costs (Orkustofnun, n.d.). Fundamentally, stepwise development entails streamlining the development of new and existing geothermal wells based on the monitoring and measured use of initial geothermal wells in a project area.

Emissions

Geothermal gases and fluids contain a range of elements and compounds. Geothermal gases, for example, may be high in sulfur, which, when emitted in an open system, can smell like rotten eggs and be toxic at certain levels. Greenhouse gases, like carbon dioxide (CO_2), can also be emitted from open geothermal systems (CO_2 arises from natural flux in geothermal processes). This release of CO_2 differs from that emitted during the combustion of traditional fossil fuels (IEA, 2011*b*). Table A-2 shows the relative releases of these emissions by various forms of energy use. Today's newer geothermal plants are typically designed to be closed-loop, with no direct operational release of carbon dioxide.

Fluid Pollutants

Geothermal fluids can also contain radon, arsenic, boron, mercury, and/or ammonia, chemicals that, if not managed properly, can pollute freshwater resources (Krater and Rose, 2009). A related concern is tied to the presence of salt in geothermal fluid, which may build up in system pipes. In such conditions, corrosion and scaling must be monitored.

Seismicity

Seismicity is another area of environmental concern with geothermal energy. As with hydraulic fracturing in the oil and gas industries, the EGS process can induce seismic activity. Induced seismicity and micro-seismicity at EGS sites are being studied and will likely continue to be factors affecting decisions about investment going forward (IEA, 2011*b*).

Table A-2. CO_2 Emissions by Energy Use

Energy Use	CO_2 Emissions
Low temperature applications of geothermal	1 g/kWh_e
High temp, hydrothermal fields, partially open-cycle, geothermal power or heat plants	0 to 740 g/kWh_e (Worldwide average: 120 g/kWh_e)
Closed–loop geothermal power plant systems*	0 g/kWh_e
Lignite/brown coal plant	940 g/kWh_e
Natural gas plant	370 g/kWh_e

SOURCE: IEA (2011*b*), citing Bertani and Thain (2002), Bloomfield et al. (2003), and IEA (2010).

* NOTE: *These assume geothermal fluids are reinjected below ground with no atmospheric loss of vapor or gas.*

Costs, Risk, Financing, and Economics

In terms of the cost competitiveness of geothermal energy use, the physical conditions of the resource can produce varying opportunities. In regions with high-temperature hydrothermal resources, geothermal-based electricity is often competitive with new power plants that use conventional fuels (IEA, 2011*b*). The use of binary plants may also be done competitively. However, costs will differ based on plant size, the temperature of the resource, and other geological conditions. Looking beyond the power sector, geothermal heating can also be competitive with district heating systems.

Broadly speaking, geothermal energy projects have high, front-loaded capital costs for exploration, drilling, and plant construction. A study by the Geothermal Energy Association (GEA, 2005) outlined the breakdown of costs for geothermal plant development (Figure A-2). Other than the construction of the plant itself, which comprises more than 50% of the costs, the GEA study found drilling to be the most expensive component at roughly 25% of the total (GEA, 2005). Such costs rose sharply after 2000, driven in large part by higher oil prices at the time (Thorsteinsson and Tester 2010, referencing Augustine et al., 2006). Inherently, drilling carries a high level of uncertainty and risk.

Uncertainty and large-scale, front-end investment needs mean that debt financing and government support can play an important role in determining whether a geothermal project occurs. In so-called open markets, limited options exist for financing (IEA, 2011*b*). Here, resource verification loans are an instrument that can assist in covering the cost of drilling and the testing of production wells. Co-financing, grants, and partnerships, in addition to multilateral/bilateral bank funding, are other key means for addressing the financing challenge.

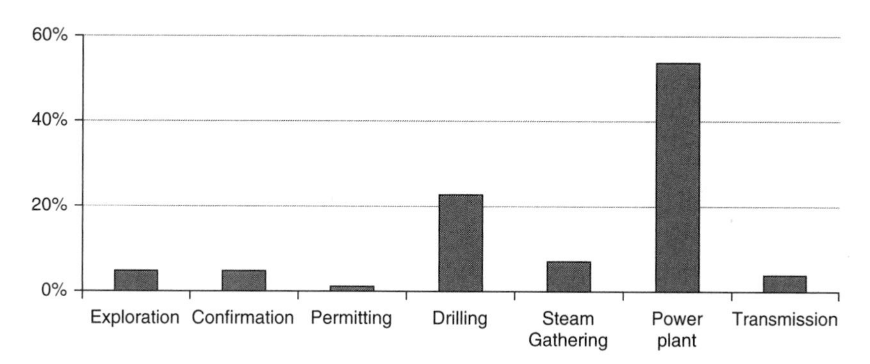

Figure A-2 Typical cost breakdown for geothermal power plants.
SOURCE: Compiled with Geothermal Energy Association data (2005).

Multidimensional and Global Aspects to Geothermal Energy

Among fuel types, geothermal energy is a prime candidate for combined heat and power plants (CHP).[3] Essentially, heat emitted in electricity production is routed for secondary use in heating. This sustainable optimization of energy embodies the application of the Second Law of Thermodynamics, which sets a limit on the overall thermal efficiency attainable by heat engines.

In 2015, the globally installed geothermal power plant capacity was estimated at 12.6 with approximately 73.5 GWh of geothermal power produced (Bertani, 2015). Global data on geothermal energy in heating are less readily available.

NUCLEAR POWER AND TECHNOLOGY

Basics

Today's use of nuclear energy in the power sector begins with the controlled and sustained release of energy in the process of nuclear fission. In an induced chain reaction, atomic nuclei are split, producing heat that, as steam, drives a turbine to provide electricity.[4] In this process, which can be generated in a way that is self-sustaining, considerable amounts of energy are produced (Tester et al., 2005).

A simple comparison of annual power plant fuel requirements for a 1,000 MW_e power plant powered by coal versus of one powered by nuclear energy illustrates the order of magnitude difference for energy releases from chemical reactions and nuclear fission. Such a plant that is powered by coal utilizes about 3,000,000 tons of the fuel, whereas, one that is powered by nuclear energy uses about 36 tons of enriched uranium (~1 ton of U-235) for a light water cooled reactor, if one factors for different fuel concentrations (Tester et al., 2005).

3. Combined heat and power plants (CHPs) can be traced to Thomas Edison's first commercial power plant (Ailworth, 2014). These plants are attractive for companies engaged in electricity and heat services as well as water services. Because of the conduciveness of geothermal energy for CHP plants, there may be an overlap in industries and utility functions by geothermal energy–producing companies.

4. The process of nuclear fission differs from that of nuclear fusion in that the latter brings atomic nuclei together to produce energy. Another distinction exists in the technological and commercial maturity of the two processes. Fusion is still very much in a precommercial stage of development after more than 50 years of study and worldwide research and development (R&D) expenditures possibly on the order of $30 billion (in 2006 $) (Holdren, 2006). By contrast, fission has been commercially used in the power sector since the 1950s (World Nuclear Association [WNA], n.d.).

For nuclear fission to occur in a nuclear power plant, basic material inputs include: (1) a fuel, (2) a moderator, and (3) a coolant. Fuel feedstock can be uranium, plutonium, or thorium, with uranium being the most common form (Barre, 2008; Sovacaool and Valentine, 2012).[5] These fuels may be utilized as an oxide, metal, carbide, or nitride, yet must be in a form where isotopes can split by fission (i.e., be fissile) (Barre, 2008). Uranium 235 (^{235}U) is the only naturally occurring fissile isotope. Alternative fissile isotopes, such as plutonium 239 (^{239}Pu) and uranium 233 (^{233}U), are produced from naturally occurring fertile isotopes, notably uranium 238 (^{238}U) and thorium 232 (^{232}Th) (Barre, 2008).[6] The determination of fuel type can be contingent on the moderator and reactor technology.[7]

The other primary input for a nuclear reactor is a coolant. The coolant absorbs the heat released from fission and produces steam, which powers a generator. Pressurized water is the most typical coolant; however, coolants can also be a gas or a liquid metal, like molten sodium (Barre, 2008).

Many combinations of fuel, moderators, and coolants in a nuclear reactor are theoretically possible; nonetheless today's commercial plants are typically represented by thermal neutron technology that can be classified in the following categories by the moderator.

Light water reactors (LWRs) account for roughly 80% of the nuclear power plants (NPPs) in use globally today (Sovacool and Valentine, 2012, citing Froggatt, 2010). The prevalence of LWRs is largely explained by their use of common water as a moderator. Water is not only a good moderator and coolant, but it is also relatively abundant, inexpensive, and easy to manage (Barre, 2008; Sovacool and Valentine, 2012). Such reactors must be sited near a river or sea, allowing the body of water to serve as a heat sink.[8]

The LWR reactor can be divided into two primary designs: boiling water reactors (BWRs) and pressurized water reactors (PWRs). With BWRs, the

5. Blending of uranium and plutonium will be discussed later. Thorium is not currently utilized at an industrial scale, but is the focus of research in areas that are rich in the resource, namely India (Barre, 2008; Pradhan, 2012).

6. For more extensive discussion of isotopes, as well as fertile and fissile materials, see Barre (2008) and Kravit, Lehr, and Kingery (2011).

7. A moderator is material that slows fast neutrons released from nuclear fission, thus enabling a nuclear chain reaction to occur (Barre, 2008). Reactors can generally be distinguished by whether they employ moderators, and of what type; for example, fast breeder reactors do not require a moderator, whereas thermal neutron reactors, such as those with light water technology, do (Barre, 2008).

8. A heat sink transfers thermal energy from a higher temperature medium to a lower one. Fourier's law of heat conduction and Newton's law of cooling form the basis for this concept.

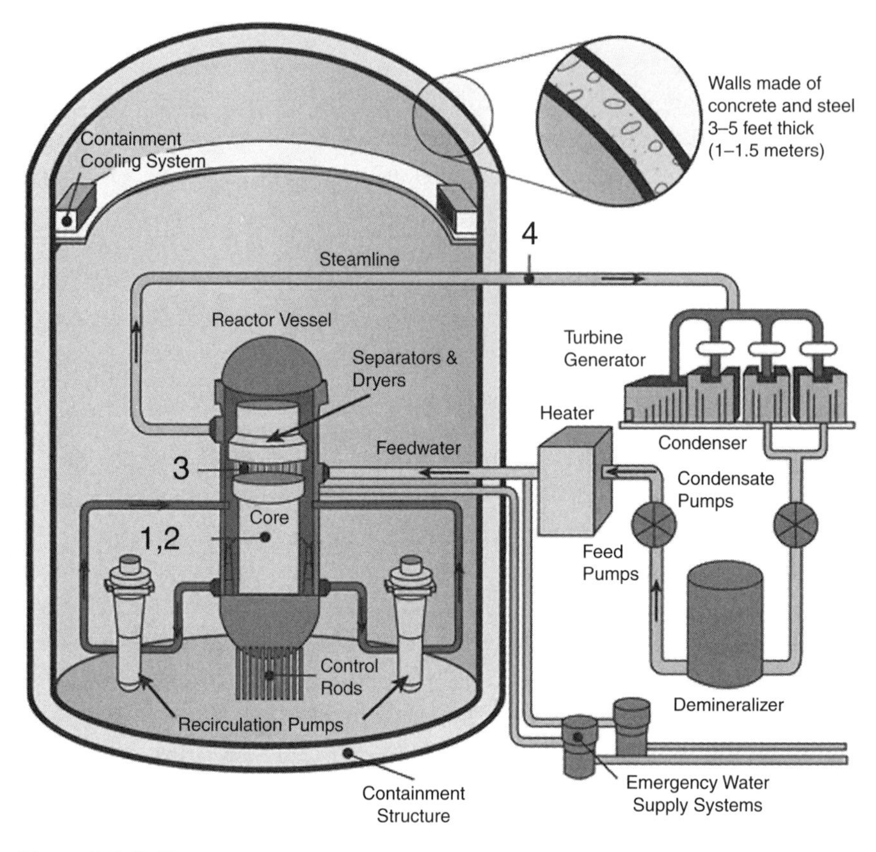

Figure A-3 Boiling water reactor.
CREDIT: US NRC (n.d.) Note: 1. Core, 2. Steam water is produced, 3. Steam water leaves the core, 4. Steamline directs steam to turbine.

coolant water is converted into steam within the reactor vessel. The steam is then conveyed via a pipe to drive a turbine (Figure A-3). In PWRs, the coolant water is circulated at high pressure between the reactor core and a steam generator, coming into contact with secondary feedwater that is heated to generate steam and power a turbine (Figure A-4). The use of common (light) water as a moderator by LWRs requires enriched, rather than natural uranium; otherwise, too many neutrons would be absorbed (Sovacool and Valentine, 2012). While enriched uranium is the standard fuel for LWRs, some PWRs use a blend of uranium and plutonium known as *mixed oxide* (MOX) (Barre, 2008).

Another category of commercial reactor is the heavy water design (HWR) that is found in Canada deuterium uranium (CANDU) models. This technology utilizes D_2O (heavy water) as the moderator and frequently as the coolant (Barre, 2008). Heavy water contains an elevated concentration of molecules with deuterium atoms (Anantharaman, Rao, Castano, and Henning, 2011),

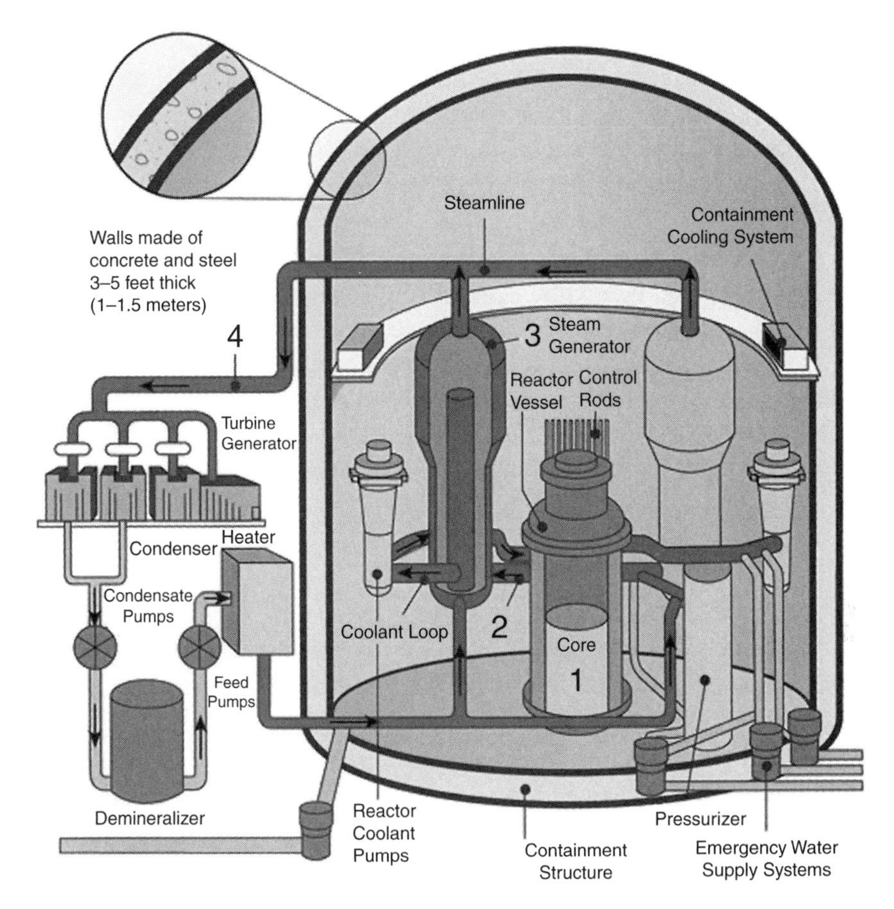

Figure A-4 Pressurized water reactor.
CREDIT: US NRC (n.d.).
NOTE: 1. Core, 2. Pressurized water in coolant loop, 3. Steam generator, 4. Steamline directs steam to turbine.

and is costly to produce (Interview, 2012; Sovacool and Valentine, 2012). Such a moderator allows the HWRs to skip the uranium enrichment process and operate with less expensive, natural uranium (Barre, 2008). Currently, about 8% of the world fleet employs this type of reactor technology (Sovacool and Valentine, 2012).

A third class of commercial reactor technology is a gas-cooled design that utilizes graphite as the moderator. Like CANDU models, this design can also run on natural uranium. Its short name in French—UNGG for *Uranium Naturel Graphite Gaz* (NUGG in English)—is associated with its feedstock and moderator. The design is considered relatively safe because of its low power density and the use of gas as a coolant (Sovacool and Valentine, 2012). An advantage to this design is in fuel loading and unloading, which can be completed while the reactor is operating (Barre, 2008; Sovacool and Valentine, 2012). A disadvantage associated with this model's early design was that fuel needed to be replaced regularly (Sovacool and Valentine, 2012).

Finally, the water-cooled, graphite-moderated design (Reaktor Bolshoi Moshchnosti Kanalnye [RBMK]) reflects a remaining category of nuclear reactor. This class shares some characteristics with the previous gas-cooled graphite-moderated design because they both use graphite for the moderator, and fuel loading/unloading may be done while the reactor operates (Barre, 2008). The RBMK differs from its gas-cooled counterpart in that conventional light water is used as the coolant, and containment buildings are not typically part of the design. RBMK technology is also subject to power surges (Barre, 2008). This reactor technology class is widely known for its association with the 1986 Chernobyl nuclear accident,[9] and was instrumental in generating weapons-grade plutonium for the Soviet Union's nuclear arsenal (Barre, 2008).

Nuclear Fuel Measurement and Uprating

The level of energy extracted from nuclear reactor fuels is typically measured in terms of a *burn-up rate*. Presented as GW days per ton of fuel, this indicator reflects the ratio of thermal energy released by nuclear fuel to the mass of fuel material consumed (Anantharaman et al., 2011).

Another concept in nuclear technology is *uprating*. If utilities want to increase the power output of a nuclear plant, they may undertake measures to uprate, which typically entails the operator refueling the plant with more enriched fuel or a higher percentage of new fuel in order to attain an increased power level (US Nuclear Regulatory Commission [NRC], n.d.). Operating at a higher power level generally produces greater amounts of steam and water flows, so components like pipes, valves, and heat exchangers will require modification (NRC, n.d).

Nuclear Technology Attributes

Nuclear power plants are generally deployed as highly centralized, stable, and somewhat inflexible technology with large-scale plants that are on the order of

9. The Chernobyl disaster began on April 26, 1986 in the Ukraine, during a systems test at reactor 4 in which an unexpected power surge occurred, followed by an attempted emergency shutdown. Deficient plant design and operation regulation together with human error factored in ensuing events, which included steam explosions and a rupture of the reactor vessel.

1,000 MW in size. In recent years, smaller modular plants have received attention (WNA, n.d.), but it is not clear that they can provide economies of scale in production or enough advantages in other areas to offset the operational economies of scale of conventional, large plants (Bunn and Malin, 2009).

Fuels and the Fuel Cycle

Uranium, the most common feedstock for nuclear fission, is found in rock and sea water (WNA, n.d,). In 2010, global production equaled 53,663 tonnes of uranium, an increase of 6% over 2009 (WNA, n.d.).[10]

When considering uranium for nuclear power plant use, the amount of fissile material matters. Natural uranium, used in NUGG reactors, contains 0.71% of the fissile isotope ^{235}U, whereas ^{235}U used in LWRs must be enriched to a concentration of 3–5% (WEC, 2010*c*). The enrichment process produces substantial quantities of depleted uranium tailings of different ^{235}U concentrations, usually in a range of 0.25–0.35% (WEC, 2010*c*). When uranium prices are high, it may be economically feasible to re-enrich the waste by-products of uranium mining, known as tailings (discussed with enrichment).

The nuclear fuel cycle involves a series of fuel processing stages.[11] At the front-end are stages for uranium mining, milling, converting, enrichment, and fabrication. These generally occur before power is generated in a reactor. The fuel cycle back-end includes reprocessing and/or the storage and disposal of spent fuel. Broadly speaking, management of fuel entails an open-cycle (i.e., once-through) or a closed-cycle approach. In open cycles, spent fuel is treated as waste with no attempt made to recover unused fissile material, whereas in closed cycles, spent fuel is treated for continued use (WEC, 2010*c*; Anantharaman et al., 2011).

Front-End Fuel Cycle

In the first stage of the fuel cycle, mining approaches range from direct excavation to in-situ extraction. With *underground mining*, long, thin shafts are dug to extract uranium from underground seams (Sovacool and Valentine, 2012). In contrast is the more prevalent *open pit mining*, which involves the removal of rock layers to extract the underlying uranium. *In-situ leaching* differs, using a process in

10. The resource availability of uranium is characterized by price bands. Total identified resources as of January 1, 2009, equaled 5.4 million tonnes in the <$130/kg U category (<$50/lb U_3O_8) and 6.3 million tonnes in the <$260/kg U category (<$100/lb U_3O_8) (NEA/OECD, 2010).

11. The focus here is principally on uranium as a feedstock.

which underground uranium deposits are immersed in acidic or alkaline solutions before the uranium ore is pumped to the surface (Sovacool, 2008*b*; WNA, n.d.). The last approach presents important tradeoffs because less time and cost are typically involved, but more water is used (up to 7–8 gallons of water per kWh of nuclear power generated) (Sovacool and Valentine, 2012, citing US DOE, 2006).

As with fossil fuel extraction, uranium mining can have a significant environmental impact. To extract and use a typical amount of uranium in a nuclear reactor for a year (25 tons), an estimated 500,000 tons of waste rock, 100,000 tons of toxic mill tailings, 144 tons of additional solid waste, and 1,343 square meters of additional liquid waste are also produced (Sovacool and Valentine, 2012, citing Thorpe, 2008).

Once uranium is extracted, it must be primed for fuel use. A milling process breaks down the mined material with an acidic or alkaline wash, leaching the uranium from the ore. The remaining powder is roughly 75% uranium oxide (U_3O_8), known as yellowcake (for more discussion, see Sovacool, 2008*b*). Residual material, namely oxide and tailings (i.e., ore with rock material), is radioactive, so it must be treated (WNA, n.d). Solutions such as the acids used in milling must also be neutralized (Sovacool, 2008*b*, citing Fleming, 2007, and Heaberlin, 2003).

The next stage of the fuel cycle entails the conversion of uranium oxide into uranium dioxide, which can then be utilized in reactors suited for natural uranium. For plants requiring enriched uranium, the feedstock must be transformed into uranium hexafluoride for added processing (UF_6) (WNA, n.d.; Sovacool, 2008*b*).

Following conversion, uranium may then be enriched, increasing the proportion of ^{235}U relative to ^{238}U material in nuclear fuel (Anantharaman et al., 2011; Sovacool and Valentine, 2012, citing Yudin, 2009). Two commercially popular approaches include diffusion and centrifuge separation (Sovacool and Valentine, 2012).[12] The former, which accounts for roughly 25% of the global enrichment capacity, entails the filtering of pressurized UF_6 through a porous membrane in a cascaded flow involving roughly 1,400 stages (WNA, n.d). The latter and more popular approach today entails the funneling of UF_6 gas through rapidly rotating vacuum tubes. Centrifugal forces separate ^{238}U and ^{235}U (Sovacool, 2008*b*, citing Uranium Information Centre, 2007).

Once enrichment occurs, uranium that is earmarked for reactor fuel generally undergoes an additional fuel fabrication process to become UO_2 where it is pressed and sintered (baked) into pellets (Sovacool and Valentine, 2012; WNA, n.d.). Pellets are incorporated into fuel rods that are then combined into fuel assemblies for use in reactors.

12. Another approach involves the repurposing of nuclear weapons materials.

Back-End Fuel Cycle

Fuel in nuclear power plants must be periodically replenished to replace built-up fission by-products with new uranium. In line with this, reprocessing may occur. Reprocessing basically entails the separation of spent fuel into three streams: uranium, plutonium, and waste.[13] Separated uranium can then be channeled through the conversion process once again on the front-end for enrichment. Plutonium can also be used. If plutonium undergoes fuel fabrication, it may be utilized in MOX fuel as an alternative to enriched uranium fuel, or in weapons (for a discussion, see Sovacool and Valentine, 2012 and WNA, n.d.). In the latter instance, weapons-grade plutonium must come from under-irradiated spent fuel (Interview, 2012). Notably, plutonium from LWR spent fuel has never been used for weapons (Interview, 2012).

Prior to reprocessing or long-term disposal, interim storage of spent fuel can be accomplished either in wet or dry form. Wet storage requires the immersion of spent fuel assemblies in water within concrete pools encased in steel (Sovacool and Valentine, 2012; WNA, n.d.). The water not only cools the fuel assembly, but also provides a shield from radiation. Dry storage, which may follow wet storage after a period of years, employs gas or air as a coolant with metal or concrete as a barrier.

Long-term storage is regularly discussed, but has not yet happened because there are no appropriate disposal facilities in place (Sovacool and Valentine, 2012; WNA, n.d.; see also section on "Societal Concerns"). An option currently receiving much scrutiny involves the use of geological repositories for sequestering waste. Such locations minimize the chance that radioactive substances will diffuse into the atmosphere or expose humans to dangerous levels of radiation.

When discussing nuclear waste management, it is useful to bear in mind that lengthier durations of interim storage generally mean that waste will be more manageable for long-term storage (WNA, 2017). This occurs because there is a progressive reduction in radioactivity and heat production.

13. Reprocessing approaches include *purex* and *pyroprocessing*, among others:

- Purex, also known as plutonium uranium extraction, entails the extraction of chemically pure plutonium. Originally, this process was utilized to prime weapons-grade material and has since been applied to the civilian power sector (Hannum, Marsh, and Stanford, 2007; Stanford, Marsh, and Hannum, 2009). It is, however, expensive and requires comparatively tighter security procedures to minimize proliferation risk.
- Pyroprocessing was developed in the 1980s and 1990s to allow for reprocessing without the risk of generating weapons-grade plutonium (Hannum et al., 2007; Stanford et al., 2009). According to some, the waste from this approach becomes "essentially harmless" within a few hundred years compared to the standard period of tens of thousands of years (Hannum et al., 2007; Stanford et al., 2009).

Decommissioning

Nuclear reactors and uranium enrichment facilities must be decommissioned as part of their closure processes. After shutting down, a plant must first cool, usually for 50–100 years, at which point it can be dismantled for final disposal (Sovacool, 2008*b*, 2010). Estimates of the energy needed for decommissioning indicate that the process may exceed that of the original construction by as much as 50% (Sovacool, 2008*b*, citing Fleming, 2007). To date, there is limited experience with decommissioning. Examples from the United States and the United Kingdom point to costs ranging from $300 million to $5.6 billion per facility (Sovacool, 2010).

Water

In addition to the use of water in milling and mining processes, substantial amounts of water are required for operating conventional nuclear reactors (Sovacool, 2010). Relative to other power generation plants, nuclear facilities are some of the most water-intensive (Sprang, Moomaw, Gallagher, Kirshen, and Marks, 2014). However, a portion of the water can be reused. This condition becomes important when droughts occur or where water is scarce because a lack of water in an area or region may limit nuclear power use as an option.

Societal Concerns

The use of nuclear power invariably raises questions about the environment, health, and societal implications. A number of aspects, including nuclear accidents, radiation exposure, waste, proliferation, and sabotage merit closer focus.

Nuclear Incidents and Accidents

When considering nuclear power's safety, it is important to understand distinctions relating to incidents and accidents. A nuclear incident is an event or technical failure during standard plant operations that does not produce off-site releases of radiation or considerable damage to equipment (for more discussion, see Sovacool, 2011*a* and 2011*b*). By contrast, a nuclear accident refers to the same operational context, yet one in which off-site releases of radiation or considerable damage occurs to plant equipment. The International Nuclear and Radiological Event Scale (INES) ranks the severity of nuclear and radiological events with a scheme in which Levels 1–3 are "incidents," and Levels

4–7 are "accidents" (IAEA, n.d.). This scheme is controversial since quantifying and ranking such impacts oversimplifies their complexity. Regardless, the indicator provides a working means to compare and discuss events.

Looking more closely at accident examples, the 2011 Fukushima accident was a Level 7 accident with an early estimate of 21 deaths and $152 billion in costs (in 2010 dollars) (Sovacool, 2011*a*).[14] This contrasts with the 1986 Chernobyl accident, which was also Level 7 and is estimated (albeit debated) to have caused 4,056 deaths and $7.2 billion (in 2010 dollars) in costs (Sovacool, 2011*a;* Chernobyl Forum, 2006). The Three Mile Island (TMI) accident was a Level 5 accident with estimates indicating no fatalities and costs of $2.6 billion (in 2010 dollars) (Sovacool, 2011*a*).[15]

Naturally, the way that nuclear safety is framed shapes discussion and decision-making. Technical failures, equipment damage, and releases of radiation can occur in the context of power plants shutting down, at research reactors, with military testing, or during different stages of the nuclear fuel cycle that are not covered within nuclear power assessments (Sovacool, 2011*a*). For instance, accident-related releases of radioiodine occurring at the Savannah River reprocessing plant in South Carolina to date have exceeded that of TMI by a factor of 10 (Sovacool, 2011*a*). There is also the case of a break in the uranium mine tailings dam in New Mexico in 1979, which produced the single largest release of radioactive material in the United States (excluding military weapons testing) (Sovacool, 2011*a*). In this example, water was left undrinkable for 2,000 people living in the nearby Navajo reservation, and farm animals were heavily contaminated with lead 210, polonium 210, thorium 230, and radium 236.

Larger system interdependencies also have relevance for nuclear safety, as was evident with the Fukushima accident. In that example, the compound occurrences of an earthquake, followed by a tsunami that flooded areas of Japan, left the back-up system for the Fukushima nuclear plant without power (WEC, 2012). In another case, the 2003 black-out in Northeastern United States and Canada revealed inadequate maintenance of back-up systems for more than a dozen nuclear power plants in the United States and Canada (Sovacool, 2011*b*).

14. The Fukushima Daiichi power plant accident began on March 11, 2011 in Japan, following a tsunami and magnitude 9.0 earthquake. The disaster was directly attributed to plant design issues and human error. Substantial amounts of radiation were released, and three of the six reactors underwent nuclear core meltdown.

15. On March 28, 1979, a partial core meltdown occurred in Unit 2 of the Three Mile Island nuclear power plant in Pennsylvania following mechanical failures in the secondary (non-nuclear) and primary systems along with human error.

When considering nuclear safety, responsibility lies with facility operators to address issues and to ensure safety under the oversight of national nuclear authorities (Bunn and Malin, 2009). Complementing this is the international safety regime, which consists of the Convention on Nuclear Safety, other safety and liability agreements, organizations focused on nuclear safety, and broad, mostly voluntary standards (Bunn and Malin, 2009).[16] In practice, the international regime adds important value to the nuclear safety landscape yet is based on limited real authority and voluntary compliance (Bunn and Heinonen, 2011). As a consequence of events like Fukushima, nuclear safety has received increased attention. This has led to additional testing and adaptations in standards, design, training, and processes (Pouget-Abadie, 2012; UCS, n.d.; WEC, 2012; WNA, n.d.). However, the changes are uneven and cannot guarantee against subsequent issues.

A global study of 279 accidents across fuel types from 1907 to 2007 found that the three accidents culminating in the most fatalities involved hydropower, nuclear power, and gasoline (Sovacool, 2008*a*). The structural failure of the Shimantan hydropower facility in China in 1975, for example, caused the greatest number of deaths with 171,000 fatalities ($9 billion in property damage). The Chernobyl accident, noted earlier, followed next in terms of fatalities. A 1998 petroleum pipeline explosion in the Niger Delta, Nigeria, ranked third for fatalities with 1,078 deaths ($54 million in property damage). Considering variables other than fatalities, the study also found that natural gas had the highest number of accidents (33%) among fuel types, whereas nuclear energy ranked highest for costs from damages, accounting for 41% of all property damage among accidents studied.

Radiation

Central to concerns over nuclear accidents is exposure to ionizing radiation. Such radiation can damage or modify living cells, leading to death (Chernobyl Forum, 2006). Living organisms may be exposed to this from natural sources, like cosmic rays, as well as anthropogenic sources, such as certain medical treatments, x-rays, and nuclear power use (Chernobyl Forum, 2006). One gauge for exposure is the *sievert*, often noted as *millisieverts*, to describe what is deemed by some to be normal exposure. The UN Scientific Committee on the Effects of Atomic Radiation (UNSCEAR) estimates that humans are annually exposed to 2.4 mSv on average through natural radiation with a typical range being 1–10 mSv (Chernobyl Forum, 2006; UNSCEAR, 2008, 2010). Low-level exposure for

16. The Convention on Nuclear Safety entered into force in 1996 and is designed for participating states with land-based, operating nuclear power plants to commit to maintaining a high degree of safety with benchmarks set for the group (IAEA, n.d.).

humans is then estimated to be a few mSv per year. When expressing exposure in population terms, a person-Sievert or person-Rem may be used. This reflects the average dosage for one individual multiplied by the number of exposed people.

To take understanding of radiation a step further, one needs to consider the half-life of radionuclides (i.e., atoms exhibiting radioactivity). Radionuclides release radiation as they decay. This decay rate is commonly described in terms of the substance's *half-life*, referring (as the name indicates) to the period required for half of a radionuclide to decay. Such rates vary widely among radionuclides. Cesium, for example, has a half-life of 30 years, whereas plutonium 239/240 has a half-life of 24,000 years. Some radionuclides also begin with a short half-life, such as plutonium 241 (14 years), but transform in the decaying process to a radionuclide with a longer half-life, such as americium 241 (430 years) (United Kingdom Environmental Agency, n.d.).

In the context of nuclear power and the environment, two isotopes of Iodine, ^{239}I and ^{131}I rank as important radioactive isotopes that are produced with the fissioning of uranium in nuclear reactors or with the fissioning of plutonium/uranium in nuclear weapons (EPA, n.d). These isotopes readily react with other chemicals and can disperse quickly in air and water. Such isotopes may cause thyroid issues, and, with long-term exposure, can lead to cancer. Notably, ^{239}I has a half-life of 15.7 million years, whereas ^{131}I has a half-life of 8 days (EPA, n.d; Tester et al., 2005, citing Parrington et al., 1996). In the event of a large release of radioactive iodine, stable iodine may be distributed by public health agencies to the population. Ingesting stable iodine improves the probability that one's thyroid will absorb the stable form instead of the radioactive form during exposure (EPA, n.d; Tester et al., 2005, citing Parrington et al., 1996).

Waste Treatment

Waste treatment for nuclear plants has evolved over time for low-, intermediate-, and high-level waste, in connection with varying degrees of risks. Low-level waste is understood to contain small amounts of radioactivity, whereas intermediate waste carries a level of radioactivity that requires protective shielding during processing and/or transport (Sovacool and Valentine, 2012). High-level waste can include reprocessing waste products and spent fuel as well as decommissioned reactors and uranium enrichment facilities (Sovacool and Valentine, 2012). Questions persist with intergenerational implications about how to manage nuclear waste on technical, ecological, societal, and economic fronts. Some may look to deep sea immersion or to outer space for long-term waste disposal. The most commonly considered option, however, is deep geological repositories.

Proliferation and Security

Among energy options, a concern that is uniquely linked to nuclear energy is the potential for proliferation of fissile material, technology, and/or knowledge to make nuclear weapons. This is underpinned by a fundamental worry that non-nuclear weapon states or radical groups might acquire tools and expertise to manufacture weapons of mass destruction.

To combat this, the Nuclear Nonproliferation Treaty (NNPT) is designed to reduce the spread of such materials and know-how, to foster disarmament, and to encourage cooperation in peaceful applications of nuclear energy.[17] By some measures, the NNPT (in conjunction with the international nonproliferation regime) is largely successful. Nearly every state worldwide is a party, and there has been no net change in the number of states with nuclear weapons for 20 years (Bunn and Malin, 2009). However, theft and trafficking of nuclear materials remain a persistent reality. Between 1993 and 2011, confirmed incidents of illicit trafficking are reported to have numbered 2,164 (International Atomic Energy Agency [IAEA], 2012).[18] Of these, 16 incidents involved unauthorized possession of highly enriched uranium or plutonium, the fundamental ingredients for nuclear weapons (IAEA, 2012; Bunn and Malin, 2009).

Closely tied to proliferation concerns is the need to safeguard nuclear reactors and the fuel cycle against sabotage. Such destructive activity can have devastating consequences, not only for people and the ecosystem in the vicinity of the targeted action, but also for those who are physically downwind or "downstream" in the ecological chain (Bunn and Malin, 2009; Dreicer and Alexakhin; 1996; National Academy of Sciences, 2002). Here, technical and procedural safeguards factor importantly as defensive measures. In conjunction with culture, practices, and leadership, such safeguarding should be highly attuned to regulatory oversight, cooperation and communication, and related security and safety concerns.[19]

Finance and Economics

Nuclear energy projects entail large-scale investments, but, once in operation, such projects tend to have low and predictable fuel, operating, and maintenance

17. The Treaty entered into force in 1970, was extended indefinitely in 1995, and currently has 190 signatories (United Nations Office for Disarmament Affairs, n.d.).

18. It is highly unlikely that these numbers include all incidents.

19. In addition to the NNPT, international conventions associated with nuclear security include the Convention on Physical Protection of Nuclear Material and Facilities and the International Convention on the Suppression of Acts of Nuclear Terrorism. Such institutional attempts at oversight, however, often lack specificity (Bunn and Malin, 2009).

costs, assuming insurance is subsidized by the government or liability is limited, and no major safety changes are needed (IEA and NEA, 2010). These cost considerations and the long operating life spans of nuclear power plants mean that high returns on investments are possible (NEA, 2007). Given such cost dynamics, existing operational nuclear power plants have traditionally been quite competitive, providing what is often the lowest-priced base-load power (WEC, 2010*c*; for discussion of base-load power, see Box A-1). However, with disruptive change in the energy playing field tied to wind power, solar power, and unconventional natural gas, among drivers, nuclear power plants are experiencing early closures (IEA, 2015*d*; Beyond Nuclear, n.d.).

Another economic aspect of nuclear power is the front-loaded nature of such costs, which serves as an investment risk and financial challenge, particularly in liberalized markets (WEC, 2010*c*). Construction costs are a major component of the final levelized costs of nuclear power, and represent a part of nuclear plant development that is subject to many overruns (Schneider and Froggatt, 2012).

Yet another economic consideration in nuclear technology is the cost tradeoff for back-end fuel cycle options. Storage of spent fuel awaiting more long-term disposal is considered to be far less costly than reprocessing (Bunn and Malin, 2009). Establishing a national repository is estimated to produce multi-billion dollar costs and requires substantial technical demands (McCombie, 2009), although limited progress in this area leaves numbers open to speculation.

Markets, Expertise, and R&D

The nuclear energy market has been described as "low volume, but high value" (NEA, 2007). It is characterized this way because there is limited turnover of equipment or technologies that are sold to large utilities or consortia of utilities, yet these involve large, financial outlays up front.

The market for nuclear power is also subject to international agreements, such as those related to export and waste transport (NEA, 2007).[20] While these agreements are implemented to address concerns about the use of nuclear technology for military purposes and safety, they can have an unintended consequence of limiting related commercial activity in civilian nuclear power.

20. Such agreements include the Joint Convention on the Safety of Spent Fuel Management and the Safety of Radioactive Waste Management, IAEA Safeguard Agreement, and the Plutonium Disposition Agreement, among others.

In terms of timescales, nuclear power projects generally require long lead times for what are often sizeable and complex systems. This necessitates the expertise of multiple disciplines.

In the 31 countries that currently use nuclear energy, a certain level of sophistication is widely recognized as necessary for managing relevant infrastructure and institutions (NEA, 2007; Interviews, 2011). To address this, countries may turn to international vendors or international collaborations.

Specific to R&D, private nuclear equipment suppliers will generally focus on incremental upgrades because the research is expensive and long-term (NEA, 2007). By contrast, governments will typically focus on more radical innovations, like the R&D that is under way with the Generation IV international partnership (NEA, 2007; Generation IV Forum, n.d.).

Nuclear Power Plants within a Grid System

When considering nuclear power plants in the context of a grid system, it is important to distinguish between base-load and load-following power plants. Base-load power plants, as noted in Box A-1, operate nearly continuously and typically produce the least-cost electricity in power markets. In the case of nuclear energy, another reason to operate them continuously is that they require time to start and shut down. By contrast, load-following plants are those that are brought online as demand increases. The rule of thumb generally in the power sector, if special priorities are excluded, is that plants with the least variable costs are brought online first. Natural gas and hydropower plants are typically used for load-following. Nuclear plants may also be used, as is the case in France. However, they are not used for rapid variations (Interview, 2012).

BIOFUELS AND RELATED TECHNOLOGIES

Biofuels

Biofuels are transport fuels produced from agricultural or other biological feedstock. The two most common forms are ethanol and biodiesel, which are often blended with petroleum-based fuels, but can also be used independently. Ethanol, also referred to as ethyl alcohol, is derived from the fermentation and distillation of sugar or starch-based biomass (i.e., plant material), including sugarcane, corn, and other cereals. By contrast, biodiesel is made by chemically reacting alcohol with lipids from vegetable oils, animal fat, or from recycled cooking oil to produce fatty acid esters.

The primary form of biofuel considered in the chapter on Brazil is ethanol, which contains the same chemical compound (C_2H_5OH) as that in alcoholic beverages (Yacobucci, 2007). Ethanol may be used in standard internal combustion engines as an anhydrous additive or with modified vehicles as a stand-alone fuel in hydrous form. Hydrous ethanol is an ethyl alcohol with a maximum water content by mass of 6.2–7.4% that differs from anhydrous ethanol, which has less than 0.6% water (Banco Nacional de Desenvolvimento Economico e Social [BNDES] and Centro de Gestao de Estudos Estrategicos [CGEE], 2008). Hydrous or hydrated ethanol is used in ethanol-only vehicles, called *neat vehicles*, or in certain *flex fuel cars*, such as those marketed in Brazil (see the section on "Automotive"). The water content of ethanol must be limited when blending with gasoline since its presence impairs the capacity of ethanol and gasoline to mix without separation, particularly in cold weather (Joseph Jr., 2010).

The production of ethanol from sugar-based feedstocks like sugarcane, sugar beets, or molasses, broadly entails (1) the processing of feedstock to separate fermentable sugars, (2) the addition of yeast to trigger fermentation, and (3) distillation of the resultant alcohol to produce a hydrous ethanol (Seelke and Yacobucci, 2007). If the ethanol product is to be used as an additive in gasoline, the hydrous form is then dehydrated with a second stage of distillation through the addition of a co-solvent to create an anhydrous ethanol (Seelke and Yacobucci, 2007; Joseph Jr., 2010). The production of ethanol from starch-based feedstocks, such as corn and cassava, requires an additional step: the hydrolysis of starch into glucose, which breaks the starch into fermentable sugar (saccharification) (Kojima and Johnson, 2006).

In contrast to ethanol, biodiesel is a fuel that can be used in diesel engines without adaptation (EIA, 2010). The most popular way to produce biodiesel is through a process of trans-esterification, which converts a base oil to an ester using an alcohol, like methanol, ethanol, or butanol, when presented with a catalyst (Coelho and Goldemberg, 2004). Feedstock can include vegetable oils, like rapeseed or soy—the two most prevalent—as well as jatropha, coconut, palm, flax, and others.[21]

One way to gauge the quality or performance of ethanol is in terms of its percent yield from final energy that is used in creating the product. The IEA has indicated that Brazilian sugarcane-based ethanol has one of the highest yields among biofuels, using only about 10–12% of ethanol's final energy to generate the ethanol. This contrasts with biodiesel and ethanol from cereals or

21. Prior to conversion, these oils may be previously unused or recycled from cooking use. Animal feedstock includes fats from cattle, poultry, and pigs, as well as fatty acids from fish oils. Other feedstocks from sewage and sea farming are also being explored.

corn, which use 30% and 60–80%, respectively (IEA, 2007). In terms of price, Brazilian sugarcane-based ethanol is competitive with oil, selling at roughly $40–50/barrel (IEA, 2007). By comparison, Marker crude oil prices from mid-2015 to mid-2016 were $24–62/barrel (IEA, 2016*c*).[22]

Biofuels can be classified in different ways. Taxonomies, for example, can be broken down by technology maturity, greenhouse gas (GHG) emission balances, or feedstock (IEA, 2011*e*). The current case focuses almost exclusively on sugarcane-based ethanol since it represents the majority of the Brazilian transport sector's low carbon energy. Other forms of biofuels, like biodiesel, are noted where relevant.

Biofuels have a mixed record in relation to the environment. Their use can reduce the amount of pollutants emitted compared to those produced by gasoline or diesel combustion, yet the types of resources and processes used, as well as safeguards, have different GHG effects. Recent studies indicate that the consumption of corn-based ethanol can result in 13–22% less GHG emissions compared to that from gasoline, whereas GHG emissions from sugarcane and cellulosic ethanol may be 50–90% less than that from gasoline (Seelke and Yacobucci, 2007, citing Farrel et al., 2006; EPA, 2011*b*).[23] Results may be fairly sensitive to the assumptions, definitions, and scoping of a specific study, including locational conditions. (For an in depth discussion of this, see Araújo, Mahajan, Kerr, and da Silva, 2017.)

In recent years, GHGs emissions from biofuels and biomass have been undergoing re-evaluation because forms of land use, agricultural practices, and feedstock vary widely with respect to the level of associated emissions (Berndes, Bird, and Cowie, 2010; European Community [EC], 2010; IEA, 2011*e*; Lee, Clark, and Devereaux, 2008; Renewable Energy Policy Network [REN21], 2010). Particular concerns exist over biofuel production that entails new land clearing, use of nitrogen-based fertilizers, slash-and-burn harvesting techniques, and substantial transport to distilleries because each of these activities entails the release of CO_2 emissions in relation to corresponding processes that do not (Berndes, Bird, and Cowie, 2010; EC, 2010; IEA, 2011*e*; Lee, Clark, and Devereaux, 2008; REN21, 2010; EPA, 2007).

The food versus fuel debate is another consideration that drew widespread attention following agricultural commodity price surges in the periods 2006 to 2008, and 2010 to 2011 (High Level Panel of Experts of Food Security and Nutrition [HLPE], 2011 and 2013; Tomei and Helliwell, 2016). Briefly sketched,

22. This covers the week of June 15, 2015 to the week of June 6, 2016 (IEA, 2016*c*). The substitution value for ethanol relative to gasoline is discussed in the "Automotive Technology" section.

23. EPA analysis considers the farm-to-tailpipe emissions plus direct and indirect land use.

the concern is that land conversion for fuel crops may displace food-based crops, in turn leading to higher food prices (Earley and McKeown, 2009; Lee et al., 2008; Runge and Senauer, 2007; Yacobucci and Schnepf, 2007). Analyses by the World Bank and Fundaçao Getulio Vargas indicate that a variety of factors contributed to grain price increases, including high oil prices, poor harvests, and financial investor speculation—more so than biofuel production (Fundaçao Getulio Vargas [FGV], 2008; World Bank, 2010*a*). However, reports also show that this may not be as relevant for Brazil since sugarcane fields for fuel production correspond to roughly 5% of cultivated land and 0.5% of the entire country's land area (BNDES and CGEE, 2008; Food and Agriculture Organization [FAO], 2008).

In line with these issues is the concern that feedstock crops may displace biodiversity-rich habitats, such as the Amazonian rain forest and savanna lands, or pastures for raising cattle (Macedo, 2005). It is worth noting that Brazilian sugarcane production occurs in the Southcentral and Northeastern regions, mostly at great distances from the rainforest and savanna (Interviews, 2011–12; World Wildlife Fund [WWF], n.d.). Nevertheless, questions remain with respect to ranching activity that is displaced from the sugarcane regions and that then migrates to new land in Northwestern forested regions. The US Environmental Protection Agency reviewed ethanol reporting and satellite imagery of land use to appraise biofuels, like Brazilian ethanol, for certification of fuels in its Renewable Fuel Standard (EPA, 2011*b*, Greene, 2009; Interview, 2011). Brazilian ethanol was classified as an advanced biofuel in 2010, implying that it met a 50% GHG emissions reduction requirement.

Automotive Technology

Automotive technology that uses ethanol can be classed by categories including standard spark-ignition internal combustion, compression ignition, neat, and flex fuel technology.

1. *Spark-ignition internal combustion (ICE) vehicles* are the most prevalent on the roads today (Milligan, 2011). Their engines are versatile and can run on fuels including oil, liquefied petroleum gas (LPG), ethanol, and natural gas. Among engine technologies, the ICE is one of the less efficient models for transforming fuel to mechanical energy, at a roughly 15–20% conversion rate (Milligan, 2001, citing Wu and Ross, 1997). Its technology can use gasoline-ethanol blends, generally up to 10% ethanol (E10), with no changes in materials, components, or engine calibration (BNDES and CGEE, 2008; Interviews 2010–12). In such technology, ethanol can serve as a

fuel octane booster and reduce pollution, such as that from tetraethyl neat and other oxidizing agents that are banned or being restricted (BNDES and CGEE, 2008; Interviews 2010–12).

2. *Compression ignition, internal combustion engine vehicles* are generally heavy-duty vehicles such as trucks, buses, and trains. The compression ignition limits the fuel types that may be used to diesel, petrol, LPG, and a few others.[24] The efficiency of these engines is generally 22–28%, with an upper limit of roughly 43% (BNDES and CGEE, 2008; Interviews 2010–12).
3. *Neat vehicles* are manufactured or adapted for hydrous ethanol and typically have a higher compression ratio of 12:1 relative to standard internal combustion engines, which have a compression ratio of 8:1 (BNDES and CGEE, 2008; Goldemberg, unpublished). This is important since the compression ratio indicates the performance level of an engine. A higher ratio is advantageous in that it allows the engine to extract more mechanical energy from the air–fuel mix. If lower octane-rated fuel is used, however, higher compression ratios can also predispose engines to knocking (i.e., detonation). Knock sensors can mitigate the reduced efficiency and engine damage associated with this. When an engine runs on a higher compression ratio, it requires a richer air–fuel mix and more fuel, producing increased horsepower and torque (Gatti, 2010). Combining the higher compression ratio and lower energy content of ethanol relative to gasoline means roughly 1.25 units of pure ethanol replaces 1 unit of gasoline (Demetrius, 1990; Goldemberg, unpublished). In terms of technical change, neat vehicles require modifications in fuel-feed systems and the ignition to account for differences in the air–fuel relationship, in addition to changes to materials that are more compatible with ethanol.
4. *Flex fuel vehicles*, also called flexible fuel (FF), use an internal combustion engine that is adapted to run with more than one fuel stored in the same

24. The characteristics that make ethanol suited to spark-ignition internal combustion engines are not as conducive to the compression ignition engines (diesel cycle) used in larger vehicles such as buses and trucks (BNDES and CGEE, 2008). However, in related research, development, and demonstration (RD&D), Sweden has adapted diesel engines with some success (BNDES and CGEE, 2008, citing Sopral, 1983). Studies have also been done in Brazil to adjust fuel systems and integrate spark ignitions into diesel vehicles in order to utilize ethanol. Such a breakthrough would be significant because about half of Brazil's transport fuel is currently diesel-based (BNDES and CGEE, 2008).

> tank. There are some indications that the Ford Model T was the first flex fuel car (Fuel Testers, n.d.).[25] In the 1980s, this technology evolved in the United States, later advancing in Brazil (Gatti, 2010, citing Pefley et al., 1980; Joseph Jr., 2010). Flex fuel technology used in Northern Hemisphere markets is primed to run with a blend of 15% gasoline and 85% anhydrous ethanol (E85) to allow for cold starts at temperatures below 11°C and to minimize low-temperature ethanol emissions (Davis, 2001; Davis, Heil, and Rust, 2000). The flex fuel technology used in Brazil allows any combination of gasoline with ethanol, as well as 100% gasoline or 100% ethanol. The technology used differs from that of standard ICEs, principally through a number of modifications to the fuel system and engine (DOE, 2012*a*). Flex fuel technology is distinguished from dual-fuel technology, which may also use more than one fuel, but the engine runs on one fuel at a time with each fuel stored independently in different tanks (DOE, 2012*b*).

Generally speaking, the utilization of ethanol as the stand-alone fuel or additive in spark-ignition ICEs has disadvantages related to ignition temperature and energy content yet also has advantages in cleaner combustion, improved engine performance, and reduced emissions (BNDES and CGEE, 2008).[26]

Octane Rating

Octane ratings measure the resistance of fuels to detonate and self-ignite. Ethanol is considered to be an excellent anti-detonating additive that can substantially improve the octane rating of a gasoline (Joseph Jr., 2009).

Cold Starts

Ethanol has a higher ignition temperature (420°C) relative to gasoline (220°C) (BNDES and CGEE, 2008, citing API, 1998 and Goldemberg and Macedo, 1994). If advanced injection systems are not used, then an auxiliary cold-start system is needed to start engines fueled by E100 in colder climates.

25. The Ford Model T had a low compression engine, adjustable carburetor, and a spark advance that enabled it to switch from gasoline to alcohol and kerosene (Kovarik, 1998). The drop in oil prices and limited alcohol supply during Prohibition enabled gasoline to emerge as the dominant fuel (English, 2008).

26. To contrast the use of E100 and E22, one study of the VW Total Flex found E100 power performance to be 3% higher and torque to be 2% higher, with corresponding fuel consumption 30% higher (Joseph Jr., 2009).

Exhaust Emissions

Combustion of hydrous ethanol or an anhydrous-gasoline blend produces less carbon monoxide (CO) and sulfuric oxides (SO_x), hydrocarbons, and other pollutants relative to gasoline due to the chemical composition of the fuels (BNDES and CGEE, 2008). By contrast, more aldehydes (R-CHO compounds) are produced and, depending on the engine attributes, nitrogen oxides (NO_x) may also be emitted in larger quantities. Concerns about aldehydes exist because they have carcinogenic potential, although catalytic converters reduce these pollutants to accepted ranges (BNDES and CGEE, 2008).

Other Attributes of Ethanol

In contrast to petroleum, ethanol is less likely to explode, is less toxic if spilled, and has a more basic physical and chemistry profile, making it easier to refine.

Bioelectricity

Bioelectricity (or biopower) refers to electricity derived from biomass. Sugarcane residuals, namely bagasse (i.e., biodegradable and compostable fiber) and straw, are examples of biomass materials that hold roughly two-thirds of the absorbed solar energy of the sugarcane plant (BNDES and CGEE, 2008; Goldemberg, 2010). Bagasse can be burned as a fuel in co-generation-based processing employed in sugar and ethanol plants, thereby enabling greater life-cycle energy efficiencies.[27] Straw is now also being utilized with high-pressure boilers.

WIND POWER AND RELATED TECHNOLOGIES

Wind is produced from the movement of air over the Earth's irregular surface. As solar radiation enters the Earth's atmosphere, wind is generated through uneven heating of the Earth's surface (Wiser, et al., 2011). Pressure differences produce air particle shifts that, in turn, generate wind flows that are shaped by the Earth's rotational flow, geography, and temperature gradients, among other factors (Wiser, et al., 2011, citing Burton, et al., 2001). In order to harness wind energy, the kinetic energy embodied in its motion must either be converted to mechanical power for processes like pumping water or be transformed into electricity with a generator.

27. *Co-generation* refers to the utilization of a power plant or engine in such a way as to generate electricity and useful heat.

Wind is prevalent in varying strengths virtually everywhere (Wiser et al., 2011, citing Burton, et al., 2001; Archer and Jacobson, 2005). Recent estimates of global wind resources indicate that a potential exists for more than 40 times the current electricity consumption, or more than five times the total use of energy worldwide (Lu, McElroy, and Kiviluoma, 2009).[28]

Resource characteristics of wind differ by location and time. Like basic topography, wind quality often varies within a given region. It can have different availability over various timescales, including subhourly, daily, by season, and in terms of interannual flux (Wiser et al., 2011, citing Van der Hoven, 1957). Given its dynamic nature, wind must be managed in a way that differs from non-intermittent energy sources.

The simple power of wind is measured by the cube of its speed; thus, each time the average speed of wind doubles, power increases by a factor of 8 (European Wind Energy Association/EWEA, n.d.). Subtle differences in speed, then, can have large effects on power output. If, for example, average wind speed increases from 6 meters per second (m/sec) to 10, wind power output could increase by more than 130% (EWEA, n.d.).

The fastest wind velocities are generally found at sea, on hilltops, and along open coastlines (EWEA, n.d.). To effectively tap its potential, utilization must account for wind strength, direction, and the frequency of flows,[29] which have a fundamental bearing on the performance and economics of wind generation at a given site (EWEA, n.d.).

Turbine Technology and Design

The basic functioning of a wind turbine mimics a fan working in reverse. Rather than utilizing electricity to produce wind, wind instead passes over the blades of a rotor that, through a shaft, spins a generator to produce electricity (DOE, n.d.). This operation occurs when wind flows cause lift to turn the blades (Figure A-5).

The torque (i.e., turning force) on the rotor blades is a function of the wind speed, density of air, and rotor area (DOE, n.d.). The density or heaviness of air is measured by the mass per unit of volume. Essentially, the heavier the air density, the more energy a turbine can capture. The rotor area determines how much energy can be harvested from the wind in a location by a given turbine.

28. This includes onshore and offshore wind, modeled at a 20% or higher capacity factor with a 100 meter hub height (Lu et al., 2009).

29. Mean wind speed directional data, gust information, and data on seasonal, annual, and height variation are typical parameters to gauge (Tiwari and Ghosal, 2007).

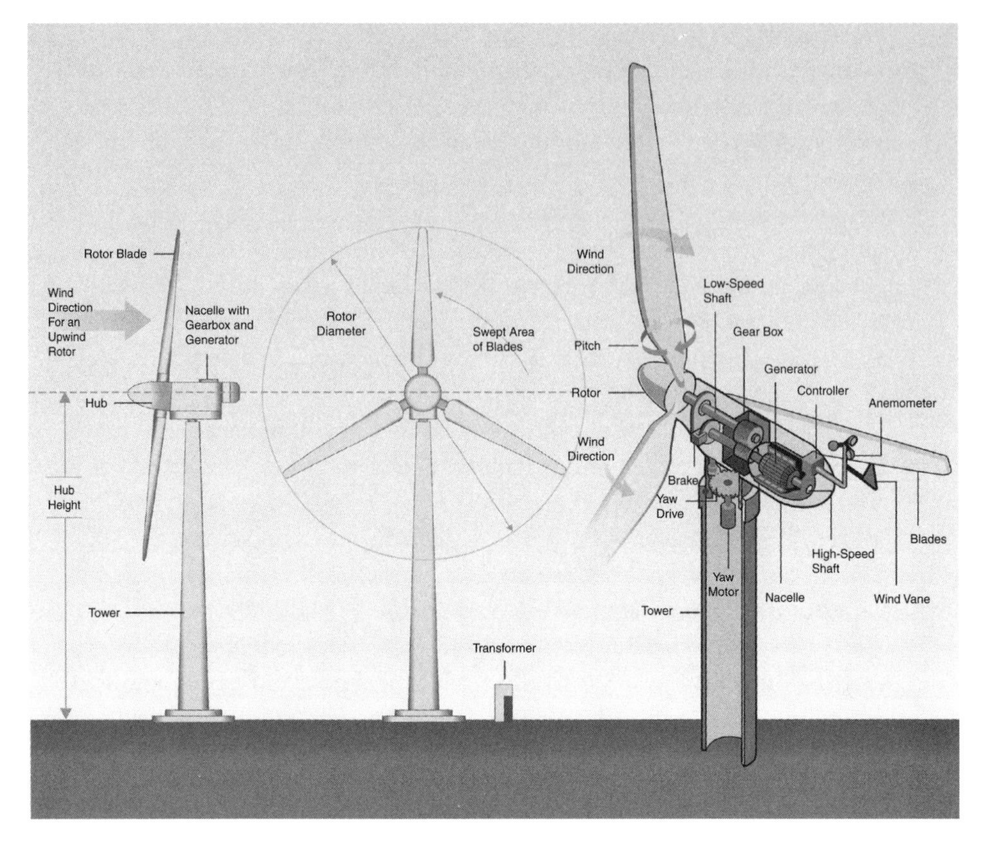

Figure A-5 Basic components of a horizontal axis wind turbine with gearbox.

- An *anemometer* gauges wind speeds.
- The *controller* starts and stops a turbine at cut-in and cut-out speeds.
- A *drive-train* generally houses a gearbox and a generator.
- The *gear box* connects low- and high-speed shafts, allowing a stepped increase of rotational speed to a range that is conducive for producing electricity.
- A *nacelle* consists of the gearbox, shafts, generator, controller, and brake.
- *Pitch* is blade positioning to control wind. Pitch and stall control are ways to control a turbine. With pitch control, an embedded anemometer tracks wind speed and conveys instructions to optimize efficiency through blade adjustment. With stall control, blades are attached in a fixed position with a design for maximal power output and equipment protection. Pitch-regulated turbines are generally more efficient than stall-regulated, though the latter, with less design complexity, are typically more reliable.
- The *rotor* consists of the blades and the hub. Upwind and downwind turbines position the rotor in relation to the wind flow. In upwind models, the rotors are located at the front of the unit and oriented by a *yaw drive* mechanism. In downwind models, the rotor is positioned at the back of the unit, so a yaw device is not necessary.

SOURCE: Wiser et al. (2011).

Basic Model Types

Two basic models of wind turbines exist: the vertical axis (VAWT) and the horizontal axis (HAWT) designs, distinguished by the direction of the rotating shaft.

Present wind turbines typically reflect the HAWT design with propeller-like blades sitting on a rotating shaft that runs parallel to the ground. The rotor in these models is often located on the front or upwind side of the tower. Such models can measure the height of a 20-story building (DOE, n.d.).

The less common VAWT design has blades oriented from base to top. A typical version is the Darrieus design, which is a two-bladed model resembling an eggbeater. VAWTs often attain a length of 100 feet and a width of 50 feet.

Most contemporary turbines operate with the three-bladed design, yet single- and two-bladed designs, as well as designs that use more than three blades, are also in use. Some of the largest blades currently used are longer than a football field (DOE, n.d.; EWEA, n.d.).

Commercial Turbine Size and Development Directions

The modern commercial manufacturing of wind turbines is often linked to the early 1980s. Since then, onshore models have increased in size from 75 kW to 2–5+ MW (2,000–5,000+ kW) (Figure A-6). Today's units frequently range from 100 kW to 3 MW (Ellenbogen et al., 2012). Depending on siting conditions, a 1 MW turbine can produce electricity for 650 households (EWEA, n.d.). Research and development focusing on wind technology size now centers on 5+ MW models (GWEC, 2016).

Wind turbines are designed to be quite reliable, with an operational availability of 98% (EWEA, n.d.).[30] During a 20-year operational life of a wind turbine, the turbine can function continuously and unattended with minimal maintenance for roughly 120,000 hours of active operation (EWEA, n.d.). By contrast, a car engine typically has a design life of roughly 6,000 hours of operation. Development efforts include material and design optimization of rotor blades, controller capabilities, alternative drive train configurations, and network operator requirements, among others.

30. The availability factor of a power plant gauges the amount of time that the plant is able to produce electricity in a given period divided by the total time of the period. Availability factors are a function of the fuel type, plant design, and operational approach so they vary widely. Thermal power plants, such as coal, nuclear, and geothermal plants, have availability factors of 70–90%. Natural gas plants tend to be in a range of 80–99%.

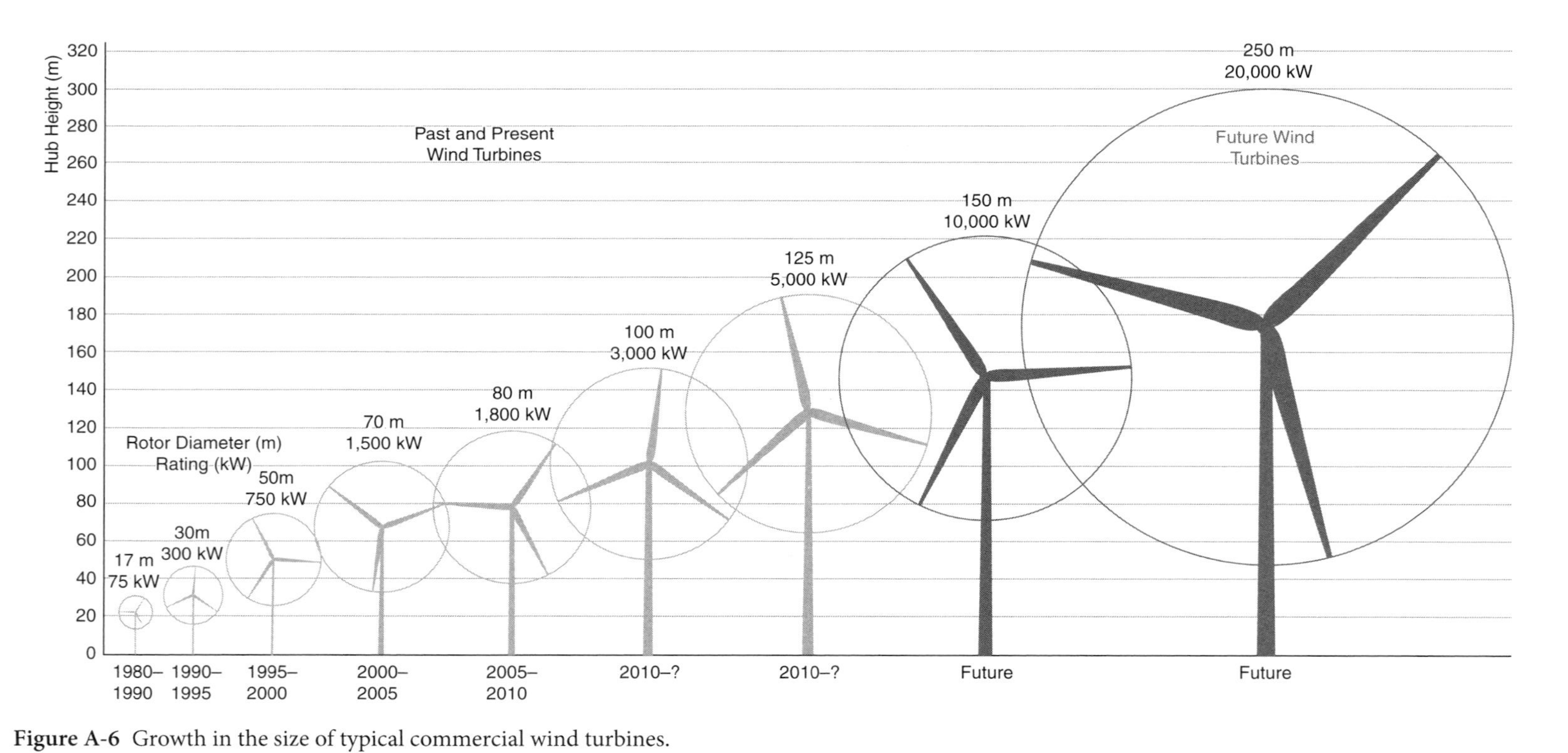

Figure A-6 Growth in the size of typical commercial wind turbines.
SOURCE: Wiser et al. (2011), design by NREL.

Modularity, Multipurposing, and Scale

Key attributes of wind power plants include modularity and the potential for multipurpose use of space. Turbine modularity enables the quick installation of single units, clusters, or large-scale wind farms. The relatively nonintrusive nature of wind turbines means that land between the sited units can be simultaneously used for other purposes, such as farming and ranching.

Contemporary wind farms can consist of 300 or more turbine units covering an expanse of several hundred square miles (EWEA, n.d. and 2009). Some of the largest onshore wind farms include the Gansu Wind Farm in China (5,000+ MW) and the Alta Wind Energy Center in the United States (1,020 MW). By comparison, fossil fuel plants are often rated at 500–1,000 MW and nuclear plants are usually rated at 1,000–1,500 MW. Currently, the largest offshore wind farm is the London Array Wind Farm in the United Kingdom (630 MW).

Onshore and Offshore Wind

Onshore wind technology has a good track record. In locations with reasonable wind resources, particularly where the cost of carbon is internalized, its generated power is competitive with newly built, conventional power plants (Arvizu, et al., 2011; IEA, 2009, 2015*c*; REN21, 2016).

Offshore wind technology is at an earlier stage in commercialization, yet the wind resource is typically much more powerful. Investment costs for offshore wind projects have been roughly twice that of onshore wind investments until recently, yet, due to strong and more frequent winds, the quality of offshore wind resources can be 50% better relative to onshore wind resources (EWEA, n.d.; IEA, 2009; IRENA, 2012; for discussion of recent cost declines in offshore wind, see Smart, et al, 2016; Wiser, et al, 2016). Recognizing the stronger power production opportunity and rugged conditions of the sea, offshore wind turbines need to be more resilient than their onshore counterparts. Currently, offshore wind represents a small fraction of the cumulative installed capacity. Nonetheless, it is key to a number of regions' energy plans, so market expectations are high (EWEA, 2009). The growing market for offshore wind raises new challenges in the logistics of manufacturing, testing, transport, installation, and maintenance of wind turbines. Special approaches and vessels for installation have been developed, and the means of access to offshore turbines in operation is an important factor in terms of cost, safety, and availability (EWEA, n.d.).

Wind Resource Assessments and Technology Indicators

Wind resource assessments are typically done to evaluate the resource potential or to forecast expected wind power production in a given location. Forecasts are used for trading, scheduling, and power dispatch in the power sector. In contrast, spatial assessments using maps or atlases are done to determine optimal locations for siting wind turbines and to estimate annual energy output (Krohn, 2002*b*).

Wind assessments are best developed through on-site measurements that utilize anemometers and weather vanes that can be augmented with data from area weather stations (EWEA, 2009). In conjunction with this, or as an alternative, computer modeling can be used, accounting for topography, ground surface cover, and elevation (EWEA, 2009). Resource classes are assigned to factor for wind speed variability, average wind speed, and average air density (Sawin, 2001). Using the resource class and other parameters, wind potential is estimated in terms of power generation. Estimates for the offshore wind potential in the European Union, for instance, indicate a resource of up to 3,000 TWh, which exceeds the total electricity consumption there (EWEA, n.d.).

When considering the power of a specific wind turbine (or power plant), the rated power output reflects the upper limit of electric power generation. This is typically represented as kW or MW. With wind turbines, this aligns with the cutout speed of a turbine, or the point at which a turbine will stop for security reasons at high wind speeds.

When evaluating wind power in regional plans, indicators often include the generated capacity or generation, the average annual capacity factor, installed capacity, and the share of total electricity. The generated capacity represents the amount of electricity produced. The average annual capacity factor is a function of the area wind resource and the way turbines are configured (see Box A-2). This indicator is dependent on the hub height of the turbine, as well as on the generator capacity divided by the swept rotor area (W/m^2) (Energinet and Energistyrelsen, 2012). Installed capacity represents the full load that a turbine, project, or region is technically able to produce. The share of total electricity derived from wind power, as the name suggests, is the ratio of wind generation for a given period divided by all electricity produced for that period.[31] Installed capacity is useful for gauging wind power infrastructure at a project or country level, whereas the share of wind power in total electricity gauges wind's contribution to the overall electricity mix.

31. The term "wind power share of electricity" is used interchangeably, here, with wind intensity or wind penetration. In some reporting, this measure is normalized against a wind index that serves as a baseline for the average wind resource over a period of time.

Box A-2

Capacity Factors

The capacity factor for wind power, as with other types of power plants, is the ratio of actual power produced in a given period relative to the plant unit's full capacity. Typical capacity factors for various types of plants are: wind 20–40%, biomass 25–80%, geothermal 45–90%, solar photovoltaic 6–20%, hydropower 20–90%, nuclear 60–100%, and large coal-fired facilities 70–90% (Kagel, Bates, and Gawell, 2007; Renewable Energy Research Laboratory, n.d.; UNDP et al., 2004).

Grid Integration

As an intermittent source of energy, wind power is managed differently from fossil fuel–powered plants. The flexibility of grid systems allows wind variability to be smoothed.

What determines a grid system's flexibility depends on factors like forecasting, spatial dispersion and aggregation, and the use of interconnections, among other elements. With forecasting, accurate predictions of expected wind power output, particularly in increments specific to generation and transmission scheduling, allow grid operators to manage wind generation fairly effectively (EWEA, 2009). If wind producers are able to provide forecasts close to real time (i.e., shorter gate closures), they are able to submit more reliable information (EWEA, 2010). In terms of spatial management of wind power, wind turbine output can be aggregated across a wide geographic expanse, allowing a natural leveling of variation. The use of interconnections (i.e., links between two networks) allows not only geographic smoothing but also trade across networks. With increased levels of interconnection, grid operators are able to leverage the aggregate potential of reserves, generation mixes, and geography across larger physical areas. This approach enhances the ease of smoothing power supply and demand (with or without wind inputs), and it can reduce balancing costs. Finally, network optimization, accomplished by using several elements: a diverse mix of energy types, including dispatchable renewables like such as hydropower, geothermal energy, biomass, waste, and solar thermal electricity; reserve availability; storage/demand-side options; and conducive market rules allows large-scale wind power outputs to be managed (EWEA, 2009; Lovins, 2011).

Not unlike other types of power generators, wind turbines must meet certain conditions to be able to connect to the grid. Output must synchronize with the power on the grid system in terms of voltage, current, frequency, and amplitude. Codes designate technical requirements for connection of a power plant (turbine unit or otherwise) to the grid. Such requirements, like those specific to tolerance, protective devices, active and reactive power, and power quality, are evolving as greater amounts of wind power are used (EWEA, 2009). These requirements can also address newer functions, like active control and provision of grid support services.

Environment

Wind power projects have a much smaller environmental footprint when compared to conventional power plants. Wind plants have no direct emissions when in use, although some emissions exist in turbine manufacture, transport, installation, and decommissioning. They also produce minimal waste, use negligible amounts of water, and do not involve mining or drilling (Wiser et al., 2011), except for instances of mining rare earths. Because of these features, the life cycle effects of wind generation relative to other types of fuel generation are quite favorable (EWEA, 2009).

Broadly speaking, the ecological considerations of wind projects focus on sound, shadow flicker, impacts on birds and bats, and, among other qualities, the effects on marine life in the case of offshore wind.[32]

Sound from wind turbines can be audible or subaudible, and, if not managed properly, may be an issue for some individuals.[33] Sound is generally a function of turbine design, distance from structures and listeners, placement of the turbine, proximate terrain, and atmospheric conditions (Ellenbogen et al., 2012).[34] When hub height wind speeds are high and ground-level wind speeds are low, sound emissions may be greater (Wiser et al., 2011). In this set of conditions, the absence of ground-level ambient sound from wind, when combined with higher sound levels of hub-height winds, can produce higher audibility (Wiser et al., 2011, citing van den Berg, 2004, 2005, 2008, and Prospathopoulos and Voutsinas, 2005).[35]

32. Because of the highly subjective nature of aesthetic considerations, such as visual preferences, this study does not cover them for any of the technologies.

33. Interestingly, sound for other power plants can be louder, but often isn't considered.

34. Upwind and downwind turbines have different sound attributes, since the blade and wind speed interaction occurs differently behind the tower as opposed to in front of it (Ellenbogen et al., 2012). Terrain and atmospheric conditions influence the sound dynamics of wind turbines because sound refraction can be shaped by hillsides, temperature gradients, and atmospheric absorption, among other factors (Ellenbogen et al., 2012).

35. Some claim that the subaudible sound of wind turbines has links to health effects (Wiser et al., 2011, citing Alves-Perreira and Branco); however, a range of studies and reports have not

Major efforts have been made over time to reduce the sound levels of wind turbines (Wiser et al., 2012).

Shadow flicker, the rapid changes between shadow and light that can occur when blades rotate between the sun and an observer, depends on the location of the observer relative to the turbine and the time of year/day (Ellenbogen et al., 2012).[36] Such flicker can exist at ranges of 1,400 meter or less from a turbine, but proper siting reduces this effect.

Bird and bat collision fatalities are a concern when siting new projects. Scientific understanding of the risks associated with this phenomenon is still evolving (Clarke and Ricci, 2003) and much depends on the species, the locational character of turbine siting, and the turbine(s) (Wiser et al., 2011). Effects can also vary across stages of installation, operation, and decommissioning. Measures to mitigate such issues might include painting turbines in colors that are more noticeable for such species and siting turbines away from migration routes.

As access to offshore wind power increases, its effects on marine life are being studied pre- and post-construction. Similar to conditions related to birds and bats, environmental effects can vary between stages and are a function of the species and the locational character of turbine siting and the turbine(s). Adverse impacts may relate to disruption of marine life activities, sounds and vibrations, and electromagnetic fields (Wiser et al., 2011). Positive effects can include the creation of new breeding grounds and shelters, as well as artificial reefs related to the physical structures of the wind turbine foundation (Wiser et al., 2011). Currently, effects do not appear large, but further study is warranted.

To balance the positive character of wind power's low environmental footprint with its negative effects, the key is in planning, permitting, production, and monitoring to minimize adverse impacts (Clarke and Ricci, 2003).

Economics of Wind Power

The economics of wind power generation, like that of other power plants, must account for the costs of investment, operation and maintenance (O&M), financing, annual energy production, and the assumed economic life of a plant,

found sufficient evidence to support this (Ellenbogen et al., 2012; Wiser et al., 2011, referencing various). Guidelines by the World Health Organization and U.S. Environmental Protection Agency are generally believed to be sufficient to avoid direct physiological health effects (Wiser et al., 2011, citing EPA, 1974 and World Health Organization 1999, 2009).

36. The shadow flicker frequencies of wind turbines are proportional to the rotational speed of the rotor multiplied by the blade number. This typically falls between 0.5 and 1.1 Hz for large turbines (Ellenbogen et al., 2012).

among other considerations (IEA, 2011*d*; Wiser et al., 2011).[37] These elements allow for the calculation of a levelized cost of energy, thus providing a form of comparison for energy projects across varied fuel types and facility life. If levelized cost is considered alongside economic support policies, power market rules, and the costs of alternatives, the economic feasibility of projects can be appraised (IEA, 2011*d*; Wiser et al., 2011).[38] A 2010 IEA study surveyed a wide range of power plant costs, ultimately concluding that hydropower reflected the lower bound of levelized costs at around $45–240/MWh, and solar photovoltaic energy reflected the upper bound at $674–1,140/MWh), with wind power (onshore and offshore) reflecting a mid-range of $91–181/MWh (IEA, 2011*d*; Wiser et al., 2011; and IEA, 2010*f*). A more recent study by the IEA (2015*c*) included fewer levelized cost estimate (LCOE) data points and different assumptions and data. LCOE for onshore wind was largely unchanged at a 10% discount rate, compared to solar photovoltaics, which had declined substantially. Recent disruptive change with offshore wind prices could alter the overall LCOE of wind power substantially going forward (IEA-RETD, 2017; IEA, 2016*a*).

From the early 1980s through 2004, investment costs for wind power underwent a steady decline (IEA, 2009). However, international cost trends increased from 2004 to 2009, due to the rising price of turbines (larger models), supply constraints on materials (i.e., copper, steel, carbon fiber, cement, etc.), pressures on labor costs, currency valuations in manufacturing countries, and the outmatching of demand with supply (Wiser et al., 2011). A study by the International Renewable Energy Agency found positive developments in wind power; namely, that the costs of wind generation in the best sites of North America were $0.04–0.05 in 2010 were "competitive with or cheaper than gas-fired generation even in the so-called 'golden age of gas'" (International Renewable Energy Agency [IRENA], 2012). In addition, the costs of wind turbines for countries like China are 50–60% cheaper than in North America (IRENA, 2012).

Another way to consider the economics of wind is in terms of payback period and the energy costs of production versus earnings from energy generation. Studies indicate that a modern wind turbine pays for itself in a very short period of time, often producing 35 times more energy than what was required to produce it (Danish Wind Turbine Owners Association [DWTO], n.d). Recent analysis of energy payback periods for wind turbines also indicates that a 3 MW offshore wind turbine is associated with a 6.8-month payback window, whereas a similar onshore turbine has a 6.6-month window (DWTO, n.d., citing Vestas, 2007). Analysis of the Horns Rev 1 wind farm in Denmark, an offshore project brought online in 2002, shows that the payback for its 2 MW turbines was 3.1 months,

37. Decommissioning is another cost but is not typically expected to be considerable for wind power plants (Wiser et al., 2011).

38. For more discussion on costs, see Araújo (2016).

similar to a 2 MW onshore wind turbine based in Western Denmark, an area with very good wind resources (DWTO, citing Vestas, 2007).

COUNTRY TIMELINES

Timeline of Iceland's Energy Transition

Late 19th century	Experiments with geothermal steam in gardening
1899	Electricity first produced in Iceland
Early 20th century	Geothermal steam applied to heat greenhouses, and buildings
1904	First hydropower turbine begins operation
1908	Stefan Jonsson pipes steam from hot springs to home
1911	Erlendur Gunnarsson harnesses geothermal energy for heating and cooking
World War I	Price increases in coal spur interest in harnessing geothermal energy for space heating and other purposes.
1930	Large-scale utilization of geothermal energy in space heating begins with a pipeline constructed in Reykjavik
Beginning of World War II	Imported coal is the key energy source followed by oil; geothermal energy and hydropower
1940	Act on Ownership and Rights of Usage of Geothermal Resources No. 98/1940
1943	First district heating company Reykjavik District Heating is now part of Reykjavik Energy
1944	Experiment in geo-steam electricity
1946	Electricity Act provides for the establishment of the State Electricity Authority to advance knowledge on geothermal resource and utilization
1956	Geothermal Department established in the State Electricity Authority
	Government launches rural electricity program that is completed in the late 1970s
1957	Reykjavík and the State buy a oil drill rig with a capacity to drill 3,000-meter deep wells.
1965	Energy company Landsvirkjun is formed.
1967	Energy Act of 1967 (in conjunction with the Water Act of 1923) indicates that the ownership of energy resource lies with ownership of the land, subject to restrictions (Energy Act 58/1967)

	State Electricity Authority becomes the NEA (Orkustofnun) and State Electricity Power Works splits off as a separate organization
1969	First program for exploration of the high-temperature fields
	Bjarnarflag Geothermal Plant, located near Námafjall and Lake Myvatn in northeast Iceland, is the first commercial geothermal power plant to go online in Iceland
1969–1970	Iceland's first aluminum smelter, ISAL, is built outside of Hafnarfjordur for a Swiss Aluminum company, now owned by Rio Tinto Alcan; and fueled by hydropower
1973–1974	Oil Embargo/first oil shock
1974	Inflation 43%
	Hitaveita Sudurnesja is formed
1977	Krafla's power production begins
	Svartsengi begins operations
Late 1970s	Power line interconnection installed
1979	Second oil shock, Iranian Revolution
	UN University established in Iceland for postgraduate training in geothermal energy
	Ferrosilicon smelter plant owned by Elkem opens at Grundartangi
Early 1980s	Fishing quotas imposed
1980	Legislation passed granting Hitaveita Suðurnesja permission to increase the electrical capacity by 6 MW
1981	Icelandic International Development Agency is established as an autonomous agency under the Ministry of Foreign Affairs
	Power Plant Act No. 60/1981
1983	Akureyri acquires a share in Landsvirkjun
Mid 1980s	Major increase in the number of fish farms
1984	All major power stations were connected to the grid; all regions have access to hydropower; diesel-based power used for reserves
	Althingi modifies HS charter to heating/water and power (No. 91/1984)
1985	85% of the population uses geothermal heating.
	Althingi lowers state ownership of HS from 40% to 20% (No. 101, 1985);
1986	Chernobyl nuclear accident occurs

1990s	Cod Wars
1990	Ministry of Environment is established
1993	Iceland joins to the EEA, effective 1994
1995–1996	New heavy industry contracts are signed
	Safety of Electricity Installations, Consumer Apparatus and Electrical Materials, Act No. 146/1996
1997	Planning and Building Act No. 73/1997
1998	Resources Act 57/1998 supersedes the Energy Act 58/1967, the Act on Ownership and Rights of Usage of Geothermal Resources No. 98/1940, and the Power Plant Act No. 60/1981
	Common Land and Boundaries of Private Land, Common Land and Highland Pasture Act, No 58/1998
1999	Reykjavik District Heating is merged with Reykjavik Electricity and Reykjavik Waterworks to form a new company Reykjavik Energy (Orkuveita Reyjavikur)
	Hygienic and Pollution Act No. 7/1998
	Monitoring Act 27/1999
	Nature Protection Act No. 44/1999
1999–2003	Master Plan, Phase 1
2000	EIA Act, Act 106/2000 to avoid or minimize environmental impacts of projects; geothermal and other thermal plants of 50 MW and more are subject to EIAs
2002	Karahnjukavirkjun hydropower project (690 MW) presented for a vote in Parliament
2003	Natl Energy Authority Act No. 87/2003;
	Electricity Act, Act 65/2003
2004–2010	Master Plan Phase 2
2005	Competition Act No. 44/2005
	Major protests occur over Karahnjukar hydro project
2006	Strategic Environmental Assessment Act No. 105/2006
2007	Emission of GHG Act No. 65/2007
	Government releases its Climate Change Strategy (No. 3)
	HS Orka is privatized
2008	Amendment to the Resources Act No. 58/2008
	Banking crisis
2009	A steering committee is appointed to advise on comprehensive energy policy for Iceland
2012	Kyoto exemption expires
	The United Kingdom and Iceland sign a deal for a subsea cable

Timeline of French Energy Transition

1945	CEA is founded
1946	France nationalizes its electricity sector and creates EDF
1957	Euratom Treaty is signed
1960s	Debate between EDF and CEA occurs over nuclear reactor technology; it ends with LWRs being chosen
1972	Committee against Atomic Pollution is established in The Hague
	Survivre et Vivre reveals cracks in nuclear waste storage barrels in Saclay
	Eurodif partnership (Belgium, France, Iran, Italy, and Spain) chooses diffusion process for its Tricastin enrichment plant that will be based in France.
1973	First oil shock
1974	Prime Minister Messmer announces what became known as the Messmer Plan
1975	Approximately 30 CNRS physicists expressed opposition to the Messmer Plan with a petition that became known as the Appeal of 400
1976	CEA creates COGEMA
1977	First criticality of Fessenheim 1 is reached. Order placed for the Superphenix 1200 MWe fast breeder reactor
	As part of the European consortium of utilities (EDF = 51%), EDF begins building the world's only commercial fast breeder, Superphenix at Creys-Melville on the Rhone River
	Major demonstrations take place near the SuperPhenix site, one demonstrator is killed
1978	Creation of EURODIF, a subsidiary of now AREVA, to exploit the French gaseous diffusion technology (with Italy, Spain, Belgium, and later Iran)
Late 1970s	Fading of the national antinuclear movement
1979	Second oil shock, Iranian Revolution
	TMI nuclear accident
	Tricastin enrichment begins
1981–1995	Major change in French politics—the Socialist party an absolute majority
1983	High inflation and economic problems; economic reform reversed to pursue fiscal and spending restraint.
Mid 1980s	EDF begins sales and services in international market
1985	Framatome, under the leadership of Jean Claude Leny, sets out to diversify

1986	Chernobyl nuclear accident
	A subsea power cable connecting France and England is completed (constructed for contingencies, but becomes an export route)
1987	France's first severity scale is developed by Conseil Supérieur de la Sûreté et de l'Information Nucléaires (French Higher Council for Nuclear Safety and Information [CSSIN])
	Brundtland Report is released
Late 1980s	Riots occur relating to waste management
1988	EDF plants operate at an average load factor of 61% (W Germany 74%, Switzerland 84%, Finland 92%)
1989	Framatome and Siemens form a subsidiary which leads to the EPR design
1990s	The INES2 International Nuclear and Radiological Event Classification scale is published by the IAEA
1990	All waste exploratory digging/processes are stopped; issue turned over to Parliament, which appoints Christian Bataille
1995	Strikes protest a move toward free market reform
	France ends its nuclear weapons tests (1960–1995)
1996	France ceases production of weapons-grade fissile materials
1997	Kyoto Protocol is adopted
	The Jospin government stops the Superphénix
1998	SPD-Green coalition in Germany decides on a nuclear phase-out (accelerated in 2011)
1999	Level 2 Event at La Blayais
2001	The last uranium mine closes in France
2003	Order for the first EPR by TVO (Olkiloto 3)
2004	Public debates begin
	Government and EDF announce Flamanville as the next EPR site
2006	Construction begins on Flamanville EPR and at Tricastin for a new enrichment facility (George Besse II) to replace Eurodif (in operation since 1978)
	Laws passed on waste management, transparency, and security in the nuclear field
2007	EDF begins construction on its first domestic EPR, Flamanville-3
2008	Flamanville-3 construction is suspended by the French nuclear safety authorities to correct safety measures
	EDF and AREVA sign a framework agreement for reprocessing all spent fuel, excluding MOX, through 2040

2010	Reprocessing—La Hague handles 1,050 tons of spent fuel/year from EDF (vs. 850 previously)
2011	Fukushima nuclear accident
	German Chancellor Merkel announces an immediate shutdown of 8/17 reactors; Legislation is passed in the Bundestag (7/31/11), commencing a phase-out of nuclear energy
	French Prime Minister asks the Cour des Comptes to prepare a report on the costs of the nuclear power sector to be submitted before January 31, 2012
2012	Two small fires are contained at water-cooled Penly PWR reactor; release of smoke spurred automatic shutdown, no environmental effects reported
	Shutdown of Eurodif and commissioning of Georges Besse 2, the new centrifuge enrichment plant
	Newly elected President François Hollande announces the shutdown of Fessenheim
2016	Flamanville EPR plant planned completion

Timeline for Brazilian Energy Transition

1970	Ethanol equals less than 1% of the total transport and automotive fuels in Brazil
1973–1974	Oil Embargo/first oil shock
1975	June—President Geisel visits Professor Stumpf's lab where tests of hydrous and anhydrous ethanol on engines were conducted
	November—*ProAlcool* Program launched
1979	Second oil shock, Iranian Revolution
	Mandated installation of 100%-only ethanol pumps at gas stations
	Price of 1 liter of ethanol set at 59% of 1 liter of gasoline
	The government provides low-interest loans to agriculture to produce ethanol
	The government and auto industry reach agreement that industry will produce mostly lead vehicles
1980	Hydrated ethanol and neat vehicle (100% ethanol) sales begin
1986	Alcohol vehicle sales peak at approximately 90% of new light vehicles
	Collapse of world oil prices
1986–2000	Flat/incremental growth or decline in ethanol production

1989–1990	Increase in ethanol prices
	Removal of most ethanol policies
	Decline in ethanol production and supply shortages resulting in ethanol imports
	Alcohol vehicle sales decline substantially
1990–2000	Deregulation of energy and sugar markets
1991	*ProAlcool* officially ended, although CINAL maintained some functions
1993	Blending set at 22%
1994	End to universally set fuel price
1997–1999	Subsidies end
1998-2002	Neat vehicle sales end
	Ethanol prices liberalized
2002–2004	Biodiesel specifications are set
2003	Flex fuel vehicle sales begin
	Ethanol blending set at 25%
	Surge in oil prices
2004	Brazilian sugarcane-based ethanol is cost-competitive with gasoline
	Biodiesel Program (CNPB) is launched
2006	Brazil attains net oil independence
2008	Biodiesel blend of 3% required
	Public auctions for biodiesel occur
2011	Ethanol equals roughly 17% of total transport fuels and 41% of automotive fuels in Brazil

Timeline for Danish Energy Transition

1962	Danish Parliament gives authority to Ministry of Education to decide on the installation of nuclear plants
1970	Ministry of Trade produces a report on large-scale wind on the Danish power grid
1971	ELSAM decides to build a nuclear power plant
1973	Denmark joins the EEC
1973–1974	Oil embargo/first oil shock
1974	ELSAM proposes a nuclear power plant in Jutland
	OOA is formed

1975	OVE is formed
1975–1978	Tvind Turbine built
1976	DEA established
	First Energy Plan
	Alternative Energy Plan put forward by ATV
	Energy Research Program launched (publically funded R&D), including funding for wind research
	Electricity Supply Act—requires cost-based prices with allowance for reserves and depreciation; no reference to competition
1977	Energy taxes introduced
1978	Test station established for small wind turbines with approval scheme
	DWTO and DWIA formed
	Riso wind research funded
1979	Ministry of Energy established
	Energipakken: Investment grants for various technologies, including wind turbines through 1989; 30% of capital cost; guarantees for grid connection and buy-backs;
	Large-scale test turbines (Nibe 1 and Nibe 2) are erected by utilities
	Second oil shock, Iranian Revolution
	TMI nuclear accident
	PURPA passed in the United States providing favorable policy conditions for wind exports
	First agreements for natural gas supply to Denmark with Maersk Oil
1981	Energy Plan 81, first with a strong wind dimension
	Production subsidy begins
	Wind Atlas released
	Gorm field production begins (oil)
Early 1980s	Exports begin to California wind market
1982	Skjold field production begins (oil)
	Renewable Energy Committee is established
1983	Alternative energy plan is put forward
	Nordic Folkcentre for Renewable Energy is established
1984	Gas production begins
	Amendment to Electricity Act of 1984 provides subsidies for grid-connected electricity from wind

	First voluntary purchase agreement—utilities and turbine owners/manufacturers
	New environmental regulations—50% reduction of sulfur emissions by 1995; limits on the burning of straw
1985	Danish turbines represent 50% of the California market
	Nuclear plans for Denmark are withdrawn
	Denmark joins the EU
	Government–utility settlement for Danish power companies to install 100 MW of wind over next 5 years
	Restrictions established for "local ownership" and investment share for turbines
1986	Subsidies in the American market are discontinued and reduced in Denmark: 20+ bankruptcies for Danish wind turbine manufacturers follow and market consolidation ensues
	Chernobyl nuclear accident occurs
	Danish Energy Agency and the Renewable Energy Committee implement programs to foster decentralized CHP/DH
	New resource study is released
1987	Brundtland Report is released
	Government committee on offshore regulatory conditions issues report
1990	Energy 2000 plan
	Government–utility deal for 100 MW of wind power
	Danish Turbine Guarantee established
	Committee is established to provide planning guidelines
1991	Energy tax reform
	National wind map is published
	Council for Renewable Energy is established, including development program for RETs
	First offshore wind farm built worldwide
	Denmark is self-sufficient in oil and gas
1992	Wind Turbine Location Committee provides recommendations on increased coordination
	Systematic planning is instituted by the national government with direction to the local authorities
	Role of co-ops shrinks with change in planning procedures
	Government rule obligates power companies to develop/enhance the grid for wind additions
	CO_2 tax package is passed
	Rio Summit

1993	Svend Auken is named Minster of the Environment
1994	Under a modified government, the Ministry of Energy and Environment are merged
	Municipal authorities are ordered to indicate wind-suitable land in their plans (location and extent)
Mid-1990s	Laws pertaining to farming sector open new prospects for farmers to own wind turbines; ownership now more mixed
1995	Second offshore wind project is brought online at Tuno Knob
	Government committee on offshore regulatory conditions reports
1996	Energy 21 Plan
	Amendment to Electricity Supply Act; transition toward competition begins
1997–1998	Offshore planning regulations are established with one-stop permitting
	Offshore plan of action prepared; mapping of potential sites identifies roughly 4,000 MW of immediate potential
	Kyoto Protocol signed
	Requirement imposed on power companies to establish an additional 750 MW of offshore wind before 2008
1998/2002	Liberalization of electricity and gas markets
1999	Electricity Reform is introduced; Electricity Supply Act (Law no. 375, 2 June) plus political agreements among parties form the cornerstone for competition in the Danish electricity market
	New, detailed Wind Atlas is released
2000	Denmark is 142% self-sufficient in energy
	Energy Savings Act is adopted and Climate 2012 is released
	EU approves electricity reform except for requirements on minimum price guarantees
	First large offshore wind project commissioned (40 MW) at Middlegrunden
	Danes reconfirm decision not to join the Euro Zone in a referendum
	Liberalization of the gas market begins; Natural Gas Supply Act, July 1, 2000
	CO_2 quotas set for power producers (CO_2 bank)
2001	Parliament approves Kyoto Protocol in May
	New Liberal-Conservative government takes office in November

	Liberalization and Renewable Energy programs abolished
	Funding for wind power shifts to PSO paid by electricity consumers
	Parliamentary hearing; government postpones green certification implementation
2002	Repowering program launched by pre-2001 administration leads to a short revitalization spurt in wind installations
	Offshore development and repowering sustain some growth in capacity
	Wind subsidies are reduced, leading to almost full stoppage in renewable energy development
	Repowering scheme ends
2003	New price system goes into effect with a cap of .36 Dkk/kWh including CO_2 compensation
	Private investment in onshore wind comes to a standstill
2004	New incentive program with market premium is adopted in lieu of green certificates; extra price paid for onshore turbine installation, if repowered
	Danish power sector is restructured: distribution, transmission, and production became independent sectors
	New Certification Scheme, based on IEC WT01 System for Conformity Testing and Cert of Wind Turbines, is introduced to replace the Type Approval Scheme
2005	Energinet.dk, the Danish transmission system operator, is formed after liberalization of the electricity market and the merger of Eltra, Elkraft System, Elkraft Transmission, and Gastra
	Kyoto Protocol comes into force
	DEA starts new plan for siting offshore wind farms from 2010–2025
	Energy Strategy 25 is published
Mid-2000s	Three working groups are formed for new wind siting: offshore, onshore, and to position new industrially developed turbines (0 series)
2006–2007	Municipal reorganization occurs with larger units
2007	Government puts forward a vision for Denmark without fossil fuels
	Government proposes increase in wind subsidies
2008	Political agreement: raising of subsidies with cap on load years
	Law No. 1392 of 27/12/2008 passed; implementation of the February 2008 Agreement

2009	EU institutions adopt the Third Liberalization Package to open power markets more and make them fairer
2010	*Green Energy,* is published by the independent Danish Commission on Climate Change Policy
2011	*Energy Strategy 2050,* government plan outlines policy instruments to transform Denmark into a low carbon society with a stable and affordable energy supply
	Fukushima nuclear accident occurs
2012	Offshore wind farm project at Kattegart (400 MW) to be implemented
	Political agreement by 95% of Parliament to advance a long-term, green growth strategy

INTERVIEWS BY COUNTRY AND AFFILIATION

*Note: If interviews were conducted with more than one person at a single organization, then the institutional name is listed mutiple times. If more than one interview occurred with an interviewee, then the entry is marked with **.*

ICELAND		PRIMARY PERIOD
Agricultural University of Iceland		2012
Almenna (formerly)	**	2012
Boston University		2012
Canadian Geothermal Energy Association; Islandsbanki (formerly)	**	2012
Cornell University		2011
Gekon Cluster (Geothermal)		2012
Gekon Cluster (Geothermal)		2012
HS Orka	**	2011–2012
Iceland Geosurvey/ISOR		2012
Iceland Geosurvey/ISOR		2012
Icelandic Nature Conservancy	**	2011–2012
Independent Author and Documentary Film Maker		2012
Innovation Centre Iceland, Technological Development		2012
Invest Iceland		2010
Landsvirkjun (National Power Company of Iceland)	**	2011–2012
Mannvit		2012
Marketing Office, Landsvirkjun-Ministry of Industry (formerly)	**	2012

Ministry for the Environment, Office of Policy and International Affairs	**	2011–2012
Ministry of Industry, Energy and Tourism	**	2011–2012
Ministry of Industry, Energy and Tourism	**	2011–2012
Ministry of Industry, Energy and Tourism	**	2011–2012
Morgunbladid Newspaper		2012
Orkustofnun / National Energy Authority		2012
Orkustofnun / National Energy Authority		2012
Orkustofnun / National Energy Authority (formerly)	**	2012
Orkustofnun / National Energy Authority	**	2011–2012
Orkustofnun / National Energy Authority, and Iceland University (retired)	**	2012
Orkustofnun / National Energy Authority, and Iceland University (retired)	**	2011–2012
Reykjavik Geothermal		2011
Reykjavik Geothermal		2011
Reykjavik Geothermal	**	2011–2012
Reykjavik University		2011
Reykjavik University		2011–2012
Saga Journal, Historical Society of Iceland, Author and Editor		2012
Samorka - Icelandic Energy and Utilities		2012
Sure Step Iceland; Fish Inspecting and Consulting (formerly)		2011
United Nations Geothermal Training Programme		2012

France

Agence Nationale Pour la Gestion des Déchets Radioactifs/National Agency for the Management of Radioactive Waste (ANDRA)	**	2011–2012
AREVA	**	2012
AREVA (retired)	**	2011–2012
AREVA (retired)	**	2011–2012
Atomic Energy and Alternative Energies Commission (CEA)	**	2011–2012
Atomic Energy and Alternative Energies Commission (CEA)		2011

Atomic Energy and Alternative Energies Commission (CEA) (fomerly)	**	2011–2012
Atomic Energy and Alternative Energies Commission (CEA) and French Nuclear Safety Authority (ANS) (formerly)		2010
Directorate General for Energy & Climate (DGEC), Ministere Ecologie, Energie, Developpement Durable et Mer		2011–2012
École Polytechnique	**	2011–2012
Électricité de France (EDF)		2011
Électricité de France (EDF)	**	2011
Électricité de France (EDF) (formerly)		2011
Environmentalists for Nuclear Energy	**	2011
Framatome (fomerly)	**	2011
Framatome (formerly)		2011
Le Centre National de la Recherche Scientifique	**	2011–2012
Massachussetts Institute of Technology (MIT)		2012
Mycle Schneider Consulting	**	2012
Nuclear Energy Academy (INEA) and European Nuclear Society (ENS)	**	2011–2012
Nuclear Energy Counsel, Embassy of France, Washington DC	**	2011–2012
Sortir du Nucleaire		2011
Université Paris X		2011

Brazil

Associação Nacional dos Fabricantes de Veículos Automotores (ANFAVEA) (formerly)		2012
Banco Nacional de Desenvolvimento Economico e Social (BNDES)		2010
Banco Nacional de Desenvolvimento Economico e Social (BNDES)		2010
Center for Strategic Studies and Management in Science, Technology and Innovation (CGEE)		2010
Center for Strategic Studies and Management in Science, Technology and Innovation (CGEE)		2010
Embrapa Agroenergia		2010
Independent energy consulting		2010

Instituto Alberto Luiz Coimbra de Pos-Graduacao e Pesquisa de Engenharia (COPPE), UFRJ		2010
Instituto Alberto Luiz Coimbra de Pos-Graduacao e Pesquisa de Engenharia (COPPE), UFRJ	**	2010–2012
Instituto Alberto Luiz Coimbra de Pos-Graduacao e Pesquisa de Engenharia (COPPE), UFRJ		2012
Inter-American Development Bank		2011
Laboratorio Nacional de Ciência e Tecnologia do Bioetanol (CTBE)		2010
Laboratorio Nacional de Ciência e Tecnologia do Bioetanol (CTBE)	**	2010
Laboratorio Nacional de Ciência e Tecnologia do Bioetanol (CTBE)	**	2010–2012
Laboratorio Nacional de Ciência e Tecnologia do Bioetanol (CTBE); Copersucar-CTC (formerly)	**	2010–2012
Ministry of Agriculture	**	2010
Ministry of Agriculture (formerly)		2012
Ministry of Mines and Energy, and Petrobras (formerly)	**	2010–2012
Ministry of External Relations		2010
Ministry of Finance and Ministry of Planning, Budget, and Management (formerly)		2010
Ministry of Mines and Energy		2010
Ministry of Mines and Energy		2011
Ministry of Mines and Energy		2012
Ministry of Mines and Energy		2013
Ministry of Science and Technology		2010
Office of the President, Undersecretariat, Government Programs and Actions; Ministry of Agriculture (formerly)	**	2010–2012
Operador Nacional do Sistema Elétrico (ONS) (National Electricity System Operator)		2010
Petrobras	**	2011
Planalto, Análise e Acompanhamento de Políticas Governamentais		2010
Planalto, Executive Interministerial Commission for Biodiesel		2010

PSR, Electricity and Natural Gas Consuting and Technical Services		2012
PSR, Electricity and Natural Gas Consuting and Technical Services	**	2010–2012
Sugar Cane Industry Association (UNICA)		2010
Sugar Cane Industry Association (UNICA)	**	2010–2012
Universidade de Campinas (UNICAMP) (retired)	**	2012
Universidade de São Paulo (USP)		2010
Universidade de São Paulo (USP); Ministry of Science and Technology (formerly); Ministry of Education (formerly); Ministry of Environment (formerly); Secretariat of Environment, State of São Paulo (formerly);	**	2010–2012
Universidade de São Paulo and Centro Nacional de Referência em Biomassa (USP and CENBIO)	**	2012
Universidade Federal de Itajubá	**	2012
Universidade Federal de São Carlos	**	2011–2012
University Federal de Pernambuco; Ministry of Science and Technology (formerly)	**	2010–2012
VW Brazil	**	2011–2012

Denmark

Aalborg University	**	2011–2012
Aalborg University	**	2011–2012
Aalborg University	**	2011–2012
Aalborg University	**	2012
Aalborg University Denmark	**	2012
BTM Consulting/Navigant	**	2011–2012
Danish Wind Secretariat/Sekretariatschef, Vindmøllesekretariatet	**	2011–2012
Danish Energy Association		2011
Danish Wind Industry Association	**	2012
Dong Energy		2011
Dong Energy		2011
Energinet		2011
Energinet	**	2011–2012
EU/Commission, Directorate General Environment (formerly)		2011

Holmgaard Consulting; Vestas and DONG Energy (formerly)	**	2011
Middelgrunden Wind Turbine Cooperative		2011
Ministry of Energy (formerly); Ministry for Development Cooperation (formerly)	**	2011–2012
Nordic Folkecenter for Renewable Energy		2011
Poul la Cour Foundation and Museum	**	2011–2012
Regional Development, Central Denmark Region	**	2012
Regional Political Directorate, Danske Regioner		2012
Ringobing Landobank		2012
Samsoe Island, Consulting		2011
Samsoe Municipal Government		2011
Soren Krohn Consulting; World Bank; Danish Wind Energy Association (formerly)		2012
Spok Aps, Middelgrunden, and Danish Wind Turbine Owners Association Board	**	2011
Technical University of Denmak (DTU)	**	2011–2012
Technical University of Denmak (DTU)	**	2012
Technical University of Denmak (DTU)	**	2011–2012
Technical University of Denmark (DTU)	**	2012
Vestas		2011
Vestas		2011

ACRONYMS AND ABBREVIATIONS

AAAS	American Academy of Arts and Sciences
ACEEE	American Council for Energy Efficient Economy
ACRO	Association pour le Contrôle de la Radioactivité dans L'Ouest (Association for the Control of Radioactivity in the West)
ANDRA	L'Agence Nationale pour la Gestion des Dechets Radioactifs (National Agency for Radioactive Waste Management)
ANEEL	Agência Nacional de Energia Elétrica (National Agency for Electric Energy)
ANFAVEA	Associação Nacional dos Fabricantes de Veículos Automotores (National Association of Automotive Vehicle Manufacturers
ANP	Agência Nacional do Petróleo, Gás Natural e Biocombustíveis (National Agency for Petroleum, Natural Gas and Biofuels)
AREVA T&D	AREVA Transmission and Distribution
ASN	Autorité de sûreté nucléaire (Authority for Nuclear Safety)
ASTRID	Advanced Sodium Technological Reactor for Industrial Demonstration
ATV	Academy of Technical Sciences
B2	2% Biodiesel (Note: This is a sample. Various numbers are used.)
BBA	Bolsa Brasileira do Alcool (Brazilian Alcohol Exchange)
BBC	British Broadcasting Company
BDKT	Bupp-Derian-Komanoff-Taylor hypothesis
BEN	Balanco de Energia Nacional

BM&F	Bolsa Mercantil e Future (Mercantile and Futures Exchange)
BNDES	Banco Nacional de Desenvolvimento Economico e Social (National Bank of Economic and Social Development)
BNEF	Bloomberg New Energy Finance
BOVESPA	Bolsa de Valores, Mercadorias and Futuros de São Paulo
BP	Formerly British Petroleum; BP is now the name
BRL	Brazilian Reais
BWR	Boling water reactor
CANDU	Canada Deuterium Uranium
CANGEA	Canadian Geothermal Energy Association
CCS	Carbon capture and sequestration
CEA	Comisariat a l'energie atomique (Atomic Energy Commission)
CEAF	Centro de Analise Funcional e Aplicações
CEIB	Interministerial Executive Commission (Brazil)
CELS	Complex established legacy sectors
CENELEC	European Committee for Electro-technical Commission
CENPES	Centro de Pesquisas Leopoldo Americo Miguez de Mello (Research Center, Petrobras)
CEO	Chief Executive Officer
CEPEA	Centro de Estudos Avancados em Economia Aplicada
CEPN	Centre d'Economie de l'université Paris
CEPOS	Center for Political Studies
CETESB	Companhia Ambiental do Estado de São Paulo (Environmental Agency of the State of São Paulo)
CFDT	Confédération Française Démocratique du Travail (French Democratic Confederation of Labor)
CGDD	French Sustainable Development Committee
CGEE	Centro de Gestão e Estudos Estratégicos (Center for Management and Estrategic Studies)
CGP	Commissariat General du Plan (Commissioner General of Planning)
CH_4	Methane
CHP	Combined heat and power
CIA	Central Intelligence Agency
CNDP	National Commission on Public Debate (La Commission nationale du débat public)
CNE	National Commission on Evaluation
CNEF	Nuclear Financing Committee

CNPB	Conselho Nacional de Petroleo Biocombustiveis (National Council for Biodiesel)
CNPE	Conselho Nacional de Politica Energetica (National Energy Policy Council)
CNRS	Centre national de la recherché scientifique (National Center for Scientific Research)
CO	Carbon monoxide
CO_2	Carbon dioxide
COD	Concerted Offshore Wind Energy Deployment
COFINS	Contribucao para o Financiamento da Seguridade Social
COP3	Conference of Parties 3
Copersucar	Cooperative of Producers of Sugarcane, Sugar, and Alcohol in the State of São Paulo
COPPE-UFRJ	Centro de Pos-graduacao e Pesquisa de Engenheiria
CP	Nuclear reactor model
CPTEC/INPE	Centro de Previsao de Tempo e Estudos Climaticos—CPTEC/INPE
CRE	Commission de régulation de l'énergie (Commission for Regulation of Energy)
CRI	Carbon Recycling International
CRIIRAD	Commission de Recherche et d'Information Independantes sur la Radioactivite (Commission for Research and Independent Information on Radioactivity)
CRS	Congressional Research Service
CSSIN	Council for Nuclear Safety and Information
CTA	Centro Tecnico Aeroespacial (Aerospace Technical Center)
CTBE	Laboratorio Nacional de Ciencia e Technologia do Bioetanol (National Laboratory for Bioethanol Science and Technology)
CTC	Centro Tecnico Canavieiro (Center for Sugarcane Technology)
DANIDA	Danish International Aid Agency
DEA	Energistyerlsen (Danish Energy Agency)
DGEC	Direction Generele de l'Energie et du Climat
DGSNR	Directorate General for Nuclear Safety and Radiation Protection
DHC	District Heating and Cooling
DIR	Directive
Dkk	Danish Krona
DMCE	Danish Ministry of Climate and Energy

DOE	Department of Energy
DOS	Department of State
DTU	Danmarks Tekniske Universitet (Technical University of Denmark)
DWIA	Danske Vindmolleindustrien (Danish Wind Industry Association)
DWMA	Danish Wind Manufacturers Association
DWTO	Danish Wind Turbine Owners Association
E100	100% Ethanol (Note: This is a sample. Various numbers are used.)
EAEM	Energy and Environmental Management
EC	European Commission
EDF	Électricité de France (Electricity of France)
EEA	European Economic Area Agreement
EEC	European Economic Community
EFTA	European Free Trade Association
EGS	Enhanced Geothermal Systems
EIA	Energy Information Administration (US) (See also next entry. These are existing conventions.)
EIA	Environmental Impact Assessment (EU) (See also above entry. These are existing conventions.)
EJ	Exajoules
EMBRAPA	Brazilian Agricultural Research Corporation
EPA	Environmental Protection Agency
EPE	Empresa de Pesquisa Energética (Energy Research Company)
EPR	European pressurized reactor
EPRI	Electric Power Research Institute
Equiv	Equivalent
ERI	Energy Research Institute
ESMAP	Energy Sector Management Assistance Program
ETIS	Energy technology innovation system
EU	European Union
EURODIF	French Nuclear Fuel Consortium
EWEA	European Wind Energy Association
EWETP	European Wind Energy Technology Platform
FAO	Food and Agriculture Organization
FAOSTAT	Food and Agriculture Organization Statistics
FAPESP	Fundacao de Amparo a Pesquisa do Estado de São Paulo
FBR	Fast breeder reactor
FCI	Framatome Connectors International

FF	Flexible fuel
FGV	Fundaçao Getulio Vargas (Getulio Vargas Foundation)
FIT	Feed-in tariff
FX	Foreign exchange
G7	Group of Seven, major advanced economies
G20	Group of twenty, major advanced economies
GDP	Gross domestic product
GEA	Geothermal Energy Association
GHG	Greenhouse gas
GJ	Gigajoules
GM	General Motors
GNI	Gross National Income
GPS	Global positioning systems
GSIEN	Groupement de Scientifiques pour l'Information sur l'Énergie Nucléaire (Scientific Group for Information on Nuclear Energy)
$GtCO_2$	Gigatons of carbon dioxide
GTOE	Gigatons of oil equivalent
GTP	Geothermal Training Programme
GTZ	Deutsche Gesellschaft für Internationale Zusammenarbeit
GW_e	Gigawatt (electric)
GWEC	Global Wind Energy Council
H_2S	Hydrogen sulfide
H/C	Hydrogen-carbon ratio
HAWT	Horizontal axis wind turbine
HDI	Human Development Index
HS	Hitaveita Sudurnesja
HTR	Heater
HVDC	High-voltage direct current
HWR	Heavy water reactor
IAA	Instituto do Açúcar e do Álcool (Institute of Sugar and Alcohol)
IAC	Instituto Agronomico de Campinas (Agronomy Institute of Campinas)
IAEA	International Atomic Energy Agency
IBGE	Instituto Brasileiro de Geografia e Estatística (Brazilian Institute of Geography and Statistics)
ICE	Internal combustion engine
ICPE	Installations classees pour la protection de l'environnement
IDDP	Iceland Deep Drilling Project

IEA	International Energy Agency
IEC	International Electrotechnical Commission
IEEE	Institute of Electrical and Electronics Engineers
IGCC	Integrated gasification combined cycle
IGO	Intergovernmental organization
IIASA	International Institute of Applied Systems Analysis
IPCC	Intergovernmental Panel on Climate Change
IPT	Instituto de Pesquisas Tecnológicas do Estado de SP
IRENA	International Renewable Energy Agency
IRSN	Institut de Radioprotection et de Suret Nucleaire (Institute for Radiation Protection and Nuclear Safety)
ISEA	International Sustainable Energy Assessment
ISAL	Icelandic Aluminum Company
ISK	Icelandic Kroner
ISOR	Iceland Geosurvey
ITA	Instituto de Tecnologia Aeronautica
ITER	International Thermonuclear Experimental Reactor
km	Kilometers
kW	Kilowatts
kWh	kilowatt-hour
LCOE	Levelized Cost of Electricity
LNG	Liquefied natural gas
LOGIT	Logistic regression
LPG	Liquefied petroleum gas
LTS	Large technical systems
LWR	Light water reactor
MAPA	Ministerio da Agrucultura Pecuaria e Abastecimento (Ministry of Agriculture, Livestock and Supply)
MBOE	Million barrels of oil equivalent
MCT	Ministerio da Ciencia e Tecnologia (Science and Technology Ministry)
MEEDDM	Ministry of Ecology, Energy, Sustainable Development and the Sea
MIC	Ministerio da Industria e Comercio (Ministry of Industry and Commerce)
MIT	Massachusetts Institute of Technology
MME	Ministerio de Minas e Energia (Ministry of Mines and Energy)
MOX	Mixed oxide
MLP	Multilevel perspective
MSR	Moisture separator reheater

Mt	Metric tons
MTBE	Methyl tert-butyl ether
MW	Megawatt
MWh	Megawatt-hour
N_2O	Nitrous oxide
NASA	National Aeronautics and Space Administration
NATO	North Atlantic Treaty Organization
NEA	Nuclear Energy Agency
NEF	New energy finance
NFCRE	Nordic Folk Center for Renewable Energy
NGO	Nongovernmental organization
NHMRC	National Health and Medical Research Council
NIC	National Innovative Capability
NIS	National Innovation Systems
NIVE	North-Western Jutland Institute for Renewable Energy
NOAA	National Oceanic and Atmospheric Administration
NOx	Nitrogen oxide
NPP	Nuclear power plant
NRC	Nuclear Regulatory Commission
NREL	National Renewable Energy Laboratory
NU	Natural uranium
NUGG	Natural uranium gas graphite
O&M	Operations and maintenance
OECD	Organization of Economic Cooperation and Development
OOA	Organization for Information on Nuclear Power
OPEC	Organization of Petroleum Exporting Countries
OVE	Organisationen for Vedvarende Energi (Organization for Renewable Energy)
P&T	Partitioning and transmutation
PBS	Public Broadcasting System
PCAST	President's Council Advisors on Science and Technology
PD	People's Daily
PEON	La Commission pour la Production d'Electricite d'Origine Nucleaire (Commission for the Production of Electricity from Nuclear Energy)
PIS/PASEP	O Programa de Integracao Sociale o Programa de Formacao do Patrimonio do Servidor Publico (PASEP)
PLANALSUCAR	Programa Nacional de Melhoramento da Cana-da-Acucar (National Program for the Improvement of Sugarcane)
PNPB	Programa Nacional do Petroleo e Biocombustiveis (National Program for Petroleum and Biofuel)

PPI	Pluri-annual Investment Plan
PPIAF	Public–Private Infrastructure Advisory Facility
PPM	Parts per million
PPP	Purchasing power parity
PRIS	Power Reactor Information System (IAEA)
ProAlcool	Programa Nacional do Álcool (National Fuel Alcohol Program)
PSO	Public shared obligation
PURPA	Public Utilities Regulatory Policy Act
PV	Photovoltaic
PWR	Pressurized water reactor
R&D	Research and development
RBMK	Reaktor Bolshoi Moshchnosti Kanalye
RCP	Reactor coolant pump
R&D	Research and development
RD&D	Research development and demonstration
REC	Renewable Energy Credit
REN21	Renewable Energy Network for the 21st Century
RET	Renewable energy technology/renewables
RETD	Renewable energy technology deployment
RFA	Renewable Fuel Association
RHR	Residual heat removal
RIDESA	Rede Intrauniversitaria para o Desenvolvimento do Setor Sucroenergetico (Interuniversity Network for the Development of the Sugar-Energy Sector)
RTE	Reseau Transport d'Electricite (Electricity Transmission System)
S/G	Steam generator
SAP	Software application program
SCPRI	French Radiation Protection Agency
SEA	Strategic environmental assessment
SITC	UN Comtrade Classification
SNM	Strategic niche management
SO_2	Sulfuric oxide
SOE	State-owned enterprise
SRREN	Special Report on Renewable Energy Sources and Climate Change Mitigation
TELESP	Telefones de São Paulo
TEP	Techno-economic paradigms
TFC	Total final consumption
TGV	Train de Grand Velocite

TIS	Technological innovation systems
TJ	Terajoule
TMI	Three Mile Island
TOE	Tons of oil equivalent
TPES	Total primary energy supply
TSO	Transmission system operator
TWh	Terawatt-hour
UCS	Union of Concerned Scientists
UFMG	Universidade Federal de Minas Gerais
UNCRET	United Nations Centre for Natural Resources, Energy and Transport
UNDP	United Nations Development Program
UNEP	United Nations Environment Programme
UNFCCC	United Nations Framework Convention on Climate Change
UNGG	Uranium naturel graphite gaz
UNICA	União da Agroindústria Canavieira de São Paulo (São Paulo Sugarcane Agroindustry Union)
UNICAMP	Universidade de Campinas
UNSCEAR	United Nations Scientific Committee on the Effects of Atomic Radiation
UNU	United Nations University
USDA	U.S. Department of Agriculture
USGS	U.S. Geological Survey
USP	Universidade de São Paulo
USSR	Union of the Soviet Socialist Republic
VAWT	Vertical axis wind turbine
VIC	Vertically integrated company
VW	Volkswagen
WCED	World Commission on the Environment and Development
WCRE	World Council for Renewable Energy
WDI	World Development Indicators
WEA	Wind Energy Association
WEC	World Energy Council
WHO	World Health Organization
WIPO	World Intellectual Property Organization
WNA	World Nuclear Association
WTI	West Texas Intermediate
WWF	World Wildlife Fund
$\mu g/m^3$	Microgram per cubic meter

BIBLIOGRAPHY

Note: Data/information for this research was often sourced from databases and websites that underwent frequent updates. In such cases, a general standard was adopted, indicating the date of publication as "n.d." in text, with the most recent access date indicated in the full reference. Specific to interviews, more than 120 were completed for this study. Information on organizational affiliations and dates are available following the Country Timelines.

ABC News. (2004, January 18). Thousands March in Paris Anti-Nuclear Protest, http://www.abc.net.au/news/2004-01-18/thousands-march-in-paris-anti-nuclear-protest/121430.

Abernathy, W., and Utterback, J. (1978). Patterns of Innovation in Technology, *Technology Review*, 80: 7, 40–47.

Abraciclo (n.d.). Information. http://www.abraciclo.com.br/, Accessed January 10, 2017.

ABS Energy Research. (2009). *Geothermal Energy Report* (6th Edition), London: ABS Energy Research, http://www.absrenewables.com.

Achen, C. (1986). *The Statistical Analysis of Quasi-Experiments*. Berkeley: University of California Press.

Achen, C., and Snidal, D. (1989). Rational Deterrence Theory and Comparative Case Studies, *World Politics*, 41, 143–169.

Ackermann, T. (Ed.). (2005). *Wind Power in Power Systems*. Chichester: Wiley.

Ackermann, T., and Soder, L. (2000). Wind Energy Technology and Current Status: A Review, *Renewable and Sustainable Energy Reviews*, 4, 315–374.

Adam, S., and Kriesi, H. (2007). The Network Approach, Chapter 5 in: P. Sabatier (Ed.), *Theories of the Policy Process*. Boulder: Westview Press.

Agência Nacional do Petróleo, Gás Natural e Biocombustíveis (ANP). (n.d). Information http://www.anp.gov.br/, Accessed January 5, 2017.

Agência Nacional do Petróleo, Gás Natural e Biocombustíveis (2008). Anuário Estatístico Brasileiro do Petróleo, Gás Natural e Biocombustíveis.

Agência Nacional do Petróleo, Gás Natural e Biocombustíveis (ANP). (2013). Leilao de Biodiesel Vai Ofrecer 700 Milhoes de Litros no Final de Fevereiro. http://www.anp.gov.br/?pg=59260&m=&t1=&t2=&t3=&t4=&ar=&ps=&cachebust=1331339359948.

Aguayo, F. (2008). *Selection Environments and Innovation Regimes*, Paper, 6th Globelics Conference, Mexico City, Mexico.

Ailworth, E. Cambridge Project Taps Excess Steam to Buildings. *Boston Globe*, May 19, 2014, https://www.bostonglobe.com/business/2014/05/18/back-future-french-company-revives-combined-heat-and-power-boston/GBv8Yqzyf33iZr51PTzoGO/story.html.

Akiyama, T., Baffes, J., Larso, D., and Varangis, P. (2001). *Commodity Market Reforms: Lessons of Two Decades, Regional and Sectoral Studies*. Washington, DC: World Bank.

Alabama Policy Institute. (n.d.). Government Intervention in the Marketplace, Guide to the Issues. http://www.alabamapolicy.org/wp-content/uploads/GTI-Brief-Gvt-Intervention.pdf. Accessed June 10, 2016.

Albertsson, A., and Jonsson, J. (2010*a*). *The Svartsengi Resource Park*, Proceedings World Geothermal Congress 2010, Bali, Indonesia, April 25–29, 2010.

Albertsson, A., and Jonsson, J. (2010*b*). *The Geothermal Installations at Svartsengi and Reykjanes*, Centre for Research and Development, Training and Educating Tourism, Proceedings World Geothermal Congress 2010, Bali, Indonesia, April 25–29, 2010.

Albertsson, A., Thorolfsson, G., and Jonsson, J. (2010). *Three Decades of Power Generation-Svartsengi Power Plant*, Proceedings World Geothermal Congress 2010, Bali, Indonesia, April 25–29, 2010.

Albright, D., Berkhout, F., and Walker, W. (1993). World Inventory of Plutonium and Highly Enriched Uranium 1992, Oxford: Oxford University Press.

Allison, G. (1969). Conceptual Models and the Cuban Missile Crisis, *The American Political Science Review*, 63: 3, 689–718.

Allison, G., and Zelikow, P. (1999). *The Essence of Decision: Explaining the Missile Cuban Missile Crisis* (2nd Edition). London: Longman.

Almeida, C. (2007, December 6). Sugarcane Ethanol: Brazil's Biofuel Success, *Science and Development Network,* http://www.scidev.net/global/policy/feature/sugarcane-ethanol-brazils-biofuel-success.html.

American Academy of Arts and Sciences (AAAS). (2011). *Beyond Technology: Strengthening Energy Policy Through Social Sciences*. Cambridge: AAAS.

American Institutes of Research. (2006). *The Contributions and the Limitations of Cross-National Comparisons in Examining Professional Development and Educational Quality*, Paper, US Agency for International Development, Cooperative Agreement No. GDG-A-00-03-00006-00.

Anadon, L., and Holdren, J. (2009). Policy for Energy Technology Innovation, in: Gallagher, K., and Ellwood, D. (Eds.), *Acting in Time on Energy Policy* (pp. 89–127). Washington, DC: Brookings Institute Press.

Anadon, L., Bunn, M., Chan, G., Chan, M., Jones, C., Kempener, R., Lee, A., Logar, N., and Narayanamurti, V. (2011). Transforming US Energy Innovation, Report, November 9, 2011, Cambridge: Belfer Center for Science and International Affairs, http://belfercenter.ksg.harvard.edu/publication/21486/background.html.

Anantharaman, K., Rao, P., Castano, C., and Henning, R. (2011). Nuclear Fission, in: Kravit, S., Lehr, J., and Kingery, T. (Eds.), *Nuclear Energy Encyclopedia: Science, Technology and Applications* (pp. xv–xviii). Hoboken: Wiley.

Anderer, J., McDonald, A., and Nakicenovic, N. (1981). *Energy in a Finite World* (Volume 1). Cambridge: Ballinger.

Andersen, P. (1998). Wind Power in Denmark. Technology, Policies and Results, Danish Energy Agency, 51171/97-0008.

Andersen, P. (2004). Sources of Experience: Theoretical Considerations and Empirical Observations from Danish Wind Energy Technology, *International Journal of Energy Technology and Policy*, 2: 1/2, 33–51.

Andersen, P. (2007, January 7). Review of Historical and Modern Utilization of Wind Power. Roskilde: Risø Laboratory.

Andersen, P., and Drejer, I. (2008). Systemic Innovation in a Distributed Network: The Case of Danish Wind Turbines, 1972–2007, *Strategic Organisation*, 6/1: 12–46.

Andersen, P., Drejer, I., and Waldstrom, C. (2006). In the Eye of the Storm, Paper presented at the DRUID Summer Conference June 18–20, 2006, Copenhagen, Denmark.

Anderson, P., and Gudmundsson, M. (1998). Inflation and Disinflation in Iceland, Report, Working Papers No. 1, Central Bank of Iceland. http://www.sedlabanki.is/uploads/files/WP-1.pdf.

Andrews, A., and Folger, P. (2012, January 6). Nuclear Power Plant Design and Seismic Safety Considerations, Washington, DC: Congressional Research Service.

Andrews-Speed, P. (2016). Applying Institutional Theory to the Low Carbon Energy Transition, *Energy Research and Social Science*, 13, 216–225.

Anex, R. (2000). Stimulating Innovation in Green Technology: Policy Alternatives and Opportunities, *American Behavioral Scientist*, 44, 188–212.

Applebome, P. (2008, August 3). They Used to Say Whale Oil Was Indispensible, Too, *New York Times*, p. A25.

Appunn, K. Germany's Power Consumption and Power Mix in Charts, *Clean Energy Wire*, June 9, 2016, https://www.cleanenergywire.org/factsheets/germanys-energy-consumption-and-power-mix-charts.

Araújo, K. (2013). Energy at the Frontier: Low Carbon Energy System Transitions and Innovation in Prime Mover Countries. PhD Dissertation, Massachusetts Institute of Technology, Cambridge, MA.

Araújo, K. (2014). The Emerging Field of Energy Transitions: Progress, Challenges, and Opportunities, *Energy Research & Social Science*, 1, 112–121, http://www.sciencedirect.com/science/article/pii/S2214629614000164.

Araújo, K. (2016). 'Truer Costs' in Energy System Change, *Papers in Energy*, 141–173, and Los 'Costes Reales' del Cambio del Sistema Energético, *Papeles de Energía*, 43–79.

Araújo, K., Mahajan, D., Kerr, R., and da Silva, M. (2017). Global Biofuels at the Crossroads: An Overview of Technical, Policy, and Investment Complexities in the Sustainability of Biofuel Development, *Agriculture*, 7: 32.

Archer, C., and Jacobsson, M. (2005). Evaluation of Global Wind Power, Stanford: Stanford University, http://www.stanford.edu/group/efmh/winds/global_winds.html.

Arctic Climate Impact Assessment. (2004). Impacts of Warming Climate. http://www.amap.no/acia/GraphicsSet1.pdf.

Arentsen, M. (2005). The Invisible Problem and How to Deal with It, in: Bemelmans-Videc, M., Rist, R., and Vedung, E. (Eds.), *Carrots, Sticks, and Sermons* (pp. 211–230). New Brunswick: Transaction Publishers.

AREVA. (n.d). Information. http://www.AREVA.com. Accessed December 10, 2016.

AREVA. (2016). Annual Report 2015. http://www.areva.com/EN/news-10717/2015-annual-results.html.

Argo, J. (2001). Unhealthy Effects of Upstream Oil and Gas Flaring, Presentation, Public Review Commission into Effects of Potential Oil and Gas Exploration, Drilling Activities within Licenses 2364, 2365, 2368. http://www.sierraclub.ca/national/oil-and-gas-exploration/soss-oil-and-gas-flaring.pdf.

Armaroli, N., and Balzani, V. (2011). *Energy for a Sustainable World.* Weinheim: Wiley VCH.

Arnorsson, S. (1975, March). *Geothermal Energy in Iceland: Utilization and Environmental Problems*, OSJHD1675, Reykjavik: Orkustofnun.

Arnorsson, S. (2008). Cooperation between the United Nations University Geothermal Training Programme and the University of Iceland, 30th Anniversary Workshop, August 26–27, 2008.

Arrhenius, S. (1896). On the Influence of Carbonic Acid in the Air upon the Temperature of the Ground, *Philosophical Magazine and Journal of Science*, 41: 5, 237–276.

Arrhenius, S. (1897). On the Influence of Carbonic Acid in the Air upon the Temperature of the Ground, *Publications of the Astronomical Society of the Pacific*, 9: 54, 14.

Arrow, K. (1962*a*). Economic Welfare and the Allocation of Resources for Invention, in: R. Nelson (Ed.) *The Rate and Direction of Inventive Activity* (pp. 609–626). Princeton: Princeton University Press.

Arrow, K. (1962*b*). The Economic Implications of Learning by Doing. *Review of Economic Studies*, 29: 3, 155–173. http://www.jstor.org/stable/2295952.

Arruda, P. (2011). Perspective of the Sugarcane Industry in Brazil, *Tropical Plant Biology*, 4, 3–8.

Arthur, W. (1988). Competing Technologies: An Overview, in: Dosi G, Freeman, C., Nelson, R., Silverberg, G., and Soete, L. (Eds.), *Technical Change and Economic Theory* (pp. 590–607). London: Pinter.

Arthur, W. (1994). *Increasing Returns and Path Dependence in the Economy.* Ann Arbor: University of Michigan Press.

Arthur, W. (2009). *The Nature of Technology: What It Is and How It Evolves.* New York: Free Press.

Arthur, W. (1989, March). Competing Technologies, Increasing Returns, and Lock-In by Historical Events, *The Economic Journal*, 99: 394, 116–131.

Arvizu, D., Bruckner, T., Chum, H., Edenhofer, O., Estefen, S., et al. (2011). Technical Summary, in: Edenhofer, O., Pichs-Madruga, R., Sokona, Y., Seyboth, K., Matschoss, P., et al. (Eds.), *IPCC Special Report on Renewable Energy Sources and Climate Change Mitigation.* Cambridge: Cambridge University Press.

Associação Nacional dos Fabricantes de Veículos Automotores (ANFAVEA). (n.d.). Information. http://www.anfavea.com.br/.html. Accessed January 3, 2017.

Associação Nacional dos Fabricantes de Veículos Automotores (ANFAVEA). (2011*a*). Indústria Automobilística Brasileira: 50 Anos. http://www.anfavea.com.br/anuario.html.

Associação Nacional dos Fabricantes de Veículos Automotores (ANFAVEA). (2011*b*). Anúario da Industria Automobilistica Brasileira. http://www.anfavea.com.br/anuario.html.

Associação Nacional dos Fabricantes de Veículos Automotores (ANFAVEA). (2012*a*). Carta da ANFAVEA, No. 308. http://www.anfavea.com.br/.

Associação Nacional dos Fabricantes de Veículos Automotores (ANFAVEA). (2012*b*). Carta da ANFAVEA, No. 309. http://www.anfavea.com.br/.

Associação Nacional dos Fabricantes de Veículos Automotores (ANFAVEA). (2017). Anuária da Indústria. http://www.anfavea.com.br/anuario.html.

Ausubel, J. (1991). Rat Race Dynamics and Crazy Companies: The Diffusion of Technologies in Social Behavior, *Technological Forecasting and Social Change*, 39, 11–22.

Ausubel, J. (2003). Decarbonization: The Next 100 Years, Alvin Weinburg Lecture, Oak Ridge National Laboratory, Oak Ridge, TN, June 5, 2003.

Autorité de Sûreté Nucléair (ASN). (1999). Press Releases. http://www.asn.fr/index.php/S-informer/Actualites/1999/COMMUNIQUE-N-3-INCIDENT-SUR-LE-SITE-DU-BLAYAIS.

Auty, R. (Ed.). (2002). *Resource Abundance and Economic Development*. Oxford: Oxford University Press.

Axelsson, G., Gunnlaugsson, E., Jonasson, T., and Olafsson, M. (2010). Low Temperature Geothermal Utilization in Iceland, *Geothermics*, 39: 4, 329–338.

Backstrand, K. (2003). Civic Science for Sustainability: Reframing the Role of Experts, Policy-makers and Citizens in Environmental Governance, *Global Environmental Politics*, 3: 4, 24–41.

Backwell, B. (2010, May 19). Spain's REE to Run TWENTIES Wind Integration Project. *Recharge News*. http://www.rechargenews.com/energy/wind/article215256.ece.

Bacon, R., and Besant-Jones, J. (2001). Global Electric Power Reform, Privatization, and Liberalization of the Electric Power Industry in Developing Countries, *Annual Review of Energy and the Environment*, 26, 331–359.

Bake, J., Junginger, M., Faaij, A., Poot, T., and Walter, A. (2009). Explaining the Experience Curve, *Biomass and Bioenergy*, 33, 644–658.

Baker, N. (2010, December 23). Iceland and China Establish Strategic Geothermal Partnership, *The Energy Collective*. http://www.theenergycollective.com/nathanael-baker/48959/iceland-and-china-establish-strategic-geothermal-partnership.

Bambrick, G. (2012, December 11). Fracking: Pro and Con, *Tufts Now*. http://now.tufts.edu/articles/fracking-pro-and-con.

Banco Nacional de Desenvolvimento Econômico e Social (BNDES) and Centro de Gestão e Estudos Estratégicos (CGEE). (2008). *Sugarcane-based Bioethanol: Energy for Sustainable Development*. Rio de Janeiro: BNDES and CGEE, https://web.bndes.gov.br/bib/jspui/handle/1408/6305.

Barbier, E. (2005). Natural Resource-Based Economic Development in History, *World Economics*, 6: 3, 103–152.

Bardach, E. (1977). *The Implementation Game: What Happens After a Bill Becomes a Law*. Cambridge: MIT Press.

Barré, B. (2008). *All about Nuclear Energy*. Paris: AREVA.

Barrow, D. (2006). Social and Cultural Factors: Constraining and Enabling, in: Moran, M., Rein, M., and Goodin, R. (Eds.), *The Oxford Handbook of Public Policy*. Oxford: Oxford University Press.

Barry, F. (2006). Foreign Direct Investment and Institutional Co-evolution in Ireland, Working Paper 200603, School of Economics. Dublin: University College.

Bartlett, A. (Winter 1997/1998). Reflections on Sustainability, Population Growth and the Environment—Revisited, *Renewable Resources Journal,* 15: 4, 6–23.

Barzelay, M. (1986). *The Politicized Market Economy*. Berkeley: University of California Press.

Bastin, C., Szloko, A., and Pinguelli Rosa, L. (2010). Diffusion of New Automotive Technologies for Improving Energy Efficiency in Brazil's Light Vehicle Fleet, *Energy Policy,* 38, 3586–3597.

Bataille, C., and Galley, R. (1999). Rapport Sur l'Aval du Cycle Nucléaire, Rapport 1359. http://www.assemblee-nationale.fr/rap-oecst/nucleaire/r1359-00.asp.

Bauen, A. (2006). Future Energy Sources and Systems: Acting on Climate Change and Energy Security, *Journal of Power Systems,* 157, 893–901.

Baumgartner, F., and Jones, B. (1993). *Agendas and Instabilities in American Politics.* Chicago: University of Chicago Press.

Baumuller, H., Donnelly, E., Vines, A., and Weimar, M. (2011). The Effects of Oil Companies' Activities on the Environment, Health and Development in Sub-Saharan Africa, Report, London: Chatham House.

Beck, F., and Martinot, E. (2004). *Renewable Energy Policies and Barriers in Encyclopedia of Energy*. San Diego: Elsevier Science.

Beck, P. (1999). Nuclear Energy in the 21st Century, *Annual Review of Energy and the Environment,* 24, 113–137.

Bell, D., Gray, T., and Haggett, C. (2005). The 'Social Gap' in Wind Farm Siting Decisions: Explanations and Policy Responses, *Environmental Politics*, 14: 4, 460–477.

Bemelmans-Videc, M. Rist, R., and Vedung (Eds.). (2005). *Carrots, Sticks, and Sermons.* New Brunswick: Transaction.

Bennett, L., and Skjoeldebrand, R. (1986). Worldwide Nuclear Power Status and Trends. Nuclear's Contribution to Electricity Supply Is Growing, *IAEA Bulletin*, 28: 3, 40–45. https://www.iaea.org/sites/default/files/publications/magazines/bulletin/bull28-3/28304784045.pdf.

Benson, T. (2007, June 4). Ethanol Boom Won't Threaten Food Supply, *Reuters*. http://www.reuters.com/article/us-brazil-ethanol-summit-food-idUSN0433058520070604.

Berg, S. (2005). Glossary for the Body of Knowledge on the Regulation of Utility Infrastructure and Services, Prepared for the World Bank et al. http://regulationbodyofknowledge.org/glossary/.

Bergek, A. (2002). Shaping and Exploiting Technological Opportunities, PhD Dissertation, Department of Industrial Dynamics, Chalmers University of Technology, Goteborg, Sweden.

Berger, P., and Luckmann, T. (1966). *The Social Construction of Reality*. Garden City: Anchor.

Berkhout, F. (2002). Technological Regimes, Path Dependency and the Environment, *Global Environmental Change*, 12: 1, 1–4.

Berman, B. (2011, June 14). History of Hybrid Vehicles, Hybrid Cars. http://www.hybridcars.com/history.html.

Berndes, G., Bird, N., and Cowie, A. (2010). Bioenergy, Land Use, and Climate Change Mitigation, IEA Bioenergy. http://www.ieabioenergy.com/publications/bioenergy-land-use-change-and-climate-change-mitigation-background-technical-report/.

Bernstein, S. (2006). *The Republic of de Gaulle, 1958–1969.* (Translated by Morris, P.). Cambridge History of Modern France. Cambridge: Cambridge University Press.

Bernstein, S., and Rioux, J. (2015). *The Pompidou Years, 1969–1974.* Cambridge History of Modern France. Cambridge: Cambridge University Press.

Bertani, R. (2003). What Is Geothermal Potential? *IGA News,* 53, 1–3. http://www.geothermal-energy.org/publications_and_services/iga_newsletter.html?no_cache=1&cid=505&did=233&sechash=21ce7403.

Bertani, R. (2005*a*). World Geothermal Generation 2001–2005: State of the Art, in: Proceedings of the 2005 World Geothermal Congress, Antalya, Turkey, April 24–29, 2005.

Bertani, R. (2005*b*). World Geothermal Power Generation in the Period 2001–2005, *Geothermics,* 34, 651–690.

Bertani, R. (2006). World Geothermal Power Generation: 2001–2005. *GRC Bulletin,* 35: 3, 89–111.

Bertani, R. (2007). World Geothermal Generation in 2007, *Proceedings of the European Geothermal Congress*, Unterhaching, Germany, April–May, 2007.

Bertani, R. (2010). Geothermal Power Generation in the World: 2005–2010 Report, *Geothermics,* 41, 1–29.

Bertani, R. (2015). Geothermal Power Generation in the World 2010–2014 Update Report, Proceedings World Geothermal Congress 2015, Melbourne, Australia, April 19–25, 2015.

Besson, S. (2003, May 8). Après vingt ans de silence, un ex-député avoue l'attaque à la roquette contre Creys-Malville, Le Temps. https://www.letemps.ch/suisse/2003/05/08/apres-vingt-ans-silence-un-ex-depute-avoue-attaque-roquette-contre-creys-malville.

Beyond Nuclear (n.d.). Reactors Are Closing, http://www.beyondnuclear.org/reactors-are-closing/. Accessed December 15, 2016.

Biello, D. 100 (2010). Percent Renewable? One Danish Island Experiments with Clean Power, *Scientific American.* http://www.scientificamerican.com/article/samso-attempts-100-percent-renewable-power/.

Biofuels Digest (2011, November 22). Vale to Export Ethanol, Exapnd Santos Port Facilities, *Biofuels Digest,* http://www.biofuelsdigest.com/bdigest/2011/11/22/vale-to-export-ethanol-expand-santos-port-facilities/.

Birkland, T., and Warnement, M. (2013). Focusing Events, Risk, and Regulation, Working paper. North Carolina State University, September 2013 revision.

Birkland, T. (1997). *After Disaster: Agenda Setting, Public Policy and Focusing Events.* Washington, DC: Georgetown University Press.

Birkland, T. (1998, April). Focusing Events, Mobilization, and Agenda Setting, *Journal of Public Policy,* 18: 1, 53–74.

Birol, F. (2016). *Global Energy Markets in Transition: Implications for the Economy, Environment & Geopolitics*, IEA presentation, Tokyo, Japan, April 21, 2016.

Bjorkman, N. (2015). Bringing the Neighbors on Board, in: State of Green (Ed.), *Wind Energy Moving Ahead.* Frederiksberg: Danish Wind Industry Association.

Bjornsson, G., and Albertsson, A. (1985). The Power Plant at Svartsengi, Development and Experience, Experience, Proceedings 1985 International Symposium on Geothermal Energy, Davis: Geothermal Resources Council.

Bjornsson, S. (1970). A Program for the Exploration of High Temperature Areas in Iceland. Proceedings of the United Nations Symposium on the Development and Utilization of Geothermal Resources, *Geothermics*, 2: 2, 1050–1054.

Bjornsson, S. (2006). Geothermal Development and Research in Iceland, Reykjavik: Orkustofnun, https://rafhladan.is/bitstream/handle/10802/6401/OS-2005-Geothermal-Development.pdf?sequence=1.

Bjornsson, S. (2010). Geothermal Research and Development in Iceland, Reykjavik: Orkustofnun, www.nea.is/media/utgafa/GD_loka.pdf.

Blegaa, S., Josephsen, L., Meyer, N., and Sorenson, B. (1977, June). Alternative Danish Energy Planning, *Energy Policy*, 5: 2, 87–94.

Blight, D. (2005). Fossilized Lies: A Reflection on Alessandro Portelli's "The Order Has Been Carried Out," *The Oral History Review*, 32: 1, 5–9.

Bloomberg. (2011, September 13). IEA Cuts Oil Demand Forecast, Sees Difficult Libya Restart, *Bloomberg*, http://www.bloomberg.com/news/articles/2011-09-13/world-oil-demand-forecasts-cut-by-iea-as-global-economic-recovery-falters.

Bloomberg. (2011, September 26). Energy and Oil Prices, *Bloomberg*, http://www.bloomberg.com/energy/.

Bloomberg. (2012, May 30). Landsvirkjun in Iceland Offers the Most Competitive Energy, *Bloomberg*, http://www.bloomberg.com/article/2012-05-30/aE6vcFvg52CM.html.

Bloomberg New Energy Finance (BNEF). (2012, January 11). Moving Toward a Next-Generation Ethanol Economy, Final Study, *BNEF*, https://www.dsm.com/content/dam/dsm/cworld/en_US/documents/bloomberg-next-generation-ethanol-economy.pdf.

Bloomberg New Energy Finance. (2016, January 14). Clean Energy Defies Fossil Fuel Crash to Attract Record $329B Global Investment in 2015, *BNEF*, http://about.bnef.com/press-releases/clean-energy-defies-fossil-fuel-price-crash-to-attract-record-329bn-global-investment-in-2015/.

Blue Lagoon (n.d). Information. http://www.bluelagoon.com/. Accessed December 10, 2016.

Bock, H. (n.d.). WWER/VVER, Presentation, Vienna University of Technology, Vienna, Austria, http://www.ati.ac.at/fileadmin/files/research_areas/ssnm/nmkt/04_WWER_Overview.pdf. Accessed June 15, 2016.

Boden, T., Marland, G., and Andres, R. (2010). Global, Regional, and National Fossil-Fuel CO_2 Emissions. Carbon Dioxide Information Analysis Center, Oak Ridge National Laboratory, Oak Ridge, TN, US Department of Energy.

Boddey, R. (1995). Biological Nitrogen Fixation in Sugarcane: A Key to Energetically Viable Bio-fuel Production, *CRC Critical Review in Plant Science*, 14: 263–279.

Bogliacino, F., and Pianta, M. (2011, February). Engines of Growth, Innovation and Productivity in Industry Groups, *Structural Change and Economic Dynamics*, 22, 41–53.

Bolton, R., and Foxon, T. (2015). A Socio-technical Perspective on Low Carbon Investment Challenges, *Environmental Innovation and Societal Transitions*, 14, 165–181.

Bondergaard, M. (2015). The Future of Wind Energy in Denmark, Presentation, Copenhagen Denmark, http://windpower.org/download/2530/7_the_future_of_wind_energy_in_denmarkpdf.

Bonvillian, W., and Weiss, C. (2015). *Technological Innovation in Legacy Sectors*. New York: Oxford University Press.

Borras, S. (2011). Policy Learning and Organizational Capacities in Innovation Policies, *Science and Public Policy*, 38: 9, 725–734.

Borrel, B., Bianco, J., and Bale, M. (1994). *Brazil's Sugarcane Sector*, Policy Research Working Paper, WP 1363, Washington, DC: World Bank.

Borup, M., Gregersen, B., and Madsen, A. (2007). *Development Dynamics and Conditions for New Energy Technology Seen in Innovation System Perspective*, Paper, Druid Conference, Copenhagen, Denmark, June 18–20, 2007.

Bosch, F. A., Volberda, H. W., and Boer, M. D. (1999, May). Coevolution of Firm Absorptive Capacity and Knowledge Environment: Organizational Forms and Combinative Capabilities, *Organization Science*, 10: 5, 551–568.

Boselli, M. (2011, November 17). France Needs to Upgrade All Nuclear Reactors, *Reuters*, http://www.reuters.com/article/us-france-nuclear-tests-idUSTRE7AG0HQ20111117.

Boston Consulting Group. (1968). *Perspectives on Experience*. Boston: Boston Consulting Group Inc.

Boulin, P., and Boiteux, M. (2000). L'aventure Nucléaire en France: Grande et Petite Histoire, *Les amis de L'Ecole de Paris*.

Bouveret, P., Barrilot, B., and Lalanne, D. (2013, January 1). Nuclear Chromosomes: The National Security Implications of a French Nucelar Exit, *Bulletin of Atomic Scientists*, http://thebulletin.org/2013/january/nuclear-chromosomes-national-security-implications-french-nuclear-exit.

Bower, J., and Christensen, C. (1995, January–February). Disruptive Technologies: Catching the Wave, *Harvard Business Review*, 73: 1, 43–53.

Boyle, M., and Robinson, M. (1981). French Nuclear Energy Policy, *Geography*, 66: 4, 300–303.

BP. (2011). Energy Outlook 2030. http://www.bp.com/content/dam/bp/pdf/energy-economics/energy-outlook-2016/bp-energy-outlook-2011.pdf.

BP. (2015, 2016, 2017). Statistical Review of World Energy. http://www.bp.com/.

Bracmort, K. (2012). Is Biopower Carbon Neutral? CRS Report R41603, Washington DC: Congressional Research Services, https://www.fas.org/sgp/crs/misc/R41603.pdf.

Bradshaw, M. (2014). *Global Energy Dilemmas*. Cambridge: Polity.

Braunbeck, O., and Magalhaes, P. (2010). Technological Evolution of Sugarcane Mechanization, in: Cortez, L. (Ed.), *Sugarcane Bioethanol*. São Paulo: Blucher.

Braunbeck, O., and Neto, A. (2010). Transport Logistics of Raw Material and Waste of Sugarcane, in: Cortez, L. (Ed.), *Sugarcane Bioethanol R&D for Productivity and Sustainability*, São Paulo: Blucher.

Brianezi, P. (2010, March 18). Brazil Exports Ethanol Production Model to Poor Countries, *Reporter Brasil*.

Brimblecombe, P. (1987). *The Big Smoke: A History of Air Pollution in London Since Medieval Times*. London: Methuen.

British Broadcasting Corporation (BBC). (2000, June 15). Nuclear Doubts Gnaw Deeper, *BBC*, http://news.bbc.co.uk/2/hi/europe/792209.stm.

British Broadcasting Corporation (BBC). (2006, May 3). Bolivia's Gas Takeover, *BBC*, http://news.bbc.co.uk/2/hi/business/4969290.stm.

British Broadcasting Corporation (BBC). (2012*a*, April 12). UK in Talks with Iceland over 'Volcanic Power Link', *BBC,* http://www.bbc.co.uk/news/uk-politics-17694215.

British Broadcasting Corporation (BBC). (2012*b*, April 5). Fire at Penly Nuclear Reactor in Northern France, *BBC,* http://www.bbc.com/news/world-europe-17630358.

British Broadcasting Corporation (BBC). (2012*c*, May 29). BP to Resume Oil Operations in Libya, *BBC,* http://www.bbc.co.uk/news/business-18256587.

British Broadcasting Corporation (BBC). (2016, May 26). US Nuclear Force Still Uses Floppy Disks, *BBC,* May 26, 2016, http://www.bbc.com/news/world-us-canada- 36385839.

Brown, B. (1979). *Disaster Preparedness and the United Nations*. New York: Pergamon.

Brown, M., Chandler, S., Lapsa, M., and Sovacool, B. (2008). Carbon Lock-in, ORNL/TM-2007/124, November 2008, Oak Ridge: Oak Ridge National Laboratory.

Brown, P., and Whitney, G. (2011). US Renewable Electricity Generation: Resources and Challenges, CRS Report, R41954, Washinton, DC: Congressional Resaerch Services, https://www.fas.org/sgp/crs/misc/R41954.pdf.

Brundtland Commission. (1987). *Our Common Future*. Oxford: Oxford University Press.

BTM. (2010). Offshore Wind Power 2010, Press release, http://btm.dk/news/offshore+wind+power+2010/?s=9&p=&n=39.

BTM. (2012). International Wind Energy Development World Market Update 2011, Forecast 2012–2016, Press Release, http://www.navigant.com/windreport.

Buch, M., and Kjaer, E. (2015, May). Offshore Wind Development, Danish Energy Agency (DEA) Copenhagen, Denmark.

Buchsbaum, L. (2016), March 1). Germany's Energiewende at a New Turning Point, *Power*, http://www.powermag.com/germanys-energiewende-new-turning-point/.

Buckley, W. (1967). *Sociology and Modern Systems Theory*. Englewood Cliffs: Prentice Hall.

Budny, D. (2005). *The Global Dynamics of Biofuels, Special Report*, No. 3, Brazilian Institute, Washington, DC: Woodrow Wilson Center. https://www.wilsoncenter.org/.

Bunn, M., and Heinonen, O. (2011, September 16). Preventing the Next Fukushima, *Science*, 333, 1580–1581.

Bunn, M., and Malin, M. (2009). Enabling a Nuclear Revival—And Managing Its Risks, *Innovations,* 4: 4, 73–191.

Buongiorno, J. (2011). *Nuclear Energy,* Presentation, Massachusetts Institute of Technology, Cambridge, MA, October 7, 2011.

Burgassi, P., and Cappetti, G. (2004, April–June). 100 Year History of Geothermal Power Production at Larderello. *International Geothermal Association (IGA) Newsletter*, 56.

Bush, V. (1945). Science: The Endless Frontier, Report to the President, Office of Scientific Research and Development. http://www.nsf.gov/od/lpa/nsf50/vbush1945.htm.

Cahn, M. (1995). Playing the Policy Game, in: Theodoulou, S., and Cahn, M. (Eds.), *Public Policy: The Essential Readings*. Englewood Cliffs: Prentice Hall.

California Energy Commission. (2011). Overview of Wind Energy in California, California Energy Commission. http://www.energy.ca.gov/wind/overview.html.

Callon, M. (2012). Society in the Making: The Study of Technology as a Tool for Sociological Analysis, in: W. Bijker, T. Hughes, and T. Pinch (Eds.), *The Social Construction of Technological Systems*. Cambridge: MIT Press.

Calvalcanti, M., Szloko, A., and Machado, G. (2012). Do Ethanol Prices in Brazil Follow Brent Price and International Gasoline Price Parity? *Renewable Energy*, 43, 423–433.

Cameron, J. (1973). Uranium Resources and Supply, *IAEA Bulletin*, 15: 5, 10–13. https://www.iaea.org/sites/default/files/18104881218.pdf.

Canada Geothermal Association (n.d.). Information. http://www.cangea.ca/what-is-geothermal/, Accessed December 31, 2016.

CanaVialis (n.d.). http://www.monsanto.com/global/br/produtos/pages/canavialis.aspx. Accessed December 31, 2016.

Canis, B. (2011). The Motor Vehicle Supply Chain: Effects of the Japanese Earthquake and Tsunami, CRS Report R41831. https://www.fas.org/sgp/crs/misc/R41831.pdf.

Cantarella, H., and Rossetto, R. (2014). Fertilizers for Sugarcane, in: Cortez, L. (Ed.), *Sugarcane Bioethanol—R&D for Productivity and Sustainability*, São Paulo: Blucher.

Carbon Recycling International (n.d.), http://carbonrecycling.is/. Accessed December 20, 2016.

Carbon Trust. (2007). Accelerating Innovation in Low Carbon Technologies, Report. http://www.carbontrust.co.uk.

Carbon Trust. (2008). Low Carbon Technology Innovation and Diffusion Centres: Accelerating Low Carbon Growth in a Developing World, Report. http://www.ccrasa.com/library_1/16515%20-%20Low%20Carbon%20Technology%20Innovation%20and%20Diffusion%20Centre%20report.pdf.

Caresche, C., Chanteguet, J., Filippetti, A., and Guibert, G. (2011, April 8). Sortons du nucléaire, *Le Monde*, http://www.lemonde.fr/idees/article/2011/04/08/sortons-du-nucleaire_1504573_3232.html.

Carle, R. (1973). Phenix Reactor, France, *IAEA Bulletin*, 15: 5, 39–41, https://www.iaea.org/sites/default/files/publications/magazines/bulletin/bull15-5/15504793941.pdf.

Carlsen, H., Zoega, H., Valdimarsdottir, U., Gíslason, T., and Hrafnkelsson, B. (2012). Hydrogen Sulfide and Particle Matter Levels Associated with Increased Dispensing of Anti-asthma Drugs in Iceland's Capital, *Enviornmental Research*, 113, 33–39. http://dx.doi.org/10.1016/j.envres.2011.10.010.

Carrington, D. (2012, April 11). Iceland's Volcanoes May Power UK, *The Guardian*. http://www.guardian.co.uk/environment/2012/apr/11/iceland-volcano-green-power.

Carrington, D. (2015, May 18). Fossil Fuels Subsidizes by $10M a Minute, Says IMF, *The Guardian*, https://www.theguardian.com/environment/2015/may/18/fossil-fuel-companies-getting-10m-a-minute-in-subsidies-says-imf.

Casti, J. (1989). *Paradigm Lost: Tackling Unanswered Mysteries of Modern Science*. New York: Avon.

Cavallo, A., Hick, S., and Smith, D. (1993). Wind Energy, in: Johannsson, J., Kelly, H., Reddy, A., and Williams, R. (Eds.), *Renewable Energy: Sources for Fuels and Electricity*. Washington, DC: Island Press.

Cenbio (2008), Here Comes the "Flex' Vehicles 3rd Generation, *Revista Brasileira de Bioenergia*, Year 2: 3, 31–34, August 2008.

Center for Energy Economics. (2007). Brief History of LNG, http://www.beg.utexas.edu/energyecon/lng/LNG_introduction_06.php.

Central Intelligence Agency (CIA) (1984). French Nuclear Reactor Fuel Reprocessing Program, Washington DC: CIA, https://www.cia.gov/library/readingroom/docs/DOC_0000832123.pdf

Central Intelligence Agency (CIA) (n.d.). World Factbook, CIA, https://www.cia.gov/library/publications/the-world-factbook/, Accessed on May 5, 2017.

Centre for European Policy Studies. (2008). Energy Policy for Europe. CEPS Task Force Report, https://www.ceps.eu/system/files/book/1623.pdf.

Chakravarty, S., and Mitra, A. (2009). Is Industry Still the Engine of Growth? *Journal of Policy Modeling*, 31: 1, 22–35.

Chamorro, C., Mondekar, M., Ramos, R., Segovia, J., Martin, M., et al. (2012). World Geothermal Power Production Status: Energy, Environmental and Economic Study of High Enthalpy Technologies, *Energy*, 42, 10–18.

Chandler, H. (2012, April). Case Study: Denmark, in: Cochran, J., Bird, L., Heeter, J., and Arent, D. (Eds.), *Integrating Variable Energy in Electric Power Markets: Best Practices from International Experience*, Report, NREL/TP-6A00-53732, Golden: NREL, https://www.nrel.gov/research/publications.html.

Channell, J., Curmi, E., Nguyen, P., Prior, E., Syme, A., et al. (2015). *Energy Darwinism II: Why a Low Carbon Future Doesn't Have to Cost the Earth*. Report, Citigroup GPS, August 2015.

Charbol, M. (2016). Re-examining Historical Energy Transitions and Urban Systems in Europe, *Energy Research and Social Science*, 13, 194–201.

Charpin J., Dessus, B., and Pellat, R. (C-D-P). (2000). Etude économique prospective de la filière nucléaire: Rapport au Premier Ministre, Paris, France, http://www.ladocumentationfrancaise.fr/rapports-publics/004001472/index.shtml.

Chatterji, M. (Ed.). (1981). *Energy and Environment in the Developing Countries*. New York: Wiley.

Chaussade, J. (1990). Public Confidence and Nuclear Energy, *IAEA Bulletin*, 32: 2, 7–10, https://www.iaea.org/sites/default/files/publications/magazines/bulletin/bull32-2/32204790709.pdf.

Cheavegatti-Gianotto, A., Abreu, H., Arruda, P., Filho, J., Burnquist, W., et al. (2011). Sugarcane: A Reference Study for the Regulation of Genetically Modified Cultivars in Brazil, *Tropical Plant Biology*, 4, 62–89.

Chen, L. (2011, December 9). What Makes Mario Run? *Shanghai Daily*, http://www.shanghaidaily.com/feature/What-makes-Mario-run-Ethanol/shdaily.shtml.

Chernobyl Forum (2006). Chernobyl's Legacy: Health, Environmental and Socio-Economic Impacts and Recommendations to the Governments of Belarus, the Russian Federation and Ukraine, The Chernobyl Forum: 2003–2005, Second revised version (INIS-XA-902), International Atomic Energy Agency (IAEA), Vienna, Austria, https://www.iaea.org/sites/default/files/chernobyl.pdf.

Cherp, A., Adenikinju, A., Goldthau, A., Hernandez, F., Hughes, L., et al. (2012). Energy and Security, in: *Global Energy Assessment—Toward a Sustainable Future* (pp. 325–384), Cambridge/Laxenburg: Cambridge University Press/International Institute for Applied Systems Analysis.

China Daily. (2012, April 22). Wen Pledges Co-operation with Iceland in Geothermal Energy, http://www.chinadaily.com.cn/china/2012-04/22/content_15108563.htm.

Christensen, C. (2003). *The Innovator's Solution: Creating and Sustaining Successful Growth*. Cambridge: Harvard Business Press.

Christensen, C., Baumann, H., Ruggles, R., and Sadtler, T. (2006, December). Disruptive Innovation for Social Change, *Harvard Business Review*, https://hbr.org/2006/12/disruptive-innovation-for-social-change.

Churchill, R. (1968). *Winston Churchill, Volume 2, Young Statesman, 1901–1914*. London: Heiman.

Churchill, W. (1928). *The World Crisis, 1911–1918*. London: Penguin.

Churchill, W., and Heath, F. (1965). *Great Destiny: Sixty Years of the Memorable Events in the Life of the Man of the Century Recounted in His Own Incomparable Words*. New York: Putnam.

Cimoli, M., and Dosi, G. (1995). Technological Paradigms, Patterns of Learning and Development: An Introductory Roadmap, *Journal of Evolutionary Economics*, 5, 243–268.

Clark, G., and Jacks, D. (2007). *Coal and the Industrial Revolution, 1700–1869*. European Review of Economic History (pp. 39–72). Cambridge: Cambridge University Press.

Clarke, J., and Ricci, H. (2003). Mass Audubon's Position Statement on Wind Energy Development. http://www.massaudubon.org/our-conservation-work/climate-change/climate-change-policy/wind/position-statement.

Clausen, N. (2008). *Environmental Impacts of Offshore Wind Farms: The Danish Experience*, IEA Task 23—Offshore Wind Energy, Ecology and Regulation, Workshop Petten, Holland, February 28–29, 2008.

Cleveland, C. (Ed.). (2007). Energy Quality, Net Energy and the Coming Energy Transition. Frontiers in Ecological Economic Theory and Application (pp. 268–284). Northampton: Edward Elgar.

Cleveland, C. (Ed.). (2008). Energy Transitions Past and Future, *Encyclopedia of the Earth*. http://www.eoearth.org/article/Energy_transitions_past_and_future.

Cleveland, C. (Ed.). (2009). Concise Encyclopedia of the History of Energy. Salt Lake City: Elsevier.

Cochran, J., Bird, L., Heeter, J., and Arent, D. (2012). Integrating Variable Energy in Electric Power Markets: Best Practices from International Experience, Report, NREL/TP-6A00-53732, Golden: NREL, April 2012.

Coelho, S., and Goldemberg, J. (2004). Alternative Transportation Fuels: Contemporary Case Studies, in: Cleveland, C. (Ed.), *Concise Encyclopedia of the History of Energy*. Salt Lake City: Elsevier.

Coelho, S., and Guardabassi, P. (2014). Brazil: Ethanol, in: Solomon, B., and Bailis, R. (Eds.), *Sustainable Development of Biofuels in Latin America and the Caribbean* (pp. 71–101). New York: Spring Science and Business Media.

Cohen W., and Levinthal, D. (1990). Absorptive Capacity: A New Perspective on Learning and Innovation. *Administrative Science Quarterly* 35: 1, 128–152.

Cole, A. (2008). *Governing and Governance in France*. Cambridge: Cambridge University Press.

Collins, C. (1983, July). Framatome: French Nuclear Monopoly Finds Fertile Ground Abroad, International Monetary Fund, *Multinational Monitor*, 4: 7.

Combs, C. (2010). French Nuclear Power: A Model for the World? *Hinckley Journal of Politics*, 11, http://epubs.utah.edu/index.php/HJP/article/view/305/0.

Coombs, R. (2016, June). China's Energy Transition: Rapid Growth on a Long Road, International Perspectives, Energy Transition, http://energytransition.de/2016/06/regional-perspective-china/.

Commissariat a l'Energie Atomique et Aux Energies Alternatives (CEA). (n.d), Information. http://ww.cea.fr. Accessed January 10, 2017.

Commisariat Général au Développement Durable (CGDD). (2015). Chiffres and Statisiques, No. 683, http://www.developpement-durable.gouv.fr/commissariat-general-au-developpement-durable-cgdd.

Commission of the European Communities (EC). (2008). Commission Staff Working Document: The Support of Electricity from Renewable Energy Sources, http://iet.jrc.ec.europa.eu/remea/commission-staff-working-document-support-electricity-renewable-energy-sources.

Congressional Budget Office (CBO). (2008, May). Nuclear Power's Role in Generating Electricity, Pub. No. 2986, Washington, DC: CBO.

Connors, S., and McGowan, J. (2000). Windpower: A Turn of the Century Review, *Annual Review of Energy and the Environment*, 25: 147–97.

Copersucar. (2012). History. http://www.copersucar.com.br/historico_en.html. Accessed May 10, 2012.

Corbett, M. Oil Shock of 1973–74, Federal Reserve History, http://www.federalreserve-history.org/Events/DetailView/36. Accessed December 10, 2016.

Cornwall, A. (2002). *Beneficiary, Consumer, Citizen: Perspectives on Participation for Poverty Reduction*. Stockholm: Sida Studies.

Cortez, L. (2010). Introduction, in: Cortez, L. (Ed.), *Sugarcane Bioethanol.* São Paolo: Blucher.

Costa, R., and Sodré, J. (2010). Hydrous Ethanol vs. Gasoline-ethanol Blend: Engine Performance and Emissions, *Fuel*, 89: 2, 287–293.

Costanza, R., and Mathias, R. (1998). Using Dynamic Modeling to Scope Environmental Problems and Build Consensus, *Environmental Management*, 22: 183–195.

Cour des Comptes, France (2012, January). Summary of the Public Thematic Report: The Costs of the Nuclear Power Sector. http://www.ccomptes.fr/index.php.

Cowan, R. (1990). Nuclear Power Reactors: A Study in Technological Lock-in, *Journal of Economic History*, 50: 541–567.

Cowen, T. (2002). Public Goods and Externalities, *The Concise Encyclopedia of Economics*. http://www.econlib.org/library/Enc1/PublicGoodsandExternalities.html.

Crease, R. (2004). Energy in the History and Philosophy of Science, in: Cleveland, C. (Ed.), *Encyclopedia of Energy.* Boston: Elsevier.

Creative Commons. (2012). Prime Minister of Iceland. http://en.Wikipedia.org/wiki/Prime_Minister_of_Iceland. Accessed June 10, 2012.

Creste, S., Pinto, L., Xavier, M., and Landell, M. (2010). The Importance of Germoplasm in Developing Agro-Energetic Profile of Sugarcane Cultivars, in: Cortez, L. (Ed.), *Sugarcane Bioethanol.* São Paulo: Blucher.

Cruz, C. (2008). Bioenergy in Brazil, Presentation, São Paulo: FAPESP, http://www.fapesp.br/pdf/5023/PFPMCG_2702_brito.pdf.

Cuff, M. (2016, February 23). Energy Executives Reveal Readiness for Low-Carbon Transition, *Business Green*, http://www.businessgreen.com/bg/analysis/2448205/energy-executives-reveal-readiness-for-low-carbon-transition.

Cummings, C., and Stikova, E. (Eds.). (2007). *Strengthening National Public Health Preparedness and Response to Chemical, Biological, and Radiological Agent Threats.* NATO Security Series, Human and Societal Dynamics, Vol. 20. Amsterdam: IOS Press.

Da Costa, A., Pereira Junior, N., and Aranda, D. (2010). The Situation of Biofuels in Brazil: New Generation Technologies, *Renewable and Sustainable Energy Reviews*, 14, 3041–3049.

Da Costa, M., Cohen, C., and Schaeffer, R. (2007). Social Features of Energy Production and Use in Brazil: Goals for a Sustainable Energy Future, *Natural Resources Forum* 31, 11–20.

Da Cunha, M. (2010). Sustainability of Sugarcane Bioenergy: Socioeconomic Impacts, Presentation, Brasilia, Brasil, November 8–9, 2010.

Da Silva, J., Serra, G., Moreira, J., Concalves, J., and Goldemberg, J. (1978). Energy Balance for Ethyl Alcohol Production from Crops, *Science*, 201, 903.

Dahl Anderson, A. (2008). *Emergence of the Biofuel Sector in Brazil and in the US—The Role of Innovation Policy with Focus on Public Procurement*, Working Paper, Globelics 2008, Tempere, Finland, June 2–13, 2008.

Dahl, C. (2004). *International Energy Markets: Understanding Pricing, Policies & Profits*. Tulsa: Pennwell Corporation.

Dal-Bianco, M., Carneiro, M., Hotta, C., Chapola, R., Hoffmann, H., Garcia, A., and Souza, G. (2012). Sugarcane Improvement: How Far Can We Go? *Biotechnology*, 23, 265–270.

Daly, H. (1996). *Beyond Growth: The Economics of Sustainable Development*. Boston: Beacon.

Dambourg, S., and Krohn, S. (2001). Public Attitudes toward Wind Power, http://ele.aut.ac.ir/~wind/en/articles/surveys.htm.

Damvad Analytics and DWIA. (2016). Branchestatistik for Vindmolleindustrien, Vindmølleindustrien. Frederiksberg: DWIA, http://www.windpower.org/da/fakta_og_analyser/statistik/branchestatistik.html.

Danish Energy Agency (DEA). (Energistyrelsen) (n.d.). Information. https://ens.dk/en. Accessed December 5, 2016.

Danish Energy Agency (DEA). (Energistyrelsen). (2005). Offshore Wind Power: Danish Experiences and Solutions. http://ec.europa.eu/ourcoast/download.cfm?fileID=984.

Danish Energy Agency (DEA). (Energistyrelsen). (2006). Export of Energy Technology and Energy Management in 2004. http://ens.netboghandel.dk/.

Danish Energy Agency (DEA) (Energistyrelsen). (2009). Wind Turbines in Denmark, http://www.ens.dk/sites/ens.dk/files/dokumenter/publikationer/downloads/wind_turbines_in_denmark.pdf.

Danish Energy Agency (DEA) (Energistyrelsen). (2011*a*). Energy Statistics 2010, http://www.ens.dk/en-US/Info/FactsAndFigures/Energy_statistics_and_indicators/Annual%20Statistics/Sider/Forside.aspx.

Danish Energy Agency (DEA) (Energistyrelsen). (2011*b*). The Wind Industry: A Historical Flagship, Memo, May 24, 2011. Reference PCJ/HW/FGN, http://www.ens.dk/sites/ens.dk/files/supply/renewable-energy/wind-power/facts-about-wind-power/key-figures-statistics/endelig_engelsk_notat.pdf.

Danish Energy Agency (DEA) (Energistyrelsen). (2011*c*). Energy in Denmark, Presentation, https://ens.dk/en.

Danish Energy Agency (DEA) (Energistyrelsen). (2012). Technology Data for Energy Plants, DEA, http://www.ens.dk/sites/ens.dk/files/dokumenter/publikationer/downloads/technology_data_for_individual_heating_plants_and_energy_transport.pdf.

Danish Ministry of Climate and Energy (DMCE). (2008). The Danish Example, http://www.ens.dk/en-US/policy/danish-climate-and-energy-policy/behind-the-policies/Thedanishexample/Sider/Forside.aspx.

Danish Ministry of Energy and the Environment. (1997). Background for Energi 21, http://193.88.185.141/Graphics/publikationer/energipolitik/e21dk/dc.htm.

Danish Social Democrats. (2005). A Long Party History, http://s-dialog.dk/A-English+version-A-long-party-history-default.aspx?site=english&func=article.view&menuAction=selectClose&menuID=123075&id=123074.

Danish Wind Energy Association. (2012). Market and Prices, http://www.windpower.org/en/policy/market_and_prices.html.

Danish Wind Industry Association (DWIA). (n.d.). Information, http://www.windpower.org/en/. Accessed December 10, 2016.

Danish Wind Industry Association (DWIA). (2011/2012). The Voice of the Danish Wind Industry. http://www.windpower.org.

Danish Wind Power Association. (2002). Fakta om Vindenergi, M 1. http://www.dkvind.dk/fakta/M1.pdf.

Danish Wind Turbine Owners Association (DWTO). (n.d.). Information. http://www.dkvind.dk/html/eng/eng.html. Accessed December 31, 2016.

Darby, S. (2006). *The Effectiveness of Feedback on Energy Consumption*, Review. Environmental Change Institute, Oxford: Oxford University.

Datamonitor. (2010*a*). *Biofuel Production in Brazil*. Industry Profile, New York: Datamonitor.

David, P. (1985). Clio and the Economics of QWERTY, *American Economic* Review (Papers and Proceedings), 75: 332–337.

David, P. (1986). Understanding the Economics of QWERTY: The Necessity of History, in: Parker, W. (Ed.), *Economic History and the Modern Economist*. Oxford: Oxford University Press.

David, P. (1997). Path Dependence and the Quest for Historical Economics: One More Chorus of the Ballad of QWERTY, University of Oxford Discussion Papers in Economic and Social History, Number 20. http://www.nuff.ox.ac.uk/economics/history/paper20/david3.pdf.

David, P. (2000). Path Dependence, Its Critics and the Quest for 'Historical Economics,' Working paper, All Souls College, Oxford University. http://www-siepr.stanford.edu/workp/swp00011.pdf.

David, P. A. (2001). Path Dependence, Its Critics and the Quest for "Historical Economics", in: Garrouste, P., and Ioannides, S. (Eds.), *Evolution and Path Dependence in Economic Ideas: Past and Present* (pp. 15–40). Cheltenham: Edward Elgar.

Davidsson, E. (1986). The World Bank's Strategy in the Electric Power Sector: Case Study Landsvirkjun, Raw Materials Report, Stockholm, Sweden, 5: 1.

Davis, G. (2001). Development of Technologies to Improve Cold Start Performance of Ethanol Vehicles, Final Report, Department of Consumer and Industry Services, State of Michigan, June 11, 2001, http://www.michigan.gov/documents/CIS_EO_coldstart_AF-E-62_87914_7.pdf.

Davis, G., Heil, E., and Rust, R. (2000). Ethanol Vehicle Cold Start Improvement when Using a Hydrogen Supplemented E85 Fuel, Energy Conversion Engineering Conference and Exhibit, IECEC, *Conference Proceedings*, 1, 303–308.

Davis, M. (2008). Nuclear France: Materials and Sites. http://www.francnuc.org.

De Almeida, E., Bomtempo, J., and de Souza e Silva, C. (2007). The Performance of Brazilian Biofuels: An Economic, Environmental, and Social Analysis, OECD Discussion Paper No. 2007-5, Paris: OECD, December 2007.

De Bruijne, R., van der Palen, L., and Snodin, H. (2005). Concerted Offshore Wind Energy Deployment (COD): Legal and Administrative Issues, SenterNovem, http://www.offshorewindenergy.org/cod/Final_COD_report_legal_frameworks.pdf.

De Gouvello, C. (2010). Brazil Low Carbon Country Case Study, World Bank Group, CEAF, CETESB, COPPE-UFRJ, CPTEC/INPE, EMBRAPA, UFMG, ICONE INICIATIVA VERDE, INT, LOGIT, PLANTAR, UNICAMP, USP, www.siteresources.worldbank.org/BRAZILEXTN/Resources/Brazil_LowcarbonStudy.pdf.

De Lima, J. (2006). A Riqueza é o Saber, Revista *Veja*, p. 96, http://www.veja.abril.com.br/010206/p_096.html.

De Malleray, A. (2011, April 6). Comment la France est Devenue Nucléaire (et Nucléocrate), *Slate*, http://www.slate.fr/story/36491/france-nucleaire-nucleocrate.

De Pracontal, M. (2012, July 7). L'industrie nucléaire n'est plus competitive, Mediapart—France. http://www.worldnuclearreport.org/Mediapart-France-L-industrie.html.

Del Giudice, M. (2008, March). Power Struggle, *National Geographic*. http://ngm.nationalgeographic.com/2008/03/iceland/del-giudice-text.

Del Rio, P. (2014). On Evaluating Success in Complex Policy Mixes: The Case of Renewable Support Schemes, *Policy Sciences* 47: 3, 1–21.

Delmas, M., and Heiman, B. (2001). Government Credible Commitment to the French and American Nuclear Power Industries, *Journal of Policy Analysis and Management*, 20, 433–456.

Demetrius, F. (1990). *Brazil's National Alcohol Program*. Westport: Praegar.

Demirbas, A. (2008). Progress and Recent Trends in Biodiesel Fuels, *Energy Conversion and Management*, 50: 1, 14–34.

Devereaux, C., and Lee, H. (2009). Biofuels and Certification, Discussion Paper 2009-07, Workshop at Harvard Kennedy School, Belfer Center for Science and International Affairs, Cambridge, MA, June 2009.

Devictor, N. (2015). Status of the ASTRID Project, Presentation, ESNII+ 1st Biennial Conference, Brussels, Belgium, March 17, 2015.

Devine-Wright, P., and Devine-Wright, H. (2006). Social Representations of Intermittency and the Shaping of Public Support for Wind Energy in the UK, *International Journal of Global Energy Issues*, 25: 3–4, 243–256.

Diamond, J., and Robinson, J. (2010). *Natural Experiments of History*. Cambridge: Belknap.

Diario Official da Uniao. (1979). Decreto/Decree 83700, May 7, 1979, Brasilia: Government of Brazil.

Dijk, M., Kemp, R., and Valkering, P. (2013). Incorporating Social Context and Co-evolution in an Innovation Diffusion Model with an Application to Cleaner Vehicles, *Journal of Evolutionary Economics*, 23: 2, 295–329.

DONG Energy (n.d.). Information. http://www.dongenergy.com/en. Accessed December 10, 2016.

DONG Energy, Energinet and Vattenfall. (2010). Livscyklusvurdering: Dansk el og kraftvarme (LCA of Danish Electricity and Cogeneration), Report, April 2010.

http://www.energinet.dk/SiteCollectionDocuments/Danske%20dokumenter/Klimaogmiljo/LCA%20rapport%20-%20Dansk%20el%20og%20kraftvarme%20 2008.pdf.

Doran, J. (2001). Agent-Based Modeling of Ecosystems for Sustainable Resource Management, in: Luck, M. et al. (Eds.), Multi-Agent Systems and Applications, Vol. 2086, Lecture Notes in Computer Science (pp. 383–403). New York: Springer.

Dorian, J., Frannsen, H., and Simbeck, D. (2006). Global Challenges in Energy, *Energy Policy*, 34: 15: 1984–1991.

Dormois, J. (2004). *The French Economy in the Twentieth Century*. Cambridge: Cambridge University Press.

Dornelles, R. (n.d). The Past and Present: Ethanol Production in Brazil, unpublished manuscript.

Dosi, G. (1982). Technological Paradigms and Technological Trajectories, *Research Policy*, 11: 3, 147–162.

Dosi, G., and Lovallo, D. (1997). Rational Entrepreneurs or Optimistic Martyrs? in: Garud, R., Nayyar, P., and Shapira, Z. (Eds.), *Technological Innovation: Oversights and Foresights* (pp. 41–68). Cambridge: Cambridge University Press.

Dosi, G., Freeman, C., Silverberg, G., and Soete, L. (Eds.). (1988). *Technical Change and Economic Theory*. London: Pinter.

Dosi, G., Malerba, F., Ramello, F., and Silva, F. (2006). Information, Appropriability and the Generation of Innovative Knowledge Four Decades after Arrow and Nelson, *Industrial and Corporate Change*, 15: 6, 891–901.

Dougherty, C. (2006). *Prometheus*. London: Taylor and Francis.

Dreicer, M., and Alexakhin, R. (1996). Post-Chernobyl Scientific Perspectives, Environmental Consequences, *IAEA Bulletin*, 38: 3, 25–26. https://www.iaea.org/sites/default/files/publications/magazines/bulletin/bull38-3/38305192426.pdf.

Du, Y., and Parsons, J. (2009, May). Update on the Cost of Nuclear Power, Center for Energy and Environmental Policy Research. http://www.web.mit.edu/jparsons/www/publications/2009-004.pd.

Duffy, R. (2004). Nuclear Power, History of, in: Cleveland, C. (Ed.), *Encyclopedia of Energy* (pp. 395–408). New York: Elsevier.

Dunn, S. (2000, November/December).The Hydrogen Experiment, *World Watch*, 13: 6.

Dykes, K. (2013). Wind Actor-Networks in Denmark, in: Maegaard, P., Krenz, A., and Palz, W. (Eds.), *Wind Power for the World, The Rise of Modern Wind Energy*. Singapore: Pan Stanford.

Earley, J., and McKeown, A. (2009). Red, White and Green, BWP 180, Washington DC: WorldWatch.

Eco-energy. (2011). Exporting Ethanol: Market Opportunities and Global Supply, Southeast Bioenergy Conference, Tiftin, Georgia, August 9, 2011.

Economist (2008, September 29). Testing Metal, *Economist*. http://www.economist.com/node/12323257.

Edquist, C. (Ed.). (1997). *Systems of Innovation: Technologies, Institutions, and Organizations*. London: Pinter.

Edquist, C. (2005). Systems of Innovation: Perspectives and Challenges, in: Fagerburg, J., and Mowery, D. (Eds.), *Oxford Handbook of Innovation*. Oxford: Oxford University Press.

Edquist, C., and Johnson, B. (1997). Institutions and Organizations in Systems of Innovation, in: Edquist, C. (Ed.), *Systems of Innovation: Technologies, Institutions and Organizations*. London: Pinter.

Edwards, P., Jackson, S. Bowker, G., and Knobel, C. (2007). Understanding Infrastructure: Dynamics, Tensions, and Design, Report, Workshop on 'History & Theory of Infrastructure: Lessons for New Cyberinfrastructures,' University of Michigan, School of Information, January 2007.

Einarsson, G. (1991). Geothermal Energy in Iceland: The Resource and Its Utilization, Presentation by the Special Advisor to the Ministry of Industry, Energy and Commerce, Reykjavik, Iceland.

Eisentraut, A. (2010). Sustainable Production of Second Generation Biofuels. Paris: OECD/IEA. https://www.iea.org/publications/freepublications/publication/second_generation_biofuels.pdf.

Eisentraut, A. (2011). Outlook for Biofuels in the Medium- and Long-Term, IEA Presentation on Medium-Term Oil and Gas Markets, Antwerp, Netherlands.

Electric Auto Association. (2011). EV History, http://www.electricauto.org/?page =EVHistory.

Electricite de France (EDF). (n.d.). Information, https://www.edf.fr/. Accessed August 1, 2015.

Elias, R., and Victor, D. (2005). Energy Transitions in Developing Countries, Working Paper 40, Program on Energy and Sustainable Development, Stanford University, CA.

Ellenbogen, J., Grace, S., Heiger-Bernays, W., Manwell, J., Mills, D., et al. (2012). Wind Turbine Health Impact Study: Report of Independent Expert Panel Prepared for the Massachusetts Department of Environmental Protection and Massachusetts Department of Public Health, January 2012, http://www.mass.gov/eea/docs/dep/energy/wind/turbine-impact-study.pdf.

Emberson, L., He, K., Rockström, J., Amann, M., Barron, J., et al. (2012). Energy and Environment, *Global Energy Assessment—Toward a Sustainable Future* (pp. 191–254). New York/Laxenburg: Cambridge University Press/International Institute for Applied Systems Analysis.

Encyclopedia Britannica. (n.d.). Sievert (Sv). http://www.britannica.com/EBchecked/topic/543504/sievert-Sv. Accessed December 31, 2016.

Energiministeriet. (1981). Energiplan 81, Energiministeriet. http://www.ens.dk/sites/ens.dk/files/dokumenter/publikationer/downloads/energiplan_81.pdf.

Energinet. (n.d.). Information, http://www.energinet.dk/. Accessed August 1, 2016.

Energinet and Energistyrelsen. (2012). Technology Data for Energy Plants, Report, http://www.energinet.dk/SiteCollectionDocuments/Danske%20dokumenter/Forskning/Technology_data_for_energy_plants.pdf.

Energinet. (2009*a*). Wind Power to Combat Climate Change. How to Integrate Wind Energy into the Power System, Fredericia: Energinet, http://www.e-pages.dk/energinet/126/html5/.

Energinet. (2009*b*). *Wind Power Integration into the Market*. Fredericia: Energinet.

Energinet. (2012). Annual Report 2011, Fredericia: Energinet, http://www.energinet.dk/EN/OM-OS/Nyheder/Sider/Energinetdk%27s-aarsrapport-2011.aspx.

Energinet. (2015). System Plan 2015. Doc. 15/02781-56, Fredericia: Energinet, www.energinet.dk/systemplan-2015.

Energy and Environmental Management (EAEM). (2011, October 25). IEA Chief Says Scrap Fossil Fuel Subsidies or Face Catastrophe, *EAEM News*. http://www.eaem.co.uk/news/iea-chief-says-scrap-fossil-fuel-subsidies-or-face-catastrophe.

Energy Information Administration (EIA). (2010). Biofuels, in: Yacobucci, B., and Schnepf, R. (Eds.), *Energy: Ethanol*. Government Series. Alexandria: The Capital.Net, Inc.

Energy Information Administration (EIA). (2011). International Energy Outlook. Washington, DC: US Department of Energy.

Energy Information Administration (EIA). (2012). Today in Energy, DOE EIA, http://www.eia.gov/todayinenergy/detail.cfm?id=4550.

Energy Information Administration (EIA). (2015, September 17). Coal Use in China is Slowing, EIA. http://www.eia.gov/todayinenergy/detail.cfm?id=22972.

Energy Transition. (2012). Origin of the term "Energiewende". The German Energiewende, http://energytransition.de/2012/09/first-subchapter-chapter-4/.

English, A. (2008, July 25). Ford Model T Reaches 100, *The Telegraph*, http://www.telegraph.co.uk/motoring/news/2753506/Ford-Model-T-reaches-100.html.

Environmental Defense Fund (EDF). (2015). Annual Report 2014, New York: EDF, https://www.edf.org/.

Environmental Protection Agency (EPA). (n.d.). Information, https://www.epa.gov/. Accessed June 30, 2016.

Environmental Protection Agency (EPA). (2011*a*). EPA Proposes Air Pollution Standards for Oil and Gas Production/Cost-Effective, Flexible Standards Rely on Operators' Ability to Capture and Sell Natural Gas that Currently Escapes, Threatens Air Quality, http://yosemite.epa.gov/opa/admpress.nsf/bd4379a92ceceeac8525735900400c27/8688682fbbb1ac65852578db00690ec5.

Environmental Protection Agency (EPA). (2011*b*). Biofuels and the Environment: The First Triennial Report to Congress, EPA/600/R-10/183F, Washington, DC: U.S. Environmental Protection Agency.

Euroelectric (2003). Efficiency in Electricity Generation, Report, Preservation of Resources Working Group's Upstream Sub-Group in Collaboration with VGB, July 2003, http://www.euroelectric.org.

Euronews. (2012, March 11). Anti-Nuclear Demos across Europe on Fukushima Anniversary, Euronews, http://www.euronews.com/2012/03/11/anti-nuclear-demos-across-europe-on-fukushima-anniversary/.

European Commission. (2007). Energy Technologies: Knowledge, Perception, Measures, *Special Eurobarometer* 262, Wave 65.3.

European Commission. (2010). Communication from the Commission to the European Parliament, The Council, The European Economic and Social Committee and the Committee of the Regions, Energy 2020: A Strategy for Competitive, Sustainable and Secure Energy, http://eur-lex.europa.eu/legal-content/EN/TXT/?uri=celex: 52010DC0639.

European Community (EC). (2010). Report from the Commission on Indirect Land Use Change Related to Biofuels and Bioliquids, http://ec.europa.eu/energy/renewables/biofuels/land_use_change_en.htm.

European Environment Agency (EEA). (n.d.). Combined Heat and Power, http://www.eea.europa.eu/data-and-maps/indicators/combined-heat-and-power-chp-1/combined-heat-and-power-chp-2. Accessed December 31, 2016.

European Wind Energy Association (EWEA). (n.d.). Information. http://www.ewea.org/. Accessed December 15, 2016.

European Wind Energy Association (EWEA). (2009). *Wind Energy—The Facts: A Guide to the Technology, Economics and Future of Wind Power*. London: Earthscan.

European Wind Energy Association (EWEA). (2010). Wind Energy Factsheets, http://www.oddzialywaniawiatrakow.pl/upload/file/414.pdf.

European Wind Energy Technology Platform (EWETP). (2008). Henning Kruse, Interview, Wind Directions, EWETP, http://www.windplatform.eu/.

Evans, P. (1979). *Dependent Development: The Alliance of Multinational, State, and Local Capital in Brazil*, Princeton: Princeton University Press.

Ezzati, M., Bailis, R., Kammen, D., Holloway, T., Price, L., et al. (2004). Energy Management and Global Health, *Annual Review of Environment and Resources*, 29: 383–419.

Fagel, M. (2014). *Crisis Management and Emergency Planning*. Boca Raton: CRC Press.

Fagerburg, J. (2005). A Guide to the Literature and Introduction, in: Fagerburg, J., Mowery, D., and Nelson, R. (Eds.), *Oxford Handbook of Innovation*. Oxford: Oxford University Press.

Fagnani, J., and Moatti, J. (1984). The Politics of French Nuclear Development, *Journal of Policy Analysis and Management*, 3: 2, 264–275.

Fargione, J., Hill, J., Tilman, D., Polasky, S., and Hawthorn, P. (2008, February 29). Land Clearing and the Biofuel Carbon Debt, *Science*, 319: 5867, 1235–1238.

Faro, A. (2013). The Legalities of Leaving Nuclear, *Bulletin of the Atomic Scientists*, 69: 1, 36–42.

Feingold, R., and Overcast, E. (2015, November 12). Utilities Look for More Nimble Regulatory Practices as Electricity Industry Evolves, *Breaking Energy*.

Felder, F. (2009). A Critical Assessment of Energy Accident Studies, *Energy Policy*, 37: 12, 5744–5751.

Feng, H. (2016, April 7). China Puts an Emergency Stop on Coal Power Construction, *The Diplomat*, http://thediplomat.com/2016/04/china-puts-an-emergency-stop-on-coal-power-construction/.

Fil, N. (2011). Status and Perspectives of VVER Nuclear Power Plants, Meeting of the TWG-LWR, LWRIAEA Headquarters, Vienna, Austria, July 26–28, 2011.

Finamore, B. (2016, January 6). What China's Second Red Alert Means for the Future of Clean Energy, *Fortune*, http://fortune.com/2016/01/06/china-red-alert-clean-energy/.

Finch, C. (2011, September 28). An Inside Look at the Current State of the Nuclear Industry, *Inc.*, http://www.inc.com/tech-blog/an-inside-look-into-the-nuclear-industry.html.

Finon, D., and Staropoli, C. (2001). Institutional and Technological Co-evolution in the French Electronuclear Industry, *Industry & Innovation*, 8: 2, 179–199.

Fischer, C. (2013). *Municipal Waste Management in Germany*, Working Paper, ETC/SCP, European Environment Agency, February 2013.

Fischer, F. (2000). *Citizens, Experts, and the Environment: The Politics of Local Knowledge*. Durham: Duke University Press.

Fisher, D. (2011, May 31). Japan Disaster Shakes up Global Supply Chain, Research and Ideas, Boston: Harvard Business School, http://hbswk.hbs.edu/item/japan-disaster-shakes-up-supply-chain-strategies.

Fisher, J., and Pry, R. (1970). *A Simple Substitution Model of Technological Change*. 70-C-215. Schenectady: General Electric Company.

Flak, A. (2011, September 15). South Africa Forges Ahead on Nuclear Energy Plan, *The Globe and Mail*, http://www.theglobeandmail.com/report-on-business/international-business/african-and-mideast-business/south-africa-forges-ahead-on-nuclear-energy-plan/article4182628/.

Flavin, C. (2008). Low-Carbon Energy: A Roadmap. *WorldWatch Report 178*. Washington, DC: WorldWatch Institute.

Fleten, S., and Ringen, G. (2009). New Renewable Electricity Capacity under Uncertainty: The Potential in Norway, *Journal of Energy Markets*, 2: 1.

Flipo, F. (2008). Energy: Prometheus Bound or Unbound? A Conceptual Approach. *Sapiens*, 1: 2. http://sapiens.revues.org/248.

Flynn, P. (1978). *Brazil: A Political Analysis*. London: Westview.

Foasso, C. (2003). Histoire de la sûreté de l'énergie nucléaire civile en France (1945–2000). Technique d'ingénieur, processus d'expertise, question de société. PhD Thesis in Contemporary History, Université Lumière-Lyon II.

Folketinget. (2011). The Parliamentary Electoral System in Denmark. Guide to the Danish Electoral System, November 2011, http://www.thedanishparliament.dk/Publications/The%20Parliamentary%20Electoral%20System%20in%20DK.aspx.

Folketinget. (n.d). The Parliamentary System of Denmark. Introduction to Danish Democracy, Folketinget, http://www.thedanishparliament.dk/Publications/The%20Parliamentary%20System%20of%20Denmark.aspx. Accessed December 1, 2015.

Food and Agriculture Organization (FAO). (2008). *The State of Food and Agriculture, Biofuels: Prospects, Risks and Opportunities*. Rome: United Nations.

Foxon, T. (2013). Transition Pathways for a UK Low Carbon Electricity Future, *Energy Policy*, 52, 10–24.

Foxon, T. (2011). A Coevolutionary Framework for Analysing a Transition to a Sustainable Low Carbon Economy, *Ecological Economics*, 70, 2258–2267.

Foxon, T., Hammond, G., Pearson, P. (2010). Developing Transition Pathways for a Low Carbon Electricity System in the UK, *Technological Forecasting and Social Change*, 77, 1203–1213.

Fouquet, R. (2010). The Slow Search for Solutions: Lessons from Historical Energy Transitions by Sector and Service, *Energy Policy*, 38: 11, 6586–6596.

Fouquet, R. (2011). Long Run Perspectives on Energy and Climate Change, Presentation, Bilbao Centre, July 22, 2011.

Fouquet, R. (2016). Historical Energy Transitions, *Energy Research and Social Science*, 22, 7–12.

Fouquet, R., and Pearson, P. (1998). A Thousand Years of Energy Use in the United Kingdom, *The Energy Journal*, International Association for Energy Economics, 19: 4, 1–42.

Fouquet, R., and Pearson, P. (2012). Past and Prospective Energy Transitions, *Energy Policy*, 50, 1–7.

Fouraustie, J. (1979). *Les Trente Glorieuses ou la révolution invisible de 1946 à 1975*. Paris: Fayard.

Fourquet, J., and Pratviel, E. (2013, June 23). Les Français et le Nucléaire, Résultats détaillés, JF/EP No. 111359, Paris: IFOP.

Framatome. (1984). Pressurized Water Reactors: Towards Improved Maneuverability, Report, Paris: Framatome.

Freeman, C. (1987). *Technology Policy and Economic Performance*. London: Pinter.

Freeman, C., and Louçã, F. (2002). *As Time Goes By: From the Industrial Revolutions to the Information Revolution*. New York: Oxford University Press.

Freeman, C., and Perez, C. (1988). Structural Crises of Adjustment, Business Cycles and Investment Behaviour, in: Dosi, G. (Ed.), *Technical Change and Economic Theory*. London: Pinter.

Freeman, G. (1985, October). National Styles and Policy Sectors: Explaining Structured Variation, *Journal of Public Policy*, 5: 4, 467–496.

Fridleifsson, I. (2010). Thirty One Years of Geothermal Training in Iceland, World Geothermal Congress 2010, Bali, Indonesia.

Frideifsson, I., Palsson, B., Albertsson, A., Stefansson, B., Gunnlaugsson, E., et al. (2015). IDDP-1 Drilled in to Magma – World's First Magma-EGS System Created, Prceedings World Geothermal Congress 2015, Melbourne, Australia, April 19–25, 2015.

Fridleifsson, I., and Elders, W. (2007, March). Progress Report on the Iceland Deep Drilling Project (IDDP), *Scientific Drilling*, 4.

Fridleifsson, I., and Ragnarsson, A. (2007). *Geothermal Energy, 2007 Survey of Energy Resources* (pp. 427–437). London: World Energy Council.

Fridley, D. (2010). *Nine Challenges of Alternative Energy*. The Post Carbon Reader Series. Santa Rosa: Post Carbon Institute.

Friedlander, B. (2016, May 12). Cornell and Iceland Team to Model Geothermal Energy, *Cornell Chronicle*, http://www.news.cornell.edu.

Fuel Testers. (n.d.). Ethanol Fuel History. http://www.fuel-testers.com/ethanol_fuel_history.html.

Fundaçao Getulio Vargas (FGV). (2008). Fatores Determinantes dos Preços dos Alimentos: O Impacto dos Biocombustíveis. http://bibliotecadigital.fgv.br/dspace/bitstream/handle/10438/6947/326.pdf?sequence=1.

Furman, J., Porter, M., and Stern, S. (2002). The Determinants of National Innovative Capacity, *Research Policy*, 31, 899–933.

Furtado, A. (2010). Sustainability of Sugarcane Bioenergy: Production Expansion and Evolution of Sustainability Indicators, Presentation, Brasilia, Brazil, November 8–9, 2010.

Furtado, A., Scandiffio, M., and Cortez, L. (2011). The Brazilian Sugarcane Innovation System, *Energy Policy*, 39, 156–166.

G8 and European Energy Commissioner. (2009, May 24-25). Joint Statement, Rome, Italy. http://www.g8.utoronto.ca/energy/090525_energy-g8+ec.pdf.

Gallagher, K., Holdren, J., and Sagar, A. (2006). Energy-Technology Innovation, *Annual Review of Environment and Resources*, 31, 193–237.

Galston, W. (2006). Political Feasibility: Interests and Power, in: Moran, M., Rein, M., and Goodin, R. (Eds.), *The Oxford Handbook of Public Policy*. Oxford: Oxford University Press.

Galucci, M. (2014, June 16). China 'Energy Revolution' Needed to Meet Soaring Energy Demand and Cut Pollution Problems, President Xi Says, *International Business Times*, http://www.ibtimes.com/china-energy-revolution-needed-meet-soaring-energy-demand-cut-pollution-problems-1602352.

Garcesz, C. (n.d.). Cultivating Sustainable Development: An Analysis of the Brazilian Public Policy for Biodiesel within the Context of Sustainable Development and Environmental Management, Paper, University of Brasilia, Brasil.

Garud, R., and Karnøe, P. (2001). Path Creation as a Process of Mindful Deviation, in: Garud, R., and Karnoe, P. (Eds.), *Path Dependence and Creation in the Danish Wind Turbine Field* (chapters 1–2). London: LEA Associates.

Garud, R., and Karnøe, P. (2003). Bricolage vs. Breakthrough: Distributed and Embedded Agency in Technology Entrepreneurship. *Research Policy*, 32, 277–300.

Garud, R., and Karnøe, P. (2005). Distributed Agency and Interactive Emergence, in: Floyd, S., Roos, J., Jacobs, C., and Kellermanns, F. (Eds.), *Innovating Strategy Process* (pp. 88–96). Malden: Blackwell.

Garud, R., Kumaraswamy, A., and Karnoe, P. (2010, June). Path Dependence or Path Creation? *Journal of Management Studies*, 47: 4, 760–774.

Gatti Jr., W. (2010). 35 Anos da Criação do ProÁlcool: Do Álcool Motor ao Veículo Flex Fuel, Seminarios em Adminstração, XIII Semead.

Gazaffi, R., Olveira, K., Souza, A., and Garcia, A. (2010). Sugarcane, in: Cortez, L. (Ed.), *Sugarcane Bioethanol*. São Paulo: Blucher.

Geddes, B. (1990). How the Cases You Choose Affect the Answers You Get: Selection Bias in Comparative Politics, *Political Analysis*, 2: 1, 131–150.

Geels, F. (2002). Technological Transitions as Evolutionary Reconfiguration Processes: A Multi-level Perspective and a Case-study, *Research Policy*, 31: 8, 1257–1274.

Geels, F. (2005). Processes and Patterns in Transitions and System Innovations: Refining the Co-evolutionary Multi-Level Perspective, *Technological Forecasting and Social Change*, 72: 6, 681–696.

Geels, F. (2006). Multi-level Perceptive on System Innovation: Relevance for Industrial Transformation, in: Olshoorn, X., and Wieczorek, A. (Eds.), *Understanding Industrial Transformation: Views from Different Disciplines*. Berlin: Springer.

Geels, F. (2011). The Multi-level Perspective on Sustainability Transitions: Responses to Seven Criticisms, *Environmental Innovation and Societal Transitions*, 1: 1, 24–40.

Geels, F., and Schot, J. (2007). Typology of Socio-technical Transition Pathways, *Research Policy*, 36: 3, 399–417.

Geeman, B. (2015, June 1). Big Oil Companies Want a Price on Carbon. Here's Why. *National Journal*, http://www.theatlantic.com/politics/archive/2015/06/big-oil-companies-want-a-price-on-carbon-heres-why/446637/.

Gekon (n.d.).Information. http://www.gekon.is/en/. Accessed November 30, 2016.

General Accounting Office (GAO) (1990, October). US-NATO Burden-Sharing, Allies Contributions to Common Defense During the 1980s, Report to the Chairman, Committee on Armed Services, House of Representatives. Washington, DC: GAO.

Generation IV Forum (GIF). (n.d.). Overview of the GIF and GEN IV, http://www.gen-4.org/Technology/evolution.htm. Accessed November 10, 2016.

Geothermal Energy Association (GEA). (2005). Factors Affecting the Costs of Geothermal Power Development, http:/geo-energy.org/reports/Factors%20Affecting %20Cost%20 of%20Geothermal%20Power%20Development%20-%20August%202005.pdf.

German Organization for Technical Cooperation (GTZ). (2007). Energy-Policy Framework Conditions for Electricity Markets and Renewable Energies: 23 Country

Analyses, Stuttgart: GTZ, September 2007, http://www.ifa.de/en/culture-and-foreign-policy/organisations/organisations/german-organisation-for-technical-cooperation-gtz.html.

Gettier, E. (1963). Is Justified True Belief Knowledge? *Analysis*, 23, 121–123. http://www.ditext.com/gettier/gettier.html.

Geysir Green Energy. (n.d.). Iceland: The Leader, http://www.geysirgreenenergy.com/geothermal/iceland-the-leader/. Accessed December 10, 2016.

Gibbons, J., and Gwin, H. (2009). History of Conservation Measures in Energy, in: Cleveland, C. (Ed.), *History of Energy*. Boston: Elsevier.

Giddens, A. (1979). *Central Problems in Social Theory*. Los Angeles: University of California Press.

Giddens, A. (1984). *The Constitution of Society: Outline of the Theory of Structuration*. Cambridge: Polity.

Giddens, A. (1989). A Reply to My Critics, in: Held, D., and Thompson, J. (Eds.), *Social Theory of Modern Societies: Anthony Giddens and His Critics* (pp. 249–301). Cambridge: Cambridge University Press.

Giebelhaus, A. (2004). History of the Oil Industry, in: C. Cleveland (Ed.), *Encyclopedia of Energy*. San Diego: Elsevier.

Gipe, P. (1995). *Wind Energy Comes of Age*. New York: Wiley.

Girard, P., Marignac,Y., Tassard, J. (G–M–T) (2000). Le Parc Nucle´aire Actuel. Groupe du travail Cycle Nucle´aire, Mission d'E´valuation E´conomique de la Filiere Nucle´aire, http://www.ladocumentationfrancaise.fr/rapports-publics/014000107/index.shtml.

Glaser, B., and Strauss, A. (1967). *Discovery of Grounded Theory*. Sociology Press, http://www.sociologypress.com/book.htm.

Glennie, K. (Ed.). (1998). *Petroleum Geology of the North Sea: Basic Concepts and Recent Advances* (4th Edition). Malden: Blackwell Science.

Global Security. (n.d.). La Hague, http://www.globalsecurity.org/wmd/world/france/la_hague.htm. Accessed December 15, 2016.

Global Wind Energy Council (GWEC). (n.d.). Information, http://www.gwec.net/. Accessed December 15, 2016.

Global Wind Energy Council (GWEC) (2015, 2016). Global Wind Report, Annual Market Update, https://www.gwec.net.

Global Wind Energy Council (GWEC) and International Renewable Energy Agency (IRENA). (2012). 30 Years of Policies for Wind Energy: Lessons from 12 Wind Energy Markets. Abu Dhabi: IRENA. http:// www.irena.org/ menu/ index.aspx? mnu =Subcat&PriMenuID=36&CatID=141&SubcatID=281.

Golay, M. (1991). New Technologies: A Policy Framework for Micro-Nuclear Technologies. Houston, TX: James Baker Institute for Public Policy, Rice University.

Golay, M., Saragossi, I., and Willefert, J. (1977). Comparative Analysis of United States and French Nuclear Power Plant Siting and Construction Regulatory Policies and Their Economic Consequences, Energy Laboratory Report No. MIT-EL 77-044-WP, December 1977, https://dspace.mit.edu/handle/1721.1/31297.

Goldberg, S., and Rosner, R. (2011). *Nuclear Reactors: Generation to Generation*. Cambridge: Academy of Arts and Sciences [AAAS].

Goldemberg, J. (unpublished). Sugarcane Ethanol: Strategies for a Successful Program in Brazil, University of São Paulo, Brazil.

Goldemberg, J. (1998). Leapfrog Energy Technologies, *Energy Policy*, 26: 10, 729–742.

Goldemberg, J. (2004*a*). Development and Energy, Overview, in: Cleveland, C. (Ed.), *Encyclopedia of Energy*. Boston: Elsevier.

Goldemberg, J. (2004b). The Case for Renewable Energies, Paper, International Conference for Renewable Energies, Bonn, Germany.

Goldemberg, J. (2006*a*). The Ethanol Program in Brazil, *Environmental Research Letters*, 1: 1.

Goldemberg, J. (2006*b*). The Promise of Clean Energy, *Energy Policy*, 34, 2185–2190.

Goldemberg, J. (2007). Ethanol for a Sustainable Energy Future, *Science*, 315, 808–810.

Goldemberg, J. (2008). The Brazilian Biofuels Industry, *Biotechnol Biofuels*, 1, 6, http://biotechnologyforbiofuels.biomedcentral.com/articles/10.1186/1754-6834-1-6.

Goldemberg, J. (2009, Fall). The Brazilian Experience with Biofuels, *Innovations*, 4: 4.

Goldemberg, J. (2010). The State of São Paulo Strategy for Fuel Ethanol, in: Cortez, L. (Ed.), *Sugarcane Bioethanol*. São Paulo: Blucher.

Goldemberg, J (2011, September 20). Perspective on Brazil's Bioethanol Program, Presentation. Harvard University, Cambridge, MA.

Goldemberg, J. (2013). Sugarcane Ethanol: Strategies for a Successful Program in Brazil, in: Lee, J. (Ed.), *Advanced Biofuels and Bioproducts* (pp. 13–20). New York: Springer.

Goldemberg, J., Johansson, T., Reddy, A., and Williams. R. (1988). *Energy for a Sustainable World*. New Delhi: Wiley Eastern Ltd.

Goldemberg, J., and Lucon, O. (2010). *Energy, Environment and Development*. London/Sterling: Earthscan.

Goldemberg, J., Monaco, L., and Macedo, I. (1993). The Brazilian Fuel-Alcohol Program, in: Johansson, T. et al. (Eds.), *Renewable Energy: Sources for Fuels and Electricity*. Washington, DC: Island Press.

Goldemberg, J., Coelho, S., Nastari, P., and Guardabassi, P. (2014). Production and Supply Logistics of Sugarcane as an Energy Feedstock, in: Lijun Wang (Ed.), *Sustainable Bioenergy Production* (pp. 213–222). Baco Raton: CRC Press.

Goldemberg, J., and Tadeo Prado, L. (2010). The 'Decarbonization' of the World's Energy Matrix, *Energy Policy*, 38: 7, 3274–3276.

Goldman, D., McKenna, J., and Murphy, L. (2005). *Financing Projects that Use Clean Energy Technologies: An Overview of Barriers and Opportunities*, Report, NREL/TP-600–38723, Golden: NREL, October 2005.

Goldschmidt, B. (1989). The Early French Program: France, in: Behrens, J., and Carlson, A. (Eds.), *50 Years with Nuclear Fission*. La Grange Park: American Nuclear Society.

Goldschmidt, B. (1962). France's Contribution to the Discovery of the Chain Reaction, *IAEA Bulletin*, 4, 21–24, https://www.iaea.org/sites/default/files/publications/magazines/bulletin/bull4-0/04004782124su.pdf.

Goldschmidt, B. (1978). International Cooperation in the Nuclear Field—Past, Present and Prospects, *IAEA Bulletin*, 20: 2, 13–24.

Goldschmidt, B. (2006). When the IAEA Was Born, *IAEA Bulletin*, 48: 1, 6–10. https://www.iaea.org/sites/default/files/publications/magazines/bulletin/bull48-1/48101280610.pdf.

Goldstein, B., Hiriart, G., Bertani, R., Bromley, C., Gutierrez-Negrin, L., et al. (2012). Geothermal Energy, in: Edenhofer, O., Pichs-Madruga, Y., Sokona, K., Seyboth, P., Matschoss, S., et al. (Eds.), *IPCC Special Report on Renewable Energy Sources and Climate Change Mitigation*. Cambridge: Cambridge University Press.

Gorbatchev, A., Mattei, J., Rebour, V., and Vial, E. (2001). Report on Flooding of Le Blayais Power Plant on 27 December 1999, Eurosafe 2000—Challenges Arising to Nuclear Safety in the Context of Liberalization of the Electricity Markets, Papers. Germany. https://www.eurosafe-forum.org/sites/default/files/pe_297_24_1_sem1_1.pdf.

Gordon, R. (2008). The Case Against Government Intervention in Energy Markets, Policy Analysis No. 628, Cato Institute, December 1, 2008.

Gouvernement (Republique Francaise) (n.d.), http://www.gouvernement.fr.

Graeber, D. (2015, October 16). Oil and Gas Companies Make Green Pledge, *UPI*. http://www.upi.com/Business_News/Energy-Industry/2015/10/16/Oil-and-gas-companies-make-green-pledge/8191444992172/.

Gravelle, H., and Rees, R. (2004). *Microeconomics*. Westwood: Prentice Hall.

Grayson, L. (1981). *National Oil Companies*. Chichester: Wiley.

Green Car Congress. (2006, May 19). Petrobras Develops Hydrogenation Process to Produce Diesel Fuel with Vegetable Oil, http://www.greencarcongress.com/2006/05/petrobras_devel.html.

Greene, N. (2009). Independent Peer Review Confirms EPA's Approach to Biofuels, NRDC, August 10, 2009, https://www.nrdc.org/experts/nathanael-greene/independent-peer-review-confirms-epas-approach-biofuels.

Greenwald, J. (1986, May 12). Deadly Meltdown, *Time Magazine*.

Greenwald, J., McWhirter, W., and Traver, N. (1986, June 2). Energy and Now, the Political Fallout, *Time Magazine*. http://content.time.com/time/magazine/article/0,9171,961509,00.html.

Grin, J., Rotmans, J., and Schot, J. (in collaboration with Geels, F., and Loorbach, D.). (2010). *Transitions to Sustainable Development: New Directions in the Study of Long Term Transformative Change*. New York/London: Routledge.

Grist. (2005, April 12). Arni Finnsson of the Iceland Nature Conservation Association Answers Questions, *Grist*, http://grist.org/article/finnsson.

Gronewold, N. (2009). One Quarter of the World's Population Lacks Electricity, *Scientific American*, http://www.scientificamerican.com/article/electricity-gap-developing-countries- energy-wood-charcoal/.

Grove-Nielsen, E. (n.d.). Winds of Change, http://www.windsofchange.dk/WOC-bladestory.php. Accessed June 10, 2016.

Grubler, A. (1996). Time for a Change: On the Patterns of Diffusion of Innovation, *Daedalus*, 125: 3, 19–42.

Grubler, A. (1997). *Time for a Change: On the Patterns of Diffusion of Innovation, Technological Trajectories and the Human Environment*. Washington, DC: National Academy Press.

Grubler, A. (1998). *Technology and Global Change*. Cambridge: Cambridge University Press.

Grubler, A. (2004). Transitions in Energy Use, in: C. Cleveland (Ed.), *Encyclopedia of Energy*. San Diego: Elsevier.

Grubler, A. (2008*a*). Data, in: C. Cleveland (Ed.), *Encyclopedia of the Earth*. Washington, DC: Environmental Information Coalition, National Council for Science and the Environment.

Grubler, A. (2008*b*). Energy Transitions, in: C. Cleveland (Ed.), *Encyclopedia of the Earth*. Washington, DC: Environmental Information Coalition, National Council for Science and the Environment.

Grubler, A. (2009, October 6). An Assessment of the Costs of the French Nuclear PWR Program: 1970–2000, Laxenburg: International Institute for Applied Systems Analysis, http://pure.iiasa.ac.at/9116/.

Grubler, A. (2010). The Costs of the French Nuclear Scale up: A Case of Negative Learning by Doing, *Energy Policy*, 38, 5174–5188.

Grubler, A. (2012). Energy Transitions Research: Insights and Cautionary Tales. *Energy Policy*, 50 (Special Section: Past and Prospective Energy Transitions—Insights from History), 8–16. doi: 10.1016/j.enpol.2012.02.070

Grubler, A. (2012). Policies for the Energy Technology Innovation System (ETIS), *Global Energy Assessment—Toward a Sustainable Future* (pp. 1665–1744). New York/Laxenburg: Cambridge University Press/International Institute for Applied Systems Analysis.

Grubler, A. (2014*a*). Grand Designs: Historical Patterns and Future Scenarios of Energy Technology Change, in: Grubler, A. and Wilson, C. (Eds.), *Energy Technology Innovation: Learning from Historical Successes and Failures*. Cambridge: Cambridge University Press.

Grubler, A. (2014*b*). The French PWR Programme, in: Grubler, A. and Wilson, C. (Eds.), *Energy Technology Innovation: Learning from Historical Successes and Failures*. Cambridge: Cambridge University Press.

Grubler, A., and Nakicenovic, N. (1996). Decarbonizing the Global Energy System, *Technological Forecasting and Social Change*, 53: 1, 97–110.

Grubler, A. Nakicenovic, N, and Victor, D. (1999*a*). Modeling Technological Change. *Annual Review of Energy and the Environment*, 24, 545–569.

Grubler, A. Nakicenovic, N., and Victor, D. (1999*b*). Dynamics of Energy Technologies and Global Change, *Energy Policy*, 27: 247–280.

Grubler, A., and Wilson, C. (Eds.). (2014). *Energy Technology Innovation: Learning from Historical Successes and Failures*. Cambridge: Cambridge University Press.

Grubler, A., Aguayo, F., Gallagher, K., Hekkert, M., Jiang, K., et al. (2012). Policies for the Energy Technology Innovation System (ETIS), *Global Energy Assessment—Toward a Sustainable Future* (pp. 1665–1744). New York/Laxenburg: Cambridge University Press/International Institute for Applied Systems Analysis.

Guardabassi, P., and Goldemberg, J. (2011). The State of the Art of Ethanol Production from Sugarcane in Brazil, Presentation, Rome, Italy, November 11, 2011.

Guerozoni, M., and Raiteri, E. (2015). Demand-side vs. Supply-side Technology Policies: Hidden Treatment and New Empirical Evidence on the Policy Mix, *Research Policy,* 44: 3, 726–747.

Gudmundsottir, M., Brynjolfsdóttir, A., and Albertsson, A. (2010). The History of the Blue Lagoon in Svartsengi, Proceedings World Geothermal Congress 2010, Bali, Indonesia, April 25–29, 2010.

Guey-Lee, L. (1998). Wind Energy Developments: Incentives in Selected Countries, EIA Renewable Energy Annual 1998, Washington, DC: Energy Information Administration, https://www.eia.gov/.

Gunnlaugsson, E., Frimannson, H., and Sverrisson, G. (2000). District Heating in Reykjavìk—70 Years of Experience, Proceedings World Geothermal Congress, 2087–2092, Kyushu—Tohuku, Japan, May 28–June 10, 2000.

Hadjilambrinos, C. (2000). Understanding Technology Choice in Electricity Industries: A Comparative Study of France and Denmark, *Energy Policy*, 28, 1111–1126.
Hafele, W. (1981). *Energy in a Finite World* (Volume 2). Cambridge: Ballinger.
Hálfdanarson, G. (2008). *Historical Dictionary of Iceland*. Lanham: Scarecrow.
Hálfdanarson, G. (2012). Icelandic Modernity and the Role of Natioanlism, in: Arnason, J., and Wittrock, B. (Eds.). *Nordic Paths to Modernity*. New York: Berghahn.
Halffman, W. (2005). Science-Policy Boundaries: National Styles? *Science and Public Policy*, 32: 6, 457–467.
Hall, J., Matos, S., and Silvestre, B. (2010). Energy Policy, Social Exclusion and Sustainable Development: The Biofuels and Oil & Gas Cases in Brazil, Congress Paper, World Energy Council, http://www.indiaenergycongress.in/montreal/library/pdf/431.pdf.
Hall, S., Foxon. T. J., Bolton, R. (2015). Investing in Low-carbon Transitions: Energy Finance as an Adaptive Market, Climate Policy. http://dx.doi.org/10.1080/14693062.2015.1094731.
Hall, S. Foxon, T., and Bolton, R. (2016). Financing the Civic Energy Sector: How Financial Institutions Affect Ownership Models in Germany and the United Kingdom, *Energy Research and Social Science*, 12, 5–15.
Hall, S., and Roelich, K. (2016). Business Model Innovation in Electricity Supply Markets: The Role of Complex Value in the United Kingdom, *Energy Policy*, 92, 286–298.
Halloway, J. (2012, April 21). Plans Afoot to Send Iceland's Geothermal Energy to Europe, Arstechnica. http://arstechnica.com/tech-policy/2012/04/plans-afoot-to-tap-icelands-geothermal-energy-with-745-mile-cable/.
Hamilton, K. (2008). Clean Energy Investment and the 'New Competitiveness', Paper, London: Chatham House, June 26, 2008.
Hamilton, K. (2009, December). *Unlocking Finance for Clean Energy*, Briefing Paper, EERG BP 2009/06. London: Chatham House.
Hammond, A. (1977). Alcohol: A Brazilian Answer to the Energy Crisis, *Science*, 196: 4278, 564–566.
Han, Y. (2008). Cash Crop: Brazil's Biofuel Leadership, Harvard International Review. https://www.questia.com/library/journal/1G1-184710757/cash-crop-brazil-s-biofuel-leadership.
Hance, C. (2005). Factors Affecting Costs of Geothermal Power Development, Report for the US Department of Energy, Geothermal Energy Association. http://geo-energy.org/reports/Factors%20Affecting%20Cost%20of%20Geothermal%20Power%20Development%20-%20August%202005.pdf.
Hand, M. (2015). Wind Technology, Cost, and Performance Trends in Denmark, Germany, Ireland, Norway, the European Union, and the United States: 2007–2012, Presentation, Copenhagen, Denmark.
Hannum, W., Marsh, G., and Stanford, G. (2007). Recycling Nuclear Waste, American Physical Society Special Session on Nuclear Reprocessing, Nuclear Proliferation, and Terrorism, April 15, 2007.
Hannibalsson, I. (2009). What Do Companies in Iceland Need to Do in Order to Succeed Following the Collapse of the Economy of that Small Country? in: Robinson, L. (Ed.), *Proceedings of the 2009 Academy of Marketing Science Annual Conference*. New York: Springer.

Hansen, L., and Andersen, P. (1999). Wind Turbines: Facts from 20 Years of Technological Progress, Paper, EWEC, March 1–5, 1999, Nice, France.
Haraldsdottir, K. (2010). Introduction to the Legal Environment in Iceland for the Utilization of Geothermal Energy, *Journal of Energy and Natural Resources Law*, 28: 2, 1–47.
Haraldsson I., Thorisdottir, T., and Ketilsson, J. (2010). Efnahagslegur samanburdur hushitunar med jardhita og oliu arin 1970–2009. Report OS 2010/04. Rejkyavik: Orkustofnun.
Hargadon, A., and Douglas, Y. (2001). When Innovations Meet Institutions: Edison and the Design of the Electric Light, *Administrative Science Quarterly*, 46: 3, 476–501.
Harris, J., and Casey-McCabe, N. (1996). Energy Efficient Product Labeling: Market Impacts on Buyers and Sellers, Proceedings of the 1996 ACEEE Summer Study on Energy Efficiency in Buildings, Washington, DC.
Harrison, S., and Mort, M. (1998). Which Champions, Which People? Public and User Involvement in Health Care as a Technology of Legitimation, *Social Policy and Administration,* 32: 1, 60–70.
Hayek, F. (1979). *The Political Order of Free People, Law, Legislation and Liberty*. Chicago: University of Chicago Press.
Hecht, G. (1991). Constructing Competitiveness, in: Aspray, W. (Ed.), *Technological Competitiveness*. Piscataway: IEEE Press.
Hecht, G. (2001). Technology, Politics, and National Identity in France, in: Allen, M., and Hecht, G. (Eds.), *Technologies of Power*. Cambridge: MIT Press.
Hecht, G. (2009). The Radiance of France: Nuclear Power and National Identity after World War II. Cambridge: MIT Press.
Hecht, J. (2007). Can Indicators and Accounts Really Measure Sustainability? Considerations for the US Environmental Protection Agency, Paper, http://www.joy-hecht.net/professional/papers/jhecht-sust-ind&accounts-may07.pdf.
Hekkert, M., Suurs, R., Negro, S., Kuhlmann, S., and Smits, R. (2007). Functions of Innovation Systems: A New Approach for Analysing Technological Change, *Technological Forecasting and Social Change,* 74: 4, 413–432.
Heilbroner, R. (1993). Technological Determinism Revisited, in: Smith, M., and Marx, L. (Eds.), *Does Technology Drive History?* Cambridge: MIT Press.
Helm, D. (2007). *The New Energy Paradigm*. Oxford: Oxford University Press.
Helm, D. (2016, June 1). Green Bonds for Natural Capital, http://www.dieterhelm.co.uk/natural-capital/environment/green-bond-for-natural-capital-some-issues/.
Henderson, R., and Newell, R. (2010/2011). *Accelerating Energy Innovation: Insights from Multiple Sectors*, Working Paper 10–067, Chicago: University of Chicago Press.
Henderson, R., and Cockburn, I. (1994). Measuring Competence? Exploring Firm Effects in Pharmaceutical Research, *Strategic Management Journal*, 15: 8, Winter Special Issue, 63–84.
Herath, A. (2011, October 5). The Climate Change Debate, *Live Science*. http://www.livescience.com/16388-climate-change-debate-man-nature.html.
Herron, J. (2011, October 19). IEA Sees Dire Future for Climate, Energy Without New Technology, *Wall Street Journal*, http://online.wsj.com/article/BT-CO-20111019-707671.html.

Hersir, G., and Bjornsson, A. (1991). *Geophysical Exploration for Geothermal Resources. Principles and Application*, Report. Reykjavik: UNU-GTP.
Hess, D. (2007). *Alternative Pathways in Science and Industry: Activism, Innovation, and the Environment in an Era of Globalization*. Cambridge: MIT Press.
Heymann, M. (1998). Signs of Hubris: The Shaping of Wind Technology in Germany, Denmark and the United States, 1940–1990, *Technology and Culture*, 39: 4, 641–670.
High Level Panel of Experts of Food Security and Nutrition (HLPE) (2013, June). Biofuels and Food Security, Report; HLPE. UN Committee on World Food Security.
High Level Panel of Experts of Food Security and Nutrition (HLPE) (2011, July). Price Volatility and Food Security, Report; HLPE, UN Committee on World Food Security.
Hira, A. (2011). Sugar Rush: Prospects for a Global Ethanol Market, *Energy Policy* 39, 6925–6935.
Hirschman, A. (1970). *Exit, Voice, and Loyalty: Responses to Decline in Firms, Organizations, and States*. Cambridge: Harvard University Press.
Hirsh, R., and Sovacool, B. (2006, Spring). Technological Systems and Momentum Change: American Electric Utilities, Restructuring, and Distributed Generation Technologies, *The Journal of Technology Studies*, 23, 2.
Hirsh, R. (1999). *Power Loss: The Orgins of Deregulation and Restructuring in the American Electric Utility System*. Cambridge: MIT Press.
Hjalmarsson, J. (2009). *History of Iceland*. Reykjavik, Iceland: Forlagid.
HM Treasury, United Kingdom. (2005). Stern Report on the Economics of Climate Change, http://webarchive.nationalarchives.gov.uk/+/http:/www.hm-treasury.gov.uk/independent_reviews/stern_review_economics_climate_change/stern_review_report.cfm.
Hobbes, T. (1996). *Leviathan: Authoritative Text, Backgrounds, Interpretations* (1st Edition), Flathman, R., and Johnston, D. (Eds.). New York: W.W. Norton.
Hoffman, A., and Hoffman H. (1994). Reliability and Validity in Oral History: The Case for Memory, in: *Memory and History: Essays on Recalling and Interpreting Experience* (pp. 107–136). Baylor University Institute for Oral History.
Hoffman, N., and Twining, J. (2009). *Profiting from the Low Carbon Economy*. McKinsey & Company. http://www.mckinsey.com/insights/financial_services/profiting_from_the_low-carbon_economy.
Hogselius, P. (2009). Spent Nuclear Fuel Policies in Historical Perspective: An International Comparison, *Energy Policy*, 37, 254–263.
Hohmeyer, O. (1988). *Social Costs of Energy Consumption. External Effects of Electricity Generation in the Federal Republic of Germany*. Berlin: Springer.
Holdren, J. (2006*a*). The Energy Innovation Imperative: Addressing Oil Dependence, Climate Change, and Other 21st Century Energy Challenges, *Innovations*, 1: 2, 3–23, Spring 2006.
Holdren, J. (2006*b*) Global Energy Challenges and the Role of Increased Energy Efficiency in Addressing Them, Remarks at the Symposium on The Rosenfeld Effect, Berkeley, CA, April 28, 2006.
Holdren, J. (2007*a*). Energy Policy in Theory and Practice, Presentation for the AGI. Leadership Forum on Communicating Geoscience to Policymakers, Washington, DC, April 20, 2007.

Holdren, J. (2007*b*). Meeting the Intertwined Challenges of Energy and Environment, 2007. Robert C. Barnard Environmental Lecture, American Association for the Advancement of Science, Washington, DC, October 18, 2007.

Holdren, J. (2008, January 25). Science and Technology for Sustainable Well-Being, *Science*, 319, 424–434.

Holdren, J. (2009). Energy for Change: Introduction to the Special Issue on Energy & Climate, *Innovations: Technology, Governance, Globalization*, 4: 4, 3–11.

Holm, A., Blodgett, L., Jennejohn, D., and Gawell, K. (2010). Geothermal Energy: International Market Update. GEA, http://geo-energy.org/pdf/reports/GEA_International_Market_Report_Final_May_2010.pdf.

Holodny, E. Political Instability Is on the Rise in One of OPEC's Strongest Members, *Business Insider*, December 8 2016.

Hood, C. (1968). *The Tools of Government*. London: Chatham House.

Horizon Wind. (2011). How Does a Wind Turbine Work? http://www.horizonwind.com/about/ftkc/howdoeswindturbinework.aspx.

Horst, D. (2008). Social Enterprise and Renewable Energy: Emerging Initiatives and Communities of Practice, *Social Enterprise Journal*, 4: 3, 171–185.

House, K. (2008, July 29). In Praise and Fear of France's Energy Policy, *Bulletin of Atomic Scientists*, http://thebulletin.org/praise-and-fear-frances-energy-policy.

Howlett, M. (2002). Understanding National Administrative Styles and Their Impact on Administration Reform: A Neo-Institutional Model and Analysis, *Policy and Society*, 21: 1, 1–24.

Howlett, M., and Ramesh, M. (1993). Patterns of Policy Instrument Choice: Policy Styles, Policy Learning and the Privatization Experience. *Review of Policy Research*, 12: 3–24.

Howlett, M., Ramesh, M., and Perl, A. (2009). *Studying Public Policy: Policy Cycles and Policy Subsystems* (3rd Ed). Oxford: Oxford University Press.

Howlett, M., and Rayner, J. (2013). Patching vs. Packaging in Policy Formulation: Assessing Policy Portfolio Design, *Politics and Governance* 1: 2, 170–182.

HS Orka HF (n.d.). Information. http://www.hsorka.is/. Accessed July 2012.

Hugh, S., and Meyer, H. (2009, September). Wind Energy: The Case of Denmark, Report, Center for Politiske Studier (CEPOS), https://www.wind-watch.org/documents/wind-energy-the-case-of-denmark/.

Hughes, T. (1983). *Networks of Power: Electrification in Western Society*. Baltimore: Johns Hopkins University Press.

Hughes, T. (1986). The Seamless Web: Technology, Science, Etcetera, Etcetera, *Social Studies of Science*, 16: 2 281–292.

Hughes, T. (1989). *American Genesis: A Century of Invention and Technological Enthusiasm, 1870–1970*. New York: Penguin.

Hughes, T. (1998). *Rescuing Prometheus: Four Monumental Projects That Changed the Modern World* (1st Edition). New York: Pantheon.

Hughes, T. (2004). *Human-Built World: How to Think About Technology and Culture*. Chicago: University of Chicago Press.

Hughes, T. (2012). The Evolution of Large Technological Systems, in: W. Bijker, Hughes, T., and Pinch, T. (Eds.), *The Social Construction of Technological Systems*. Cambridge: MIT Press.

Hughes, T., and Mayntz, R. (Eds.). (1988). *The Development of Large Technical Systems*. Frankfurt/Boulder: Verlag/Westview.

Huitema, D., and Meijerink, S. (2009). Transitions in Water Management, in: Huitema, D., and Meijerink, S. (Eds.), *Water Policy Entrepreneurs: A Research Companion to the Water Transitions around the Globe*. Cheltenham: Edward Elgar.

Hulac, B. (2016). Utility Companies Could Fail a "Climate Stress Test", *Scientific American*. http://www.scientificamerican.com/article/utility-companies-could-fail-a-climate-stress-test/.

Hultman, N., Malone, E., Runci., P., Carlock, G., and Anderson, K. (2012). Factors in Low Carbon Energy Transformations; Comparing Nuclear and Bioenergy in Brazil, Sweden, and the United States, *Energy Policy*, 40: 131–146.

Humphreys, H., and McClain, K. (1998). Reducing the Impacts of Energy Price Volatility Through Dynamic Portfolio Selection, *The Energy Journal*, 19: 3, 107–134.

Hvelplund, F. (2011). Innovative Democracy and Renewable Energy Strategies: A Full-Scale Experiment in Denmark, 1976–2010, in: M. Jarvela and S. Juhola (Eds.), *Energy, Policy, and the Environment*. New York: Springer.

Hylleberg, J. (2015). There Is No Status Quo in the Wind Industry, in: State of Green (Ed.), *Wind Energy Moving Ahead*. Frederiksberg: Danish Wind Industry Association, http://www.windpower.org/en/.

Iceland Deep Drilling Project [IDDP], The Drilling of the Iceland Deep Drilling Project Geothermal Well at Reykjanes has been Successfully Completed, February 1, 2017, https://iddp.is/2017/02/01/the-drilling-of-the-iceland-deep-drilling-project-geothermal-well-at-reykjanes-has-been-successfully-completed-2/

Iceland Monitor. (2014, November 2). The Little Country that Roars, http://icelandmonitor.mbl.is/news/politics_and_society/2014/11/02/the_little_country_that_roars/.

Iceland Review. (2012, January 25). Iceland Makes Energy Deal with World Bank. http://www.icelandreview.com/icelandreview/daily_news/Iceland_Makes_Energy_Deal_with_World_Bank_0_386715.news.aspx.

Icelandic Sagas. (2012). Information. http://sagadb.org/about. Accessed July 10, 2012.

IFOP (2013, June.). JF/EP No. 111359 IFOP pour Dimanche, Les Français et le nucléaire—Résultats détaillés-Juin 2013, Département Opinion et Stratégies d'Entreprise, Ouest France.

Immergut, E. (2006). Institutional Constraints on Policy, in: Moran, M., Rein, M., and Goodin, R. (Eds.), *The Oxford Handbook of Public Policy*. Oxford: Oxford University Press.

Inamasu, R., and Neto, L. (2010). Instrumentation and Automation in the Sugarcane Ethanol Chain, in: Cortez, L. (Ed.), *Sugarcane Bioethanol*. São Paulo: Blucher.

Ingham, J. (2007, May 15). Nationalisation Sweeps Venezuela, *BBC News*. http://news.bbc.co.uk/2/hi/business/6646335.stm.

Ingimarsson, J. (1996). Geothermal Energy in Iceland, Presentation by Director for Energy Affairs, Ministries of Energy and Commerce to Japan Geothermal Energy Association, August 29, 1996.

Ingram, H., Schneider, A., and Deleon, P. (2007). Social Construction and Policy Design, in: P. Sabatier (Ed.) *Theories of the Policy Process*. Boulder: Westview.

Institute for Energy Research (IER). (2011, April 14). The Significance of Spare Oil Capacity, IER, http://instituteforenergyresearch.org/analysis/the-significance-of-spare-oil-capacity/.

Institute for Energy Research (IER). (2012, September 7). Obama's Energy Tax Proposals: Wind vs. Oil and Gas, http://www.instituteforenergyresearch.org/2012/09/07/obamas-energy-tax-proposals-wind-vs-oil-and-gas/.

Intergovernmental Panel on Climate Change (IPCC). (2011). Summary for Policymakers, in: Edenhofer, O., Pichs-Madruga, R., Sokona, Y., Seyboth, K., Matschoss, P., et al. (Eds.), *IPCC Special Report on Renewable Energy Sources and Climate Change Mitigation*. New York: Cambridge University Press.

Intergovernmental Panel on Climate Change (IPCC). (1990, 1995, 2001*a*, 2007*a*). Assessment Reports, http://ipcc.ch/publications_and_data/publications_and_data_reports.shtml#1.

Intergovernmental Panel on Climate Change (IPCC). (2000*a*). Methodological and Technological Issues in Technology Transfer. http://www.ipcc.ch/ipccreports/sres/tectran/index.htm.

Intergovernmental Panel on Climate Change (IPCC). (2000*b*). Special Report on Emissions Scenarios. http://www.ipcc.ch/pdf/special-reports/spm/sres-en.pdf.

Intergovernmental Panel on Climate Change (IPCC). (2001*b*). *Climate Change 2001: Working Group III: Mitigation, Third Assessment Report*. Cambridge: Cambridge University Press.

Intergovernmental Panel on Climate Change (IPCC). (2007*b*). Summary for Policymakers, Climate Change 2007: Mitigation, in: Metz, B., Davidson, O., Bosch, P., Dave, R., and Meyer, L. (Eds), *Contribution of Working Group III to the Fourth Assessment Report of the Intergovernmental Panel on Climate Change*. Cambridge: Cambridge University Press.

Intergovernmental Panel on Climate Change (IPCC). (2011). Special Report on Renewable Energy Sources and Climate Change Mitigation, http://www.ipcc.ch/report/srren/.

Intergovernmental Panel on Climate Change (IPCC). (2013). Climate Change 2013: The Physical Science Basis, in: Stocker, T., Qin, D., PLattner, K., Tignor, M., Allen, K., et al. (Eds.), *Contribution of Working Group I to the Fifth Assessment Report of the Intergovernmental Panel on Climate Change*. Cambridge: Cambridge University Press.

Intergovernmental Panel on Climate Change (IPCC). (2014). Climate Change 2014: Synthesis Report, in: Core Writing Team, Pachauri, R. and Meyer, L. (Eds.), *Contribution of Working Groups I, II and III to the Fifth Assessment Report of the Intergovernmental Panel on Climate Change*. Geneva: IPCC.

Internal Revenue Service (IRS). (2016). Yearly Average Currency Exchange Rates, http://www.irs.gov/.

International Association of Oil and Gas Producers (OGP). (2010). Risk Assessment Data Directory, Report 434-17, London: OGP, http://www.iogp.org.

International Atomic Energy Agency (IAEA). (n.d.*a*). General information, http://www.iaea.org/. Accessed January 5, 2017

International Atomic Energy Agency (IAEA). (n.d.*b*). PRIS Database. http://www.iaea.org/. Accessed January 5, 2017.

International Atomic Energy Agency (IAEA). (1998, 2001, 2002, 2003, 2011, 2016). Country Nuclear Power Profiles: France. Vienna, Austria: IAEA. http://www.iaea.org/.

International Atomic Energy Association (IAEA). (2004). From Obrinsk Beyond: Nuclear Power Conference Looks to the Future, IAEA, https://www.iaea.org/newscenter/news/obninsk-beyond-nuclear-power-conference-looks-future.

International Atomic Energy Agency (IAEA). (2009). World Distribution of Uranium Deposits (UDEPO) with Uranium Deposit Classification, IAEA-TECDOC-1629. Vienna, Austria: IAEA. http://www-pub.iaea.org/MTCD/Publications/PDF/TE_1629_web.pdf

International Atomic Energy Agency (IAEA). (2012). Illicit Trafficking Database http://www.iaea.org/.

International Atomic Energy Agency (IAEA). (2015). Preparedness and Response for a Nuclear or a Radiological Emergency. General Safety Requirements No. GSR Part 7. Vienna: IAEA.

International Atomic Energy Agency (IAEA), COPPE, CENBIO and UN Department of Economics and Social Affairs. (2006). Brazil: A Country Profile on Sustainable Energy Development, Report, http://www-pub.iaea.org/mtcd/publications/PubDetails.asp?pubId=7490.

International CHP/DHC Collaborative. (2005). CHP/DHC Country Scorecard: Denmark. Paris: OECD/IEA.

International Directory of Company Histories. (1998). Framatome SA History, Volume 19. St. James: St. James Press.

International Energy Agency (IEA). (n.d.). Data. Accessed January 5, 2017.

International Energy Agency (IEA). (1980, 1981, 1982, 1988, 1989, 1993, 1994, 1995, 1998*a*, 1999, 2005). Energy Policies of IEA Countries. Paris: IEA/OECD.

International Energy Agency (IEA). (1998*b*, 2000*c*, 2006*a*, 2011*a*). Energy Policies of IEA Countries: Denmark. Paris: IEA/OECD.

International Energy Agency (IEA). (2000*a*, 2004*a*, 2010*a*). Energy Policies of IEA Countries: France. Paris: OECD/IEA.

International Energy Agency (IEA). (2000*b*). Experience Curves for Energy Technology Policy. Paris: IEA/OECD.

International Energy Agency (IEA). (2004*b*). Biofuels for Transport: An International Perspective. Paris: OECD/IEA.

International Energy Agency (IEA). (2004*c*). Renewable Energy—Markets and Policy Trends in IEA Countries. Paris: IEA/OECD.

International Energy Agency (IEA). (2008*a*). Deploying Renewables: Principles for Effective Policies. Paris: OECD/IEA.

International Energy Agency (IEA). (2010*b*, 2015*g*). Technology Roadmap: Nuclear Energy. Paris: IEA/OECD.

International Energy Agency (IEA). (2008*b*, 2010*c*). Energy Technology Perspectives. Paris: OECD/IEA.

International Energy Agency (IEA). (2006*b*, 2008*c*, 2010*d*, 2011*a*, 2015*b*). World Energy Outlook. Paris: OECD/IEA.

International Energy Agency (IEA). (2007). Biofuel Production, IEA Energy Technology Essentials. Paris, IEA.

International Energy Agency (IEA). (2009, 2015*f*). Technology Roadmap: Wind Energy. Paris: OECD/IEA.
International Energy Agency (IEA). (2010*e*). Sustainable Production of Second-Generation Biofuels. Paris: OECD/IEA.
International Energy Agency (IEA). (2010*f*). Projected Costs of Generating Electricity. Paris: IEA/OECD.
International Energy Agency (IEA). (2011*b*). Technology Roadmap: Geothermal, Heat and Power. Paris: IEA/OECD.
International Energy Agency (IEA). (2011*c*). Accelerating the Deployment of Offshore Renewable Energy Technologies, Report IEA-RETD. Paris: IEA/OECD.
International Energy Agency (IEA). (2011*d*). Cost of Wind Energy, IEA Task Force 26, Report. http://www.ieawind.org/task_26.html.
International Energy Agency (IEA). (2011*e*) Technology Roadmap: Biofuels for Transport. Paris: IEA/OECD.
International Energy Agency/IEA (2012*a*). Overview and History of the IEA. Paris: IEA/OECD.
International Energy Agency (IEA). (2012*b*). Key World Energy Statistics. Paris: IEA/OECD.
International Energy Agency (IEA). (2014). World Energy Outlook. Paris: IEA/OECD.
International Energy Agency (IEA). (2015*a*). CO_2 Emissions from Fuel Combustion. Paris: IEA/OECD.
International Energy Agency (IEA). (2015*b*). Energy and Climate Change: World Energy Outlook, Special Report. Paris: IEA/OECD.
International Energy Agency (IEA). (2015*c*). Projected Costs of Generating Electricity. Paris: IEA/OECD.
International Energy Agency (IEA). (2015*d*). World Energy Outlook. Paris: IEA/OECD.
International Energy Agency (IEA). (2015*e*). Electric Power Auctions, Biomass. Paris: IEA/OECD.
International Energy Agency (IEA). (2016*a*). Next Generation Wind and Solar Power. Paris: IEA/OECD.
International Energy Agency (IEA). (2016*b*). Decoupling of Global Emissions and Economic Growth Confirmed. http://www.iea.org/newsroomandevents/press-releases/2016/march/decoupling-of-global-emissions-and-economic-growth-confirmed.html?platform=hootsuite.
International Energy Agency (IEA). (2016*c*) World Energy Outlook Special Report: Energy and Air Pollution. Paris: IEA/OECD.
International Energy Agency (IEA). (2016*d*). Note on Electricity Security for the G7, IEA/GCP. Paris: IEA/OECD.
International Energy Agency (IEA). (2016*e*). Oil Market Report. https://www.iea.org/oilmarketreport/omrpublic/.
International Energy Agency (IEA). (2016*f*). World Energy Outlook. Paris: IEA/OECD.
International Energy Agency (IEA). (2016*g*). Tracking Clean Energy Progress. Paris: IEA/ OECD.
International Energy Agency-Renewable Energy Technology Development (IEA-RETD) (2017). Comparative Analysis of International Offshore Wind Energy Development. Paris: IEA/OECD.

International Energy Agency (IEA) and International Renewable Energy Agency (IRENA). (n.d.). Global Renewable Energy Policies and Measures Database. https://www.iea.org/policiesandmeasures/. Accessed December 5, 2016

International Energy Agency (IEA) and Nuclear Energy Agency (NEA). (2010). Nuclear Energy, Technology Roadmap. Paris: IEA/OECD and NEA/OECD. https://www.iea.org/media/freepublications/technologyroadmaps/nuclear_roadmap2010.pdf.

International Energy Agency Wind. (1978, 1979, 1995–2014). IEA Wind Annual Report, 1978, 1979, 1995–2014. Paris, France: IEA/OECD.

International Geothermal Association (IGA). (2011). Electricity Generation, Iceland, http://www.geothermal-energy.org/electricity_generation/iceland.html.

International Institute for Sustainable Development (IISD). (1997). Sustainable Development Timeline, http://www.iisd.org/rio+5/timeline/sdtimeline.htm.

International Renewable Energy Agency (IRENA). (2012). Renewable Energy Technology: Cost Analysis Series, Wind Power, 1:5/5, Abu Dhabi: IRENA.

International Renewable Energy Agency (IRENA). (2016). Unlocking Renewable Energy Investment. Abu Dhabi: IRENA.

International Sugar Organization (ISO). (2009). Outlook on Brazil´s Competitiveness in Sugar and Ethanol, *International Sugar Journal*, 111: 1327, 422–427.

International Sustainable Energy Assessment (ISEA). (n.d.). International Energy Treaties, http://lawweb.colorado.edu/eesi/.

International Thermonuclear Experimental Reactor (ITER). (n.d.). Information, https://www.iter.org/.

Institut de Radioprotection et de Suret Nucleaire (IRSN). (n.d.). Le Groupe Radioécologie Nord-Cotentin (GRNC), http://www.irsn.fr/FR/connaissances/Environnement/expertises-locales/GRNC/Pages/sommaire.aspx#.V4XyO7iLSM8. Accessed July 10, 2016.

Islandsbanki. (2010, April). Iceland Geothermal Energy Market Report, Islandsbanki. http://www.islandsbanki.is/english/industry-focus/sustainable-energy/research-and-publications/.

ÍSOR. (2016). European Union Grant Applications – ÍSOR's Success, January 20, 2016, http://www.geothermal.is/news/european-union-grant-applications-isors-success.

ITER. (n.d.), Information. https://www.iter.org/. Accessed December 10, 2016.

Jaccard, M., Agbenmabiese, L., Azar, C., de Oliveira, A., Fischer, C., et al. (2012). Policies for Energy System Transformations: Objectives and Instruments, *Global Energy Assessment—Toward a Sustainable Future* (pp. 1549–1602). Cambridge/Laxenburg: Cambridge University Press/International Institute for Applied Systems Analysis.

Jackson, T. (2005). Motivating Sustainable Consumption: A Review of Evidence on Consumer Behaviour and Behavioural Change. A Report to the Sustainable Development Research Network. Surrey: Center for Environmental Strategy, http://www.surrey.ac.uk/CES.

Jacobson, M. (2009). Review of Solutions to Global Warming, Air Pollution and Energy Security, *Energy and Environment Science*, 2, 148–173.

Jacobson, M., and Delucchi, M. (2009). A Plan to Power the Planet with 100 Percent with Renewables, *Scientific American*. http://www.scientificamerican.com/article/a-path-to-sustainable-energy-by-2030/.

Jacobsson, S., and Bergek, A. (2004). Transforming the Energy Sector: The Evolution of Technological Systems in Renewable Energy Technology, *Industrial and Corporate Change*, 13: 5, 815–849.

Jacobsson, S., and Johnson, A. (2000). The Diffusion of Renewable Energy Technology: An Analytical Framework and Key Issues for Research, *Energy Policy*, 28: 9, 625–640.

Jacobsson, S., and Lauber, V. (2006). The Politics and Policy of Energy System Transformation—Explaining the German Diffusion of Renewable Energy Technology, *Energy Policy*, 34: 3, 256–276.

Jaffee, A., Newell, R., and Stavins, R. (2005). A Tale of Two Market Failures, *Ecological Economics*, 54, 164–174.

James, W. (Ed.). (1989). *The Energy-Economy Link: New Strategies for the Asia-Pacific Region*. New York: Praegar.

Jamison, A., Eyerman, R., and Cramer, J. (1990). *The Making of the New Environmental Consciousness*. Edinburgh: Edinburgh University Press.

Jamison, M., and Berg, S. (2008). Annotated Reading List for a Body of Knowledge on Regulation. http://researchictafrica.net/PGCICTPR/PGCICTPR/Module_5_files/Chapter%20IV.%20Regulating%20Overall%20Price%20Level.pdf. Accessed October 30, 2010.

Jank, M. (2009, June 25). Sugarcane: Historic Advances in Labor Relations, reprinted from *O Estado de S Paolo*. http://www.unica.com.br/presidents-desk/17830428920319033149/sugarcane-por-cento3A-historic-advances-in-labor-relations/.

Janssens, T., Nyquist, S., and Roelofsen, O. (2011, November). Another Oil Shock? McKinsey Quarterly, 4, http://www.mckinsey.com/industries/oil-and-gas/our-insights/another-oil-shock.

Jasanoff, S., and Wynne, B. (1998). Science and Decision-making, in: Rayner, S., and Malone, E. (Eds.), *Human Choice and Climate Change, Volume 1: The Societal Framework*. Devon: Batelle Press.

Jasper, J. (1988). The Politics of Nuclear Energy in France, Sweden, and the United States, PhD Dissertation, University of California Berkeley, Berkeley, CA.

Jasper, J. (1992, July). Gods, Titans, and Mortals: Patterns of State Involvement in Nuclear Development, *Energy Policy* 20: 7, 653–659.

Jefferson, M. (2008). Accelerating the Transition to Sustainable Energy Systems, *Energy Policy*, 36: 4116–4125.

Jefferson, M. (2014). Closing the Gap Between Energy Research and Modeling, the Social Sciences, and Modern Realities, *Energy Research and Social Science*, 4, 42–52.

Jeffery, J. (1986). The Subsidized 'Cheapness' of French Nuclear Electricity, *Energy Policy*, 14: 4, 384–385.

Jianxiang, Y. (2016, January 13). China's 2015 Energy Mix Beats Targets, *Wind Power Monthly*. http://www.windpowermonthly.com/article/1379220/chinas-2015-energy-mix-beats-targets.

Johannesson, H., de Roo, C., and Robaey, Z. (2011). Sustainable Planning of Megaprojects in the Circumpolar North, Project Report, University of Akureyri Research Centre, Iceland.

Johansson, T., McCormick, K., Neij, L., Turkenburg, W. (2004). The Potentials of Renewable Energy, Paper, International Conference for Renewable Energies, Bonn, Germany.

Johnson, B., Lorenz, E., and Lundvall, B. (2002). Why All This Fuss About Codified and Tacit Knowledge? *Industrial and Corporate Change* 11: 2, 245–262.

Johnson, L. (2008). The Impact of Volume Risk on Hedge Effectiveness: The Case of a Natural Gas Independent Power Producer Operation, *Journal of Energy Markets*, 1: 1.

Johnson, T. (2007, August 13). *The Return of Resource Nationalism*. Analysis Brief. New York: Council on Foreign Relations, http://www.cfr.org/venezuela/return-resource-nationalism/p13989.

Johnson, T., Alatorre, C., Romo, Z., and Lie, F. (2009). *Low Carbon Development for Mexico, Report*. Washington, DC: World Bank.

Jonasson, T., and Thordarson, S. (2007). Geothermal District Heating in Iceland: Its Development and Benefits, Paper Presented at the 26th Nordic History Congress, August 8–12, 2007.

Jones, B., and Baumgartner, F. (2005). *The Politics of Attention: How Government Prioritizes Problems*. Chicago: University of Chicago Press.

Jones, C. (2009). Energy Landscapes: Coal Canals, Oil Pipelines, and Electricity Transmission Wires in the Mid-Atlantic, 1820–1930, PhD Dissertation, History of Sociology of Science, University of Pennsylvania.

Jones, R. (2009). Energy Security: The IEA Perspective, Presentation, Istanbul, Turkey, April 28, 2009.

Jorgensen, U., and Karnoe, P. (1991). The Danish Wind Turbine Story: Technical Solutions to Political Visions? in: Rip, A., Misa, T., and Schot, J. (Eds.), *Managing Technology in Society*. London: Pinter.

Jorgenson, D. (1984). The Role of Energy in Productivity Growth, *Energy Journal*, 5: 3, 11–26.

Joseph Jr., H. (2009). New Advances on (*sic*) Flex Fuel Technology, Presentation, ANFAVEA Ethanol Summit, São Paulo, Brasil, June 1–3, 2009. http://2009.ethanol-summit.com.br/.

Joseph Jr., H. (2013). Flex Fuel Vehicles in Brazil, GBEP, March 22, 2013, Brasilia, Brazil.

Joseph Jr., H. (2010). Issues Related to the Final Use of Ethanol—The Ethanol-Powered Motor: Past, Present and Future, in: Cortez, L. (Ed.), *Sugarcane Bioethanol*. São Paulo: Blucher.

Juhn, P., Kupitz, J., and Cleveland, J. (1997). Advanced Nuclear Power Plants: Highlights of Global Development, *IAEA Bulletin*, 39: 2, 13–20, https://www.iaea.org/sites/default/files/publications/magazines/bulletin/bull39-2/39204781320.pdf.

Juliusdottir, K. (2011). Skyrsla: Idnadarradherra um Raforkumalefni, Logd Fyrir Alþingi a 140. Loggjafarþingi 2011–2012, http://www.idnadarraduneyti.is/utgefid-efni/skyrslur/nr/3460.

Juma, C. (2005). Biotechnology in a Globalizing World: The Co-evolution of Technology and Social Institutions, *Bioscience*, 55: 3, 265–272.

Juma, C. (2011). *The New Harvest: Agricultural Innovation in Africa*. Oxford: Oxford University Press.

Juma, C. (2016). *Innovation and Its Enemies*. Oxford: Oxford University Press.

Junginger, H., Jonker, J., Faaij, A., Cocchi, M., Hektor, B., et al. (2011, April). IEA Bioenergy Task Force 40: Summary, Synthesis and Conclusions from IEA Bioenergy Task 40 Country Reports on International Bioenergy Trade, http://www.bioenergy-trade.org/downloads/summary-synthesis-and-conclusions-from-iea-bio.pdf.

Kagel, A., Bates, D., and Gawell, K. (2007). A Guide to Geothermal Energy and the Environment, Geothermal Energy Association, http://geo-energy.org/reports/environmental%20guide.pdf.

Kahn Ribeiro, S., Kobayashi, S., Beuthe, M., Gasca, J., Greene, D., et al, (2007). Transport and its Infrastructure, in: Metz, B., Davidson, O., Bosch, P., Dave, R., Meyer, L. (Eds.), *Climate Change 2007: Mitigation*. Contribution of Working Group III to the Fourth Assessment Report of the Intergovernmental Panel on Climate Change. Cambridge/New York: Cambridge University Press.

Kamimura, A., and Sauer, I. (2008). The Effect of Flex Fuel Vehicles in Brazilian Light Road Transportation, *Energy Policy*, 36, 1574–1576.

Kammen, D. (2004). Renewable Energy: Taxonomic Overview, in: Cleveland, C. (Ed.), *Encyclopedia of Energy*. Boston: Elsevier.

Kammen, D., and Nijkamp, P. (1991). Technogenesis: Origins and Diffusion in a Turbulent Environment, *Technological Forecasting and Societal Change*, 39, 45–66.

Kamp, L., Smits, R., and Andriesse, C. (2004). Notions on Learning Applied to Wind Turbine Development in the Netherlands and Denmark, *Energy Policy*, 32: 14, 1625–1637.

Karam, P. (n.d.). How Do Fast Breeder Reactors Differ from Regular Nuclear Power Plants? *Scientific American*, http://www.scientificamerican.com/article/how-do-fast-breeder-react/.

Kaplan, U., and Schochet, D. (2005). Improving Geothermal Power Plant Performance by Repowering with Bottoming Cycles. Proceedings World Getothermal Congress 2005, Antalya, Turkey, April 4–29, 2005.

Karnøe, P. (1990). Technological Innovation and Industrial Organization in the Danish Wind Industry, *Entrepreneurship & Regional Development*, 2, 105–123.

Karnøe, P. (1999). When Low-Tech Becomes High-Tech: Social Construction of Learning Processes in Danish and American Wind Turbine Firms, in: Karnøe, P., Kristensen, P., and Andersen, P. (Eds.), *Mobilizing Resources and Generating Competencies*. Copenhagen: Copenhagen Business School Press.

Karnøe, P., and Buchorn, A. (2008). Denmark: Path-creation Dynamics and Winds of Change, in: Lafferty, W., and Ruud, A. (Eds.), *Promoting Sustainable Electricity in Europe: Challenging the Path Dependence of Dominant Energy Systems*. Cheltenham: Edward Elgar.

Katakey, R. (2015, June 10). US Ousts Russia as Top World Oil, Gas Producer in BP Data, *Bloomberg Business*. http://www.bloomberg.com/news/articles/2015-06-10/u-s-ousts-russia-as-world-s-top-oil-gas-producer-in-bp-report.

Kaya, Y. (1990). *Impact of CO_2 Control on GNP Growth*, Paper Presented to Energy and Industry IPCC Subgroup.

Kaya, Y., Nakicenovic, N., Nordhaus, W., and Toth, F. (Eds.). (1992/1993). Costs, Impacts and Benefits of Carbon Mitigation. CP 93 2, Workshop Proceedings from Meeting September 28–30, 1992, IIASA.

Kaya, Y., and Yokobori, K. (1997). *Environment, Energy, and Economy: Strategies for Sustainability*. Tokyo: United Nations University Press.

Kemp, R. and Loorbach, D. (2006). Transition Management: A Reflexive Governance Approach, in: Voß, J.P., Bauknecht, D. and Kemp, R. (Eds.) *Reflexive Governance for Sustainable Development*. Cheltenham/Northampton: Edward Elgar.

Kemp, R., Schot, J., and Hoogma, R. (1998). Regime Shifts to Sustainability through Processes of Niche Formation: The Approach of Strategic Niche Management, *Technology Analysis and Strategic Management,* 10, 175–196.

Kent, S., and Harder, A. (2015, October 7). Oil CEOs Differ on Carbon Strategy, Highlighting Industry Divide, *The Wall Street Journal*, http://www.wsj.com/articles/oil-ceos-differ-on-carbon-strategy-highlighting-industry-divide-1444252378.

Kenward, A. (2011, April 14). In Southeast, Extreme Heat Is a Growing Concern for Nuclear Power Operators, *Inside Climate News*, http://insideclimatenews.org/news/20110414/southeast-extreme-heat-growing-concern-nuclear-power-operators?page=3.

Kern, F., and Rogge, K. (2016). The Pace of Governed Energy Transitions, *Energy Research and Social Science*, 22: 13–17.

Kern, F., Kivimaa, P., and Martiskainen, M. (2017). Policy Packaging or Policy Patching? *Energy Research & Social Science*, 23, 11–25.

Ketilsson, J., Olafsson, L., Steinsdottir, G., and Johannesson, G. (2010). Legal Framework and National Policy for Geothermal Development in Iceland, Proceedings World Geothermal Congress 2010, Bali, Indonesia, April 25–29, 2010.

Ketilsson, J., Petursdottir, H., Thoroddsen, S., Oddsdottir, A., Bragadottir, E., et al. (2015). Legal Framework and National Policy for Geothermal Development in Iceland. *Proceedings World Geothermal Congress 2015*, Melbourne, Australia, 19–25, April 2015.

Khan, H. (2002). Innovation and Growth: A Schumpeterian Model of Innovation Applied to Taiwan. *Oxford Development Studies,* 30: 3, 289–306.

Kheshgi, H., Prince, R., and Marland, G. (2000). The Potential of Biomass Fuels in the Context of Global Climate Change: Focus on Transport Fuels. *Annual Review of Energy and the Environment*, 25, 1999–1244.

Kiesling, L. (2011, August 29). Whales and Electricity, and Sustainability, Word Press, http://knowledgeproblem.com/2011/08/29/whales-and-electricity-and-sustainability/,

Kileri, S. and Kolo, A. Privatization, Washington DC, World Bank, Working paper 3765, November 2005.

Ki-moon, B. (2011). Sustainable Energy for All: A Vision Statement, Secretary-General of the United Nations, http://www.un.org/wcm/webdav/site/sustainableenergyforall/shared/Documents/SG_Sustainable_Energy_for_All_vision_final_clean.pdf.

King, G. (1989). *Unifying Political Methodology: The Likelihood Theory of Statistical Inference*, New York: Cambridge University Press.

Kingdon, J. (1995). *Agendas, Alternatives and Public Policies* (2nd Ed.). New York: Harper Collins.

Kintisch, E. (2016). Underground Injections Turn Carbon Dioxide to Stone, *Science,* June 10, 2016, http://www.sciencemag.org/news/2016/06/underground-injections-turn-carbon-dioxide-stone.

Kiruja, J. (2011). Direct Utilization of Geothermal Energy. Presented at Short Course VI on Exploration for Geothermal Resources, organized by UNU-GTP, GDC, and KenGen, at Lake Bogoria, and Lake Naivasha, Kenya, October 27–November 18, 2011.

Kiss, B., and Neij, L. (2011). The Importance of Learning when Supporting Emergent Technologies for Energy Efficiency: A Case Study on Policy Intervention for Learning for the Development of Energy Efficient Windows in Sweden, *Energy Policy*, 39: 10, 6514–6524.

Kitschelt, H. (1986). Political Opportunity Structures and Political Protests: Anti-Nuclear Movements in Four Democracies, *British Journal of Political Science*, 16: 1, 57–85.

Kjaer, C. (2001). Green Certificates in Denmark, http://www.ens.dk/en/info/publications/green-certificate-market-denmark.

Kline, S. (1985). Research, Invention, Innovation and Production: Models and Reality, Report INN-1, March 1985. Stanford: Mechanical Engineering Department, Stanford University.

Kline, S., and Rosenberg, N. (1986). An Overview of Innovation, in: Landau, R., and Rosenberg, N. (Eds.), *The Positive Sum Strategy: Harnessing Technology for Economic Growth* (pp. 275–305). Washington, DC: National Academy Press.

Knebel, J. (2009). Partitioning and Transmutation of High Level Nuclear Waste in an Accelerator Driven System, Presentation, FISA Prague, Czech Republic, June 22, 2009.

Knodt, M. Piefer, N., and Mueller, F. (2015). *Challenges of European External Energy Governance with Emerging Powers*. Oxon: Routledge.

Kojima, M., and Johnson, T. (2006). Potential for Biofuels in Transport in Developing Countries, ESMAP Paper, Knowledge Exchange Series No. 4, Washington, DC: World Bank.

Komanoff, C. (2010). Cost Escalation of France's Nuclear Reactors. http://www.komanoff.net/nuclear_power/Cost_Escalation_in_France's_Nuclear_Reactors.pdf.

Kondratiev, N. (1935). The Long Waves in Economic Life, *Review of Economic Statistics*, 17, 105–115.

Koonin, S., and Gopstein, A. (2011, Winter). Accelerating the Pace of Energy Change, *Issues in Science and Technology*, http://issues.org/27-2/koonin/.

Koplow, D., and Dernbach, J. (2001). Federal Fossil Fuel Subsides and Greenhouse Gas Emissions: A Case Study of Increasing Transparency for Fiscal Policy, *Annual Review of Energy and the Environment*, 26, 361–389.

Korns, L. (1920). *Thoughts*. Boston: The Cornhill Company.

Kovarik, W. (1998, Spring). Henry Ford, Charles Kettering and the 'Fuel of the Future,' *Automotive History Review*, 32, 7–27, http://www.radford.edu/~wkovarik/papers/fuel.html. Originally from a paper of the same name at the Proceedings of the 1996 Automotive History Conference, Henry Ford Museum, Dearborn, Michigan, September 1996.

Kovarik, W. (n.d.). Ethanol's First Century, Paper, International Symposium on Alcohol Fuels, Revised Version. http://www.radford.edu/wkovarik/papers/International.History.Ethanol.Fuel.html. Accessed June 10, 2016.

Krater, J., and Rose, M. (2009). Development of Iceland's Geothermal Energy Potential, in: Abrahamsky, K. (Ed.), *Sparking a World-Wide Energy Revolution*. Edinburgh: AK Press.

Krause, F., Bossel, H., Muller-Reißmann, K-F. (1982). Energiewende (Energy Transition: Growth and Prosperity without Oil and Uranium). Frankfurt: Fischer Verlag.

Kravit, S., Lehr, J., and Kingery, T. (Eds.). (2011). *Nuclear Energy Encyclopedia: Science, Technology and Applications*. Hoboken: Wiley.

Kristnjansson, I. (1992). Commercial Production of Salt from Geothermal Brine at Reykjanes, Iceland, *Geothermics*, 21: 5–6, 765–771.

Krohn, S. (1998, May 4). Creating a Local Wind Industry, Regroupement National des Conseils Régionaux de l'environnement du Québec.
Krohn, S. (2000). The Wind Turbine Market in Denmark. http://ele.aut.ac.ir/~wind/en/articles/wtmindk.htm.
Krohn, S. (2001). Offshore Wind Energy: Full Speed Ahead. http://ele.aut.ac.ir/~wind/en/articles/offshore.htm.
Krohn, S. (2002*a*). Danish Wind Turbines: An Industrial Success Story, Frederiksberg: Danish Wind Industry Association. http://www.ingdemurtas.it/wp-content/uploads/page/eolico/normativa-danimarca/Danish_Wind_Turbine_Industry-an_industrial_succes_story.pdf.
Krohn, S. (2002*b*). Wind Energy Policy in Denmark: 25 Years of Success—What Now? Frederiksberg: Danish Wind Industry Association, February 2002, http://www.windpower.org/en/.
Krohn, S. (2002*c*). Wind Energy Policy in Denmark, Status 2002, Frederiksberg: Danish Wind Industry Association, February 2002, http://www.windpower.org/en/.
Kuhlmann, S., Shapira, P., and Smits, R. (2010). *The Theory and Practice of Innovation Policy: An International Research Handbook.* Cheltenham: Edward Elgar.
Kuhn, T. (1996). *The Structure of Scientific Revolutions* (3rd Ed.). Chicago: University of Chicago Press.
Laabs, E., and Groteke, F. (2010). The Brazilian Biodiesel Program—A Socially Acceptable Approach in Biofuel Production? http://www.eisa-net.org/be-bruga/eisa/files/events/stockholm/Laabs-The%20Brazilian%20biodiesel%20program%20-%20a%20socially%20acceptable%20approach%20in%20biofuel%20production.pdf.
Lacey, S. (2009, December 11). Geothermal Industry Seeing New Investments, *Renewable Energy World.* http://www.renewableenergyworld.com/articles/2009/12/geothermal-industry-seeing-new-investments.html.
Lacey, S. (2010, November 17). Icelandic Backlash Weighing in on Magma Energy, *Renewable Energy World.* http://www.renewableenergyworld.com/articles/2010/11/icelandic-backlash-weighing-on-magma-energy.html.
Laird, F. (2013). Against Transitions? Uncovering Conflicts in Changing Energy Systems, *Science as Culture*, 22: 2, 149–156.
Landes, D. (1969). *The Unbound Prometheus: Technological Change and Industrial Development in Western Europe from 1750 to the Present.* Cambridge: Cambridge University Press.
Landsvirkjun (n.d.*a*). Information. http://www.landsvirkjun.com/. Accessed December 31, 2016.
Landsvirkjun. (n.d.*b*). Electricity in Iceland, Presentation. http://www.landsvirkjun.com/ . . . us/Landsvirkjun_general_presentation.pdf.
Landsvirkjun (2009, March 6). Landvirkjun's History. http://www.landsvirkjun.com/company/history.
Landsvirkjun. (2011, November 15). The Road Ahead, Autumn Meeting, Company Presentation. http://www.landsvirkjun.com/.
Lane, J. (2012, March 2). UNICA Says Brazilian Government Financing Support for Ethanol Is Not a Subsidy, *Biofuels Digest*, http://www.biofuelsdigest.com/bdigest/2012/03/02/unica-says-brazilian-government-financing-support-for-ethanol-is-not-a-subsidy/.

Lane, P., Koka, B., and Pathak, S. (2006). The Reification of Absorptive Capacity: A Critical Review and Rejuvenation of the Construct, *Academy of Management Review*, 31: 4, 833–863.

Langley, A. (2007). Process Thinking in Strategic Organization, *Strategic Organization*, 5: 3, 271–282.

Laponche, B. (2002, October). Nucléaire: l'Exception Française? EcoRev, *Revue Critique d'Ecologie Politique*, 10.

Larsen, J. (2010). Danish Experiences with Offshore Wind Farms and Wildlife, Presentation, Vattenfall, Wind Wildlife Research Meeting VIII, Colorado, October 19–21, 2010.

Law, A. (2005). How to Build Valid and Credible Simulation Models, in: M. Kuhl, N. Steiger, F. Armstrong, and J. Jones (Eds.), Proceedings of the 2005 Winter Simulation Conference. December 4–7, 2005, Orlando, FL.

Laws, D., and Hajer, M. (2006). Policy in Practice, in: Moran, M., Rein, M., and Goodin, R. (Eds.), *The Oxford Handbook of Public Policy*. Oxford: Oxford University Press.

Leclercq, J. (1986). *The Nuclear Age: The World of Nuclear Power Plants*. Paris: Sodel.

Lee, H., Clark, W., and Devereaux, C. (2008). Biofuels and Sustainable Development, Summary Report, Executive Session on Grand Challenges of the Sustainability Transition, Venice, May 19–20, 2008.

Lehtonen, M. (2007). Biofuel Transitions and Global Governance: Lessons from Brazil, Paper, Sussex Energy Group, University of Sussex, UK.

Lehtonen, M. (2010*a*). Deliberative Decision-Making on Radioactive Waste Management in Finland, France, and the UK, *Journal of Integrative Environmental Sciences*, 7: 3, 175–196.

Lehtonen, M. (2010*b*). Opening Up or Closing Down Radioactive Waste Management Policy? Debates on Reversibility and Retrievability in Finland, France, and the United Kingdom. *Risk, Hazards, and Crisis in Public Policy*, 1: 4, 6.1.

Lehtonen, M. (2015). Megaproject Underway: Governance of Nuclear Waste Management in France, in: Brunnengraber, A., Di Nucci, M., Isidoro, A., Mez L., and, Schreurs, M. (Eds.), *Nuclear Waste Governance*. New York: Springer.

Leite, R. (2010). The Brazilian Strategy for Bioethanol, in: Cortez, L. (Ed.), *Sugarcane Bioethanol*. São Paulo: Blucher.

Lester, R., and Hart, D. (2012). *Unlocking Energy Innovation: How America Can Build a Low-Cost, Low-Carbon Energy System*. Cambridge: MIT Press.

Lester, R., and Piore, M. (2004). *Innovation: The Missing Dimension*. Cambridge: Harvard University Press.

Levi, M. (2010). *Energy Security: An Agenda for Research*. New York: Council of Foreign Relations.

Levring, P. (2017, January 24). Home of Biggest Wind Turbine Maker Bypasses Trump Sphere, *Bloomberg*.

Lewis, J. (2007). A Comparison of Wind Power Industry Development Strategies in Spain, India and China, Paper, San Francisco: Center for Resource Solutions, July 19, 2007, https://resource-solutions.org/document/a-comparison-of-wind-power-industry-development-strategies-in-spain-india-and-china/.

Lewis, J., and Wiser, R. (2007). Fostering a Renewable Energy Technology Industry: An International Comparison of Wind Industry Policy Support Mechanisms, *Energy Policy*, 35, 1844–1857.

Licht, F. O. (2016). Data (permission for reporting), World Ethanol & Biofuels Report, Agribusiness Intelligence/Informa, https://www.agra-net.com/agra/world-ethanol-and-biofuels-report/.

Lilleholt, L. (2015). A Future of Wind Energy, in: State of Green (Ed.), *Wind Energy Moving Ahead*. Frederiksberg: Danish Wind Industry Association, http://www.wind-power.org/en/.

Lindberg, L. (1977). *The Energy Syndrome: Comparing National Responses to the Energy Crisis*. Lexington: Lexington Books.

Lindblom, C. (1977). *Politics and Markets: The World's Political Economic Systems*. New York: Basic Books.

Lindblom, C., and Cohen, D. (1979). *Usable Knowledge: Social Science and Social Problem Solving*. New Haven: Yale University Press.

Lindenberg, N. (2014). Definition of Green Finance, German Development Institute, April 2014. https://www.die-gdi.de/uploads/media/Lindenberg_Definition_green_finance.pdf.

Linder, S., and Peters, B. (1989, Jan–Mar). Instruments of Government: Perceptions and Contexts, *Journal of Public Policy* 9: 1, 35–58.

Ling, K. (2009, May 18). Is the Solution to the US Nuclear Waste Problem in France? *Climate Wire*. Reprinted http://www.nytimes.com/cwire/2009/05/18/18climatewire-is-the-solution-to-the-us-nuclear-waste-prob-12208.html?pagewanted=all.

Little, D. (1995). Objectivity, Truth and Method—A Philosopher's Perspective on the Social Sciences—Commentary, *Anthropology Newsletter*. https://www.researchgate.net/publication/240657347_Objectivity_Truth_and_Method_A_Philosopher's_Perspective_on_the_Social_Sciences.

Litvin, D. (2009). *Oil, Gas and International Insecurity*. Briefing Paper. London: Chatham House, https://www.chathamhouse.org/.

Loftsdottir, A., and Thorarinsdottir, R. (2006). *Energy in Iceland* (2nd Edition). Reykjavik: Orkustofnun and Ministries of Industry and Commerce.

Logadottir, H., and Lee, H. (2012). *Iceland's Energy Policy: Finding the Right Path Forward*. Cambridge: Harvard Kennedy School.

Lohner, F. (2011, July 13). Critique du Nucléaire et Gouvernement de l'Opinion, Interview with Sezin Topcu, Contretemps. http://www.contretemps.eu/interviews/critique-nucl%C3%A9aire-gouvernement-opinion.

Loorbach, D., Frantzeskaki, N., and Thissen, W. (2010). Introduction to the Special Section: Infrastructures and Transitions. *Technological Forecasting and Social Change*, 77: 8, 1195–1202.

Louvat, D., and Metcalf, P. (2010). Closing the Cycle, *IAEA Bulletin*, 51: 2, 20–23.

Lovins, A. (1976, October). Energy Strategy: The Road Not Taken, *Foreign Affairs*. https://www.foreignaffairs.com/articles/united-states/1976-10-01/energy-strategy-road-not-taken.

Lovins, A. (1986). The Origins of the Nuclear Fiasco, in: Byrne, J., and Rich, D. (Eds.), *The Politics of Energy Research and Development*. New Brunswick: Transaction.

Lovins, A. (2011, April 8). Would the World Be Better Off Without Nuclear Power? *The Economist*, www.Economist.com/debate/days/view/685
Lovins, A., and Lovins, H. (1977). *Soft Energy Paths: Toward a Durable Peace*. San Francisco: Friends of the Earth International.
Lovins, A., and Lovins, H. (1982/2001). *Brittle Power: Energy Strategy and National Security*. Baltimore: Brick House Publishing.
Lowell, F. (2016, April 13) Outgoing President of Shell Talks Energy Shifts, Scaling Green. http://scalinggreen.tigercomm.us/2016/04/shell-world-energy/.
Lowi, T. (1979). *The End of Liberalism: The Second Republic of the United States* (2nd Edition). New York: W.W. Norton.
Lu, X., McElroy, and M., Kiviluoma, J. (2009). Global Potential for Wind-Generated Electricity, *Proceedings of the National Academy of Sciences*, 106: 27, 10933–10938.
Luft, G. Energy Self-sufficiency, UN University, Oct 24, 2012, https://ourworld.unu.edu/en/energy-self-sufficiency-a-realistic-goal-or-a-pipe-dream.
Lugar, R., and Woolsey, R. (1999, January–February). The New Petroleum, *Foreign Affairs*, 78. https://www.foreignaffairs.com/articles/middle-east/1999-01-01/new-petroleum.
Lund, H., Hvelplund, F., Alberg Østergaard, P., Möller, B., Vad Mathiesen, B., et al. (2010). Danish Wind Power Export and Cost. http://vbn.aau.dk/en/publications/danish-wind-power(71a8b090-31b0-11df-b84e-000ea68e967b).html.
Lund, J., Freeston, D., and Boyd, T. (2011). Direct Utilization of Geothermal Energy 2010 Worldwide Review, *Geothermics*, 40: 3, 159–180.
Lund, P. (2006). Market Penetration Rates of New Energy Technologies, *Energy Policy*, 34, 3317–3326.
Lundvall, B. (1988). Innovation as an Interactive Process: From User–Producer Interaction to the National System of Innovation, in: Dosi, G., Freeman, C., Nelson, R, Silverberg, G., and Soete, L. (Eds.), *Technical Change and Economic Theory*. London: Pinter.
Lundvall, B. (1985). Product Innovation and User-Producer Interaction. Paper, Aalborg, University, Aalborg, Denmark.
Lundvall, B. (Ed.). (1992). *National Systems of Innovation. Towards a Theory of Innovation and Interactive Learning*. London: Pinter.
Lundvall, B. (2010). User-Producer Relationships, National Systems of Innovation and Internationalization, in: Lundvall, B. (Ed.), *National Systems of Innovation: Towards a Theory of Innovation and Interactive Learning*. London/New York: St. Anthem Press.
Lyall, S. (2007, February 4). Smokestacks in a White Wilderness Divide Iceland, *New York Times*. http://www.nytimes.com/2007/02/04/world/europe/04iceland.html.
Macalister, T. (2007, February 15). BP Offers Gazprom a Stake after Kremlin Sets Impossible Target, *The Guardian*, https://www.theguardian.com/business/2007/feb/16/russia.oilandpetrol.
Macalister, T. (2011, September 28). Shale Boom in Bath Could Pollute Water Supplies, Warn Council Leaders, *The Guardian*, https://www.theguardian.com/environment/2011/sep/28/shale-boom-bath-pollute-water.
Macedo, I. (2005). *Sugar Cane's Energy*. São Paulo: UNICA.
Macedo, I., Leal, M., and da Silva, J. (2004). Assessment of Greenhouse Gas Emissions in the Production and Use of Fuel Ethanol in Brazil, Report, São Paulo: Government of São Paulo, March 2004.

Mackay, R., and Probert, S. (1996). Iceland's Energy and Environmental Strategy, *Applied Energy*, 53: 245–281.

Madsen, B. (n.d.). Two Oil Crisis [sic] in the 70'ies: Kick off for a New Era in Wind Power Development in Denmark, *Artikel til Poul la Cour Museets publikation om Vindkrafthistorien i Danmark*.

Maegaard, P. (2009). Danish Renewable Energy Policy. World Council of Renewable Energy/WCRE. http://www.wcre.de/en/ . . . /WCRE_Maegaard_Danish%20RE%20 Policy.pdf.

Magalhaes, P., and Braunbeck, O. (2010). Introduction, and Sugarcane and Straw Harvesting from Ethanol Production, in: Cortez, L. (Ed.), *Sugarcane Bioethanol*. São Paulo: Blucher.

Magnason, A. (2008). *Dreamland: A Self-Help Manual for a Frightened Nation*. London: Citizens Press.

Malthus, T. (1999). An Essay on the Principle of Population, in: G. Gilbert (Ed.), *Oxford World's Classics*. Oxford: Oxford University Press.

Mann, M. (1984). The Autonomous Power of the State, *European Journal of Sociology*, 25, 185–213.

Mann, M., and Gaudet, B. (2015). The "Kaya Identity," Penn State Department of Meteorology https://www.e-education.psu.edu/meteo469/node/213.

Mannvit (n.d.). Bjarnarflag Geothermal Power Plant. http://www.mannvit.com/ GeothermalEnergy/ProjectExampleinfo/bjarnarflag-geothermal-power-plant. Accessed July 15, 2016.

Marcel, V. (2005). Good Governance of the National Oil Company, Position Paper for the Workshop on Good Governance in the Petroleum Sector, London: Chatham House, February 24–25, 2005.

March, D. (1985). France Tests Its Atomic Might, *New Scientist*, 1443, 18.

March, J., and Olson, J. (1989). *Rediscovering Institutions*. New York: Free Press.

March, J., and Olson, J. (1998). The Institutional Dynamics of International Political Orders, *International Organization*, 52: 4, 943–969.

Marchetti, C. (1977). Primary Energy Substitution Models, *Technological Forecasting and Societal Change*, 10, 345–356.

Marchetti, C. (1988). Kondratiev Revisited -After One Cycle, Paper, IIASA. http://www. cesaremarchetti.org/archive/scan/MARCHETTI-037.pdf.

Marchetti, C., and Nakicenovic, N. (1979). *The Dynamics of Energy Systems and the Logistic Substitution Model*. RR-79-13, Laxenburg: International Institute for Applied Systems Analysis.

Marcus, G. (2010). *Nuclear Firsts: Milestones on the Road to Nuclear Power Development*. La Grange Park: American Nuclear Society.

Marignac, Y. (2015*a*). Fabrication Flaws in the Pressure Vessel of the EPR Flamanville-3, Consolidated Version, 12 April 2015, https://www.dropbox.com/s/njavhw7ihvk-byeu/WISE-Paris-Fabrication-Flaws-EPR-Flamanville-Latest.pdf.

Marignac, Y. (2015*b*). French Nuclear Issues: A Focus on Some Recent Developments, Presentation, NIRS Meeting, Washington DC, May 11, 2015.

Markard, J., and Truffer, B. (2008). Technological Innovation Systems and the Multi-level Perspective: Towards an Integrated Framework, *Research Policy*, 37: 596–615.

Markard, J., Raven, R., and Truffer, B. (2012). Sustainability Transitions: An Emerging Field of Research and its Prospects, *Research Policy*, 41: 955–967.

Marshall, E. (1982). Super Phenix Unscathed in Rocket Attack, *Science*, 215: 4533, 641.

Martin, P., and Turner, B. (1986). Grounded Theory and Organizational Research, *Journal of Applied Behavioral Science*, 22: 2, 141.

Martinot, E, Chaurey, A., Lew, D., Moreira, J., and Wamukonya, N. (2002). Renewable Energy Markets in Developing Countries, *Annual Review of Energy and the Environment*, 27, 309–348.

Martinot, E. (2002*a*). Grid-based Renewable Energy in Developing Countries, World Renewable Energy Policy and Strategy Forum, June 13–15, 2002, Berlin, Germany.

Martinot, E. (2002*b*). Power Sector Restructuring and Environment: Trends, Policies and GEF Experience, Discussion Paper, May 21–22, 2002, Paris, France.

Martinot, E., and Junfeng, L. (2007). *Powering China's Development: The Role of Renewable Energy*, Report. Washington, DC: WorldWatch Institute.

Martinot, E., Chaurey, A., Lew, D., Moreira, J., and Wamukonya, N. (2002). Renewable Energy Markets in Developing Countries, *Annual Review of Energy and the Environment*, 27, 309–348.

Martinot, E., Dienst, C., Weiliang, L., and Qimin, C. (2007). Renewable Energy Futures: Targets, Scenarios and Pathways, *Annual Review of Environmental Resources*, 32, 205–239.

Marx, E. (2015). How Would a Low Carbon Economy Work? *Scientific American*, http://www.scientificamerican.com/article/how-would-a-low-carbon-economy-work/.

Massachusetts Institute of Technology (MIT). (2003). The Future of Nuclear Power, http://web.mit.edu/nuclearpower/.

Massachusetts Institute of Technology (MIT). (2006). The Future of Geothermal Energy, http://energy.mit.edu/system/files/geothermal-energy-1-3.pdf.

Massachusetts Institute of Technology (MIT). (2009). Update of the MIT 2003 Future of Nuclear Power, MIT Energy Initiative, http://web.mit.edu/nuclearpower/.

Massachusetts Institute of Technology (MIT). (2011). The Future of Natural Gas, http://web.mit.edu/mitei/research/studies/documents/natural-gas-2011/NaturalGas_Report.pdf.

Matsuoka, S., Ferro, J., and Arruda, P. (2009). The Brazilian Experience of Sugarcane Ethanol Industry, *In Vitro Cell Developmental Biology, Plant* 45: 3, 372–381.

Mattei, J., Vial, E., Liemrsdorf, H., and Turschmann, M. (2001). Generic Results and Conclusions of Re-Evaluating the Flooding Protection in French and German Nuclear Power Plants, *Eurosafe Forum*.

Matter, J., Stute, M., Snæbjörnsdottir, S., Oelkers, E., Gislason, S., Aradottir, E., et al. (10 June 2016). Rapid Carbon Mineralization for Permanent Disposal of Anthropogenic Carbon Dioxide Emissions, *Science*, http://science.sciencemag.org/content/352/6291/1312.

Mattsson, N., and Wene, C. (1997). Assessing New Energy Technologies Using an Energy System Model with Endogenized Experience Curves, *International Journal of Energy Research*, 21.

McCombie, C. (2009). Responsible Expansion of Nuclear Power Requires Global Cooperation on Spent-Fuel Management, *Innovations*, 4: 4, 209–212.

McKinsey & Company. (2013, September). Pathways to a Low-Carbon Economy: Version 2 of the Global Greenhouse Gas Abatement Cost Curve, http://www.mckinsey.com/business-functions/sustainability-and-resource-productivity/our-insights/pathways-to-a-low-carbon-economy.

McMakin, A., Malone, E., and Lundgren, R. (2002). Motivating Residents to Conserve Energy without Financial Incentives, Paper, Pacific Northwest National Laboratory. http://eab.sagepub.com/content/34/6/848.abstract.

McPherson, C. (2003). National Oil Companies: Evolution, Issues, Outlook, Paper, National Oil Companies Workshop, World Bank, Washington DC, May 27, 2003.

Meadows, D., Meadows, D., Randers, J., and Behrens III, W. (1972). *The Limits to Growth*. New York: Universe Books.

Medema, S. (2007). The Hesitant Hand: Mill, Sidgwick, and the Evolution of the Theory of Market Failure, *History of Political Economy*, 39: 3, 331–358.

Megavind. (2013). The Danish Wind Power Hub, May 2013, http://www.windpower.org/da/forskning/megavind.html.

Mendonca, M., Lacey, S., and Hvelplund, F. (2009). Stability, Participation and Transparency in Renewable Energy Policy: Lessons from Denmark and the United States, *Policy and Society*, doi: 10.1016/j.polsoc.2009.01.007

Meyer, D., Mytelka, L., Press, R., Dall'Oglio, L., de Sousa Jr., P., et al. (2014). Brazilian Ethanol: Unpacking a Success Story of Energy Technology Innovation, in: Grubler, A. and Wilson, C. (Eds.), *Energy Technology Innovation: Learning from Historical Successes and Failures*. Cambridge: Cambridge University Press.

Meyer, N. (1995). Danish Wind Power Development, *Energy for Sustainable Development*, 2: 1, 18–26.

Meyer, N. (2003). European Schemes for Promoting Renewables in Liberalised Markets, *Energy Policy*, 31, 665–676.

Meyer, N. (2004). Development of the Danish Wind Power Market, *Energy and Environment*, 15: 4.

Meyer, N. (2011). Learning from Danish Experiences with Modern Wind Power, Background Notes, Achieving Decarbonization, Workshop, Fletcher School, Tufts University, March 10–11, 2011, Medford, MA.

Meyer, N. (2013). Danish Pioneering of Modern Wind Power, in: Maegaard, P., Krenz, A., and Palz, W. (Eds.), *Power for the World: the Emergence of Wind Energy* (pp. 163–176). Singapore: Pan Stanford.

Meyer, P. (2011). Brazil-US Relations, CRS Report RL33456, Washington, DC: Congressional Research Services, http://fpc.state.gov/documents/organization/157069.pdf.

Meyer, P., and Gruner, W. (1989). Improvement in Current Light Water Reactors, *IAEA Bulletin*, 31: 3, 18–21.

Middelgrundens Vindmollelaug. (n.d.), About Middlegrunden Wind Cooperative, http://www.middelgrunden.dk/middelgrunden/?q=en/node/35. Accessed May 15, 2011.

Miljø & Energi Ministeriet. (1996). Energy 21, Copenhagen, Denmark.

Miljoministeriet Naturstyrelsen. (2012). Vindmoller, http://www.naturstyrelsen.dk/Planlaegning/Planlaegning_i_det_aabne_land/Vindmoeller/.

Milligan, B. (2011). Vehicle and Fuel Technologies, in: UNEP Risoe Centre (Ed.), *Technologies for Climate Mitigation, Transport Sector*. Roskilde: UNEP Risoe Centre on Energy, Climate, and Sustainable Development, Risoe DTU National Laboratory for Sustainable Energy.

Mills, D., and Manwell, J. (2012). A Brief Review of Wind Power in Denmark, Germany, Sweden, Vermont, and Maine: Possible Lessons for Massachusetts, Report, January 11 2012, http://www.mass.gov/eea/docs/dep/energy/wind/briefreview.pdf.

Mims, C. (2008). One Hot Island, Iceland's Renewable Geothermal Power, *Scientific American*, http://www.scientificamerican.com/slideshow/iceland-geothermal-power/.

Mims, C. (2009). Can Geothermal Power in Iceland Thaw a Frozen Economy? *Scientific American*, http://www.scientificamerican.com/article/iceland-geothermal-to-thaw-frozen-economy/.

Minister for Economic and Business Affairs, Denmark. (2003, October) Report to the Danish Parliament on the North Sea, October 2003, http://www.ens.dk/sites/ens.dk/files/oil-gas/licences/agreement-ap-moller-maersk/Report_parliament.pdf.

Ministry of Industry and Commerce (MIC). (1981). Secretariat of Industrial Technology, Assessment of Brazil's National Alcohol Program, Report, Brasilia: MIC.

Ministry of Mines and Energy/Ministerio de Mines e Energia (MME). (n.d). Information, Brasilia, Brazil, http://www.mme.gov.br/mme, Accessed January 3, 2017.

Ministry of Mines and Energy/Ministerio de Mines e Energia (MME). (2016). National Energy Balance, Brasilia, Brazil, http://www.mme.gov.br/web/guest/publicacoes-e-indicadores/balanco-energetico-nacional.

Ministry of the Environment, Iceland. (2010). Iceland´s Fifth National Communication on Climate Change. http://unfccc.int/resource/docs/natc/isl_nc5_resubmit.pdf.

Minkel, J. R. (2008). The 2003 NE Blackout—Five Years Later, *Scientific American*. http://www.scientificamerican.com/article/2003-blackout-five-years-later/.

Mitchell, C., Sawin, J., Pokharel, G., Kammen, D., Wang, Z., et al. (2011). Policy, Financing and Implementation, in: Edenhofer, O., Pichs-Madruga, R., Sokona, Y., Seyboth, K., Matschoss, P., et al. (Eds.), *IPCC Special Report on Renewable Energy Sources and Climate Change Mitigation*. Cambridge: Cambridge University Press.

Moe, E. (2010). Energy, Industry and Politics: Energy, Vested Interests and Long-term Economic Growth and Development, *Energy*, 35: 4, 1730–1740.

Moe, E. (2015). *Renewable Energy Transformation or Fossil Fuel Backlash?* New York: Palgrave MacMillan.

Moe, T. (2015). Vested Interests and Political Institutions. *Political Science Quarterly*, 130: 277–318.

Mokyr, J. (1990). *The Lever of Riches: Technological Creativity and Economic Progress*. New York: Oxford University Press.

Mokyr, J. (2002). *Gifts of Athena: Historical Origins of the Knowledge Economy*. Princeton: Princeton University Press.

Montavalli, J. (2008, March/April). Iceland's Abundance of Energy, *Environmental Magazine*, https://emagazine.com/icelands-abundance-of-energy/.

Montebourg, A. (n.d.). Dépasser le Nucléaire. http://www.arnaudmontebourg2012.fr/content/depasser-le-nucleaire. Acessed June 10, 2016.

Montpetit, E. (2008). Policy Design for Legitimacy: Expert Knowledge, Citizens, Time and Inclusion in the United Kingdom's Biotechnology Sector. *Public Administration*, 86: 1, 259–277.

Montroll, E. (1978). Social Dynamics and the Quantifying of Social Forces, *Proceedings of the National Academy of Sciences in the USA*, 75, 4633–4637.

Moomaw, W., Yamba, F., Kamimoto, M., Maurice, L., Nyboer, J., et al. (2011). Introduction, in: Edenhofer, O., Pichs-Madruga, R., Sokona, Y., Seyboth, K., Matschoss, P., et al. (Eds.), *IPCC Special Report on Renewable Energy Sources and Climate Change Mitigation*. Cambridge: Cambridge University Press.

Moreira, A. (2011, February 13). Frota de Veículos Cresce 119% em Dez Anos no Brasil, aponta Denatran, *O Globo*, http://g1.globo.com/carros/noticia/2011/02/frota-de-veiculos-cresce-119-em-dez-anos-no-brasil-aponta-denatran.html.

Moreira, J. (2007). Water Use and Impacts Due to Ethanol Production in Brazil, Presentation, International Water Management Institute and Food and Ag Org Conference, Hyderabad, India, January 29–30, 2007.

Moreira, J., and Goldemberg, J. (1999). The Alcohol Program, *Energy Policy*, 27, 229–245.

Mormann, F. (2011). Requirements for a Renewables Revolution, *Ecology Law Quarterly*, 38: 4 903–966.

Morris, C., and Pehnt, M. (2015, July 15). Energy Transition: The German Energiewende, Heinrich Boll Stiftung, http://www.energytransition.de.

Moton, M. (2000). *The Delafield Commission and the American Military Profession*. College Station: Texas A&M University Press.

Mowery, D., and Rosenberg, N. (1979). The Influence of Market Demand upon Innovation, *Research Policy*, 8, 102–153.

Mowery, D., Oxley, J., and Silverman, B. (1996). Strategic Alliances and Interfirm Knowledge Transfer, *Strategic Management Journal*, Winter Special Issue: 17, 77–92.

Mowery, D., and Nelson, R. (1999). Conclusion: Explaining Industrial Leadership, in: D. Mowery and R. Nelson (Eds.), *Sources of Industrial Leadership*. Cambridge: Cambridge University Press.

Mowery, D., Nelson, R., and Martin, B. (2010). Technology Policy and Global Warming: Why New Policy Models are Needed (Or Why Putting New Wine in Old Bottles Won't Work). *Research Policy*, 39: 8, 1011–1023.

Munasinghe, M. (2004). Sustainable Development: Basic Concepts and Application to Energy, in: Cleveland, C. (Ed.), *Encyclopedia of Energy*. Boston: Elsevier.

Munoz, J., Sanchez de la Nieta, A., Contreras, J., and Bernal-Augustin, J. (2009). Optimal Investment Portfolio in Renewable Energy: The Spanish Case, *Energy Policy*, 37, 5273–5284.

Murmann, J. (2003). *Knowledge and Competitive Advantage*. Cambridge: Cambridge University Press.

Murphy, R (2011, April 14). The Significance of Spare Capacity, Energy Research Institute/ERI. http://www.instituteforenergyresearch.org/2011/04/14/the-significance-of-spare-oil-capacity/.

Mutton, M., Rosetto, R., and Mutton, M. J. (2010). Agricultural Use of Stillage, in: Cortez, L. (Ed.), *Sugarcane Bioethanol*. São Paulo: Blucher.

Mytelka, L. (2003). The Dynamics of Catching Up: The Relevance of an Innovation System Approach in Africa, in: Muchie, M., Gammeltoft, P., and Lundvall, B. (Eds.), *Putting Africa First: The Making of African Innovation Systems*. Aalborg: Aalborg University Press.

Mytelka, L., and Farinelli, F. (2003). From Local Clusters to Innovation Systems, in: Casssiolato, J., Lastres, H., Maciel, M. (Eds.), *Systems of Innovation and Development—Evidence from Brazil*. London: Edward Elgar.

Mytelka, L., and Goertzen, H. (2004). Learning, Innovation and Cluster Growth: A Study of Two Inherited Organizations in the Niagara Peninsula Wine Cluster, in: Wolfe, D., and Lucas, M. (Eds.), *Clusters in a Cold Climate*. Montreal: McGill-Queen's University Press.

Mytelka, L., and Smith, K. (2001). Innovation Theory and Innovation Policy: Bridging the Gap, Paper, DRUID Conference, Aalborg, Denmark, June 12–15, 2001.

Mytelka, L., and Smith, K. (2002). Policy Learning and Innovation Theory: An Interactive and Co-Evolving Process, *Research Policy*, 14: 23, 1–13.

Mytelka, L., and Smith, K. (2003). Interactions between Policy Learning and Innovation Theory, in: Conceiçao, P., Heitor, M., and Lundvall, B. (Eds.), *Innovation, Competence Building and Social Cohesion in Europe*. Cheltenham: Edward Elgar.

Mytelka, L., Aguayo, F., Boyle, G., Breukers, S., de Scheemaker, G., et al. (2012). Policies for Capacity Development, *Global Energy Assessment—Toward a Sustainable Future* (pp. 1745–1802). Cambridge/Laxenburg: Cambridge University Press/International Institute for Applied Systems Analysis.

Nadal, C. (2007). History of Nuclear Energy in France, Presentation, June 22, 2007, EDF.

Nakicenovic, N., and Grubler, A. (1990). *Diffusion of Technologies and Social Behavior*. Berlin/Laxenburg: Springer/IIASA.

Nakicenovic, N. (1991). Diffusion of Pervasive Systems: A Case of Transport Infrastructure, *Technological Forecasting and Societal Change* 39, 181–200.

Nakicenovic, N. (1996). Decarbonization: Doing More with Less, *Technological Forecasting and Social Change*, 51: 1–17.

Nakicenovic, N. (1997). Decarbonization as a Long-term Energy Strategy, in: Kaya, Y., and Yokobiri, K (Eds.), *Environment, Energy, and Economy: Strategies for Sustainability*. Tokyo: United Nations University Press.

Nakicenovic, N., Alcamo, J., Davis, G., de Vries, B. Fenham, J., et al. (2000). Scenario Driving Forces, in: Nakicenovic, N., and Swart, R. (Eds.), *IPCC Special Report on Emissions Scenarios* (pp. 103–166). Cambridge: Cambridge University Press.

Nakicenovic, N., Davidson, O., Davis, G., Grubler, A., Kram, T., et al. (2000). *Emissions Scenarios, Special Report of Working Group III of the Intergovernmental Panel on Climate Change, IPCC*. Cambridge: Cambridge University Press.

Nakicenovic, N., Grubler, A., and Macdonald, A. (1998/1999). *Global Energy Perspectives*. Cambridge: Cambridge University Press.

Narayanamurti, V., Anadon, L., and Sagar, A. (2009*a*). Institutions for Energy Innovation: A Transformational Challenge, Energy Technology Innovation Policy Group, Belfer Center for Science and International Affairs, Harvard Kennedy School, http://belfercenter.ksg.harvard.edu/publication/19572/institutions_for_energy_innovation.html.

Narayanamurti, V., Anadon, L., and Sagar, A. (2009*b*). Transforming Energy Innovation, *Issues in Science and Technology*, Fall 2009, 57–64.

NASA Earth Observatory. (n.d.). Svante Arrhenius (1859–1927). http://earthobservatory.nasa.gov/Features/Arrhenius/arrhenius_2.php. Accessed June 10, 2016.

Nasdaq, Petrobras - PBR Company Financials, http://www.nasdaq.com/symbol/pbr/financials?query=income-statement, Accessed May 15, 2017.

National Academy of Sciences. (2010). *Hidden Cost of Energy*. Washington, DC: National Academies Press.

National Academy of Sciences, Committee on Science and Technology for Countering Terrorism. (2002). *Making the Nation Safer: The Role of Science and Technology in Countering Terrorism*. Washington, DC: National Academies Press.

National Oceanic and Atmospheric Administration (NOAA). Trends in Atmospheric Carbon Dioxide, http://www.esrl.noaa.gov/gmd/ccgg/trends/global.html. Accessed July 29, 2017.

National Renewable Energy Laboratory (NREL). (2004). Renewable Energy Policy in China: Finantial Incentives, NREL/FS -710–35786, Golden: NREL, http://www.nrel.gov/docs/fy04osti/36045.pdf.

National Renewable Energy Laboratory (NREL), Department of Energy (n.d.). Classes of Wind Power Density at 10 m and 50 m, http://rredc.nrel.gov/wind/pubs/atlas/tables/A-8T.html, Accessed June 15, 2016.

Naval History and Heritage. (n.d.). Biographies in Naval History: Admiral Hyman G. Rickover, http://www.history.navy.mil/bios/rickover.htm, Accessed June 10, 2016.

Negro, S. (2007). *Dynamics of Technological Innovation Systems – The Case of Biomass Energy*. Innovation Studies Series. Utrecht: Utrecht University Press.

Negro, S., Hekkert, M., and Smits, R. (2007). Explaining the Failure of the Dutch Innovation System for Biomass Digestion: A Functional Analysis, *Energy Policy*, 35: 925–938.

Neij, L. (1997). Use of Experience Curves to Analyze the Prospects for Diffusion and Adoption of Renewable Energy Technology, *Energy Policy*, 23: 13, 1099–1107.

Neij, L. (2014). A Comparative Assessment of Wind Turbine Innovation and Diffusion Policies, in: Grubler, A. and Wilson, C. (Eds.), *Energy Technology Innovation: Learning from Historical Successes and Failures*. Cambridge: Cambridge University Press.

Neij, L., Andersen, P., and Durstewitz, M. (2004). Experience Curves for Wind Power, *International Journal of Energy Technology and Policy*, 2: 1/2, 15–32.

Nelkin, D., and Pollak, M. (1980, Winter). Ideology as Strategy: The Discourse of the Anti-Nuclear Movement in France and Germany, *Science, Technology & Human Values*, 5: 30, 3–13.

Nelkin, D., and Pollak, M. (1982). *The Atom Besieged*. Cambridge: MIT Press.

Nellis, J., and Kikeri, S. (2002). *Privatization in Competitive Sectors: The Record to Date, World Bank Policy Research Working Paper No. 2860*. Washington, DC: World Bank.

Nellis, J., Menezes, R., and Lucas, S. (2004). Privatization in Latin America, Policy Brief, London: Center for Global Development, https://www.cgdev.org/.

Nelson, R. (1988). Institutions Supporting Technical Change in the United States, in: Dosi, G., Freeman, C., Silverberg, G., and Soete, L. (Eds.), *Technical Change and Economic Theory*. London: Pinter.

Nelson, R. (Ed.). (1993). *National Innovation Systems: A Comparative Analysis*. New York: Oxford University Press.

Nelson, R., and Sampatt, B. (2001). Making Sense of Institutions as a Factor Shaping Economic Performance, *Journal of Economic Behavior & Organization*, 44: 1, 31–54.

Nelson, R., and Winter, S. (1982). *An Evolutionary Theory of Economic Change*. Cambridge: Belknap/Harvard University Press.

Neslen, A. (2015, July 10). Wind Power Generates 140% of Denmark's Electricity Demand, *The Guardian*. https://www.theguardian.com/environment/2015/jul/10/denmark-wind-windfarm-power-exceed-electricity-demand.

Neuenschwander, J. (1978). Remembrance of Things Past: Oral Historians and Long-Term Memory, *The Oral History Review*, 6: 1, 45–53.

Neustadt, R., and May, E. (1986). *Thinking in Time: The Uses of History for Decision-makers*. New York: Free Press.

Newton, J. (1987). *Uncommon Friends: Life with Thomas Edison, Henry Ford, Harvey Firestone, Alexis Carrel, and Charles Lindbergh*. San Diego: Harcourt.

Nielsen, F. B. (2002). A Formula for Success in Denmark, in: Pasqueletti, M., Gipe, P., and Righter, R. (Eds.), *Wind Power in View*. San Diego: Academic Press.

Nielsen, K. (2001). Tilting at Windmills, PhD Dissertation, Department of History of Science, University of Aarhus and Department of Organization and Industrial Sociology, Copenhagen Business School, September 2001.

Nielsen, K. (2005). Danish Wind Power Policies from 1976–2004: A Survey of Policy Making and Techno-economic Innovation, in: V. Lauber (Ed.), *Switching to Renewable Power: A Framework for the 21st Century*. London: Earthscan.

Nielsen, K. (2010). Technological Trajectories in the Making, *Centaurus*, 52, 175–205.

Nielsen, K., and Heymann, M. (2012). Winds of Change: Communication and Wind Power Technology Development from 1973 to ca. 1985, *Engineering Studies*, 4: 1, 11–31.

Nielsen, L. (2011). Classifications of Countries Based on Their Level of Development: How It Is Done and How It Could Be Done, Working Paper WP/11/31. Washington, DC: International Monetary Fund (IMF).

Nielsen, P., Lemming, J., Morthost, P., Clausen, N., Lawetz, H., et al. (2010). Vindmollers Okonomi, UEDP projekt 33033-0196, February 2010, http://www.emd.dk/files/Vindmøllers%20økonomi_EMD-Feb2010.pdf.

Nielsen, S. (n.d.). Offshore Wind Power in Denmark: Experience and Solutions, Presentation, Copenhagen: Danish Energy Authority.

NOAA. (2016). Annual Mean CO2 Data. ftp://aftp.cmdl.noaa.gov/products/trends/co2/co2_annmean_mlo.txt.

Nogueira, L. (2005). Biodiesel in Brazil, Presentation, IEA Bioenergy Task Force 40, Campinas, December 2005.

Nonaka, I. (1994). A Dynamic Theory of Organizational Knowledge Creation, *Organization Science*, 5: 1, 14–37.

Nordic Council of Ministers. (2011). Nordic Energy Solutions 2011, http://www.nordicenergysolutions.org/solutions/geothermal.

Nordic Folkcenter for Renewable Energy (NFCRE). (n.d.). Overview, http://www.folkecenter.net/gb/overview/. Accessed June 10, 2012.

Nord Pool. (2016). Information, http://www.nordpoolspot.com. Accessed July 10, 2016.

North, D. (1990). *Institutions, Institutional Change and Economic Performance*. New York: Cambridge University Press.

Northern Territory of Australia. (2012). Development of One Stop Shop, http://www.nt.gov.au/lands/planning/onestopshop/index.shtml.

Nuclear Energy Agency (NEA). (n.d.*a*). Radiation Protection Today and Tomorrow. https://www.oecd-nea.org/rp/reports/1994/rp.html. Accessed July 10, 2016.

Nuclear Energy Agency (NEA). (n.d.*b*.), World Energy Needs and Nuclear Power, http://www.world-nuclear.org/information-library/current-and-future-generation/world-energy-needs-and-nuclear-power.aspx. Accessed July 10, 2016.

Nuclear Energy Agency (NEA). (n.d.*c*). Data. Paris: NEA/OECD. Accessed December 15, 2016.

Nuclear Energy Agency (NEA). (1992). Principles for Intervention for Protection of the Public in a Radiological Emergency, ICRP Publication 63, Ann. ICRP 22:4.

Nuclear Energy Agency (NEA). (2006). Forty Years of Uranium Resources, Production and Demand in Perspective. Paris: NEA/OECD. https://www.oecd-nea.org/ndd/pubs/2006/6096-40-years-uranium.pdf.

Nuclear Energy Agency (NEA). (2007). Innovation in Nuclear Energy Technology, Nuclear Development, NEA No. 6103. Paris: NEA/OECD. https://www.oecd-nea.org/ndd/pubs/2007/6103-innovation-technology.pdf.

Nuclear Energy Agency (NEA). (2009). Application of the Commission's Recommendations for the Protection of People in Emergency Exposure Situations, ICRP Publication 109, Ann ICRP 39:1.

Nuclear Energy Agency (NEA). (2011, 2010, 2009, 2004, 2003). Nuclear Energy Data, French Country Report. Paris: OECD/IEA.

Nuclear Energy Agency (NEA). (2015). Nuclear Energy: Combatting Climate Change, Report. Paris: NEA/OECD.

Nuclear Energy Agency (NEA) and International Atomic Energy Association (IAEA). (2010). Uranium 2009: Resources, Production, and Demand, NEA No. 6891. Paris: OECD. https://www.oecd-nea.org/ndd/pubs/2010/6891-uranium-2009.pdf.

Nuclear Energy Agency (NEA) and OECD. (2010). Technology Roadmap: Nuclear Energy. Paris: OECD/IEA. https://www.iea.org/media/freepublications/technology-roadmaps/TechnologyRoadmapNuclearEnergy.pdf.

Nunberg, B. (1979). State Intervention in the Sugar Sector in Brazil, PhD Dissertation, Stanford University, Stanford, CA.

Nurse, K. (2007, February 12). Creative Industries as Growth Engine, Policy Innovations, Carnegie Council. http://www.policyinnovations.org/ideas/innovations/data/creative_cultural.

Nye, D. (1999). *Consuming Power: A Social History of American Energies*. Cambridge: MIT Press.

Nye, D. (1992). *Electrifying America: Social Meanings of a New Technology, 1880–1940*. Cambridge: MIT Press.

Nye, D. (2009). History of Electricity Use, in: Cleveland, C. (Ed.), *History of Energy*. New York: Elsevier.

O'Connor, P. (2010). Energy Transitions, Pardee Working Papers, No. 12, Boston University, Boston, MA, http://pardee.du.edu/working-papers.

Odgaard, O. (2000). The Green Electricity Market in Denmark: Quotas, Certificates and International Trade, Workshop on Best Practices in Policies and Measures, April 11–13, 2000, Copenhagen, Denmark.

Office of Electricity Delivery and Energy Reability (OEDER). (2004, April). US–Canada Power System Outage Task Force, Final Report on the August 14, 2003 Blackout in the US and Canada, Washington, DC: OEDER, https://energy.gov/oe/downloads/us-canada-power-system-outage-task-force-final-report-implementation-task-force.

Official Website of Denmark (OWD). (n.d.). Government and Politics, http://denmark.dk/en/society/government-and-politics/. Accessed June 25, 2015.

Ogden, P., Podesta, J., and Deutch, J. (2008, Winter). A New Strategy to Spur Energy Innovation, *Issues in Science and Technology*, http://issues.org/24-2/ogden/.

Olafsson, H. (1981). A True Environmental Parable: The Laxa-Myvatn Conflict in Iceland, 1965–1973, *Environmental Review*, 5: 2–38.

Olafsson, L., Thorsteinsson, I., Petursdottir, H., and Eggersson, H. (2011). A Report on Regulation and the Electricity Market 2010, NEA OS-2011/01, Reykjavik: Orkustofnun.

Olleros, F. (1986). Emergent Industries and the "Burnout of Pioneers," *Journal of Product Innovation Mangement,* 1, 5–8.

Olson, M. (1965). *The Logic of Collective Action: Public Goods and the Theory of Groups*. Cambridge, MA: Harvard University Press.

Olson, M. (1982). *The Rise and Decline of Nations: Economic Growth, Stagflation and Social Rigidities*. New Haven, CT: Yale University Press.

Olz, S., Sims, R., and Kirchner, N. (2007). Contribution of Renewable Energy to Energy Security, IEA Paper, Paris: IEA/OECD.

Organization of Economic Cooperation and Development (OECD). (n.d.). http://www.oecd.org/. Accessed July 10, 2016.

Organization of Economic Cooperation and Development (OECD). (2015*a*). *Overcoming Barriers to International Investment in Clean Energy, Green Finance and Investment*. Paris: OECD Publishing.

Organization of Economic Cooperation and Development (OECD). (2015*b*). *Aligning Policies for the Transition to a Low-Carbon Economy*. Paris: OECD Publishing.

Organization of Economic Cooperation and Development (OECD) and Food and Agriculture Organization (FAO). (2011). OECD-FAO Agricultural Outlook 2011–2020. http://dx.doi.org/10.1787/agr_outlook-2011-en.

Organization of Economic Cooperation and Development (OECD).and Nuclear Energy Agency (NEA). (2007). Regulatory and Institutional Framework for Nuclear Activities: Denmark, Report. http://www.oecd-nea.org/law/legislation/denmark.pdf.

Orihata, M., and Watanabe, C. (2000). Evolutional Dynamics of Product Innovation: The Case of Consumer Electronics, *Technovation*, 20: 8, 437–449.

Orkustofnun. (n.d.). Information. http://www.nea.is. Accessed January 10, 2017.

Orkustofnun. (2011). Report on Regulation and the Electricity Market. http://www.ceer.eu/portal/page/portal/EER_HOME/EER_PUBLICATIONS/NATIONAL_REPORTS/National_Reporting_2011/NR_En/C11_NR_Iceland-EN.pdf.

Orkustofnun. (2016*a*). First Steps towards a North Atlantic Energy Network. January 26, 2016, http://www.nea.is/the-national-energy-authority/news/2016/01.

Orkustofnun. (2016*b*), Norges Arktiske Universitet, Energy Styrelsen, Jarðfeingi, Shetland Islands Council, Greenland Innovation Centre, North Atlantic Energy Network, January 2016.

Ornetzeder, M., and Rohracher, H. (2006). User-led Innovations and Participation Processes: Lessons from Sustainable Energy Technologies, *Energy Policy*, 34: 2, 138–150.

Osbourne, S. Vengosh, A., Warner, N., and Jackson, R. (2011). Methane Contamination of Drinking Water Accompanying Gas-well Drilling and Hydraulic Fracturing, *Proceedings of the National Academy of Sciences*, 108: 20, 8172–8176.www.pnas.org/cgi/doi/10.1073/pnas.1100682108

Osmani, S. (2008). Participatory Governance: An Overview of Issues and Evidence, in: United Nations Department of Economic and Social Affairs (UNDESA). (Ed.), *Participatory Governance and the Millennium Development Goals (MDGs).* New York: United Nations.

Ostrom, E. (1990). *Governing the Commons: The Evolution of Institutions for Collective Action*. New York: Cambridge University Press.

Ottersen, O. (2003). The Institutional Co-evolution in Diffusion of Hydroelectric Power Technology, *International Journal of Global Energy Issues*, 20: 2,168–190.

Owen, A. (2004). Environmental Externalities, Market Distortions and the Economics of Renewable Energy Technologies, *Energy Journal*, 25: 3, 127–156.

Owen, A. (2006). Renewable Energy: Externality Costs as Market Barriers, *Energy Policy*, 34, 632–642.

Ozorio del Almeida, M. (1972). *Environment and Development: The Founex Report on Development and Environment.* New York: Carnegie Endowment for International Peace.

Pacala, S., and Socolow, R. (2004). Stabilization Wedges Solving the Climate Problem for the Next 50 Years with Current Technologies, *Science*, 305: 968.

Pachauri, S., Brew-Hammond, A., Barnes, D., Bouille, D., Gitonga, S., et al. (2012). Energy Access for Development, *Global Energy Assessment—Toward a Sustainable Future* (pp. 1401–1458). Cambridge/Laxenburg: Cambridge University Press/ International Institute for Applied Systems Analysis.

Pacheco, D., York, J., Dean, T., and Sarasvathy, S. (2010). The Co-evolution of Institutional Entrepreneurship, Research Paper, No. 2010, Charlottesville: Batten Institute, https://papers.ssrn.com/sol3/papers.cfm?abstract_id=1615032.

Pacific Biodiesel. (n.d.). History of Biodiesel Fuel, http://www.biodiesel.com/biodiesel/history/. Accessed June 10, 2016.

Palfreman, J. (n.d.). Why the French Like Nuclear Power, *PBS*. http://www.pbs.org/wgbh/pages/frontline/shows/reaction/readings/french.html. Accessed June 15, 2016.

Palsdottir, S. (Ed). (2005). *Landsvirkjun, 1965–2005*. Reykjavik, Iceland: Hid Islenska Bokmenntafelag.

Paris, L., Zini, G., Valtorta, M., Manzoni, G., Invernizzi, A., et al. (1984). Present Limits of Very Long Transmission Systems, CIGRE International Conference on Large High Voltage Electric Systems, Paris, France, August 29–September 1984.

Parrish, B. (2008). Sustainability-driven Entrepreneurship: A Literature Review. Working Paper, Leeds: Sustainability Research Institute, https://www.see.leeds.ac.uk/fileadmin/Documents/research/sri/ . . . /SRIPs-09_01.pdf.

Parrish, B., and Foxon, T. (2009). Sustainability Entrepreneurship and Equitable Transitions to a Low-Carbon Economy. *Greener Management International*, 55, 47–62.

Patterson, W. (2005, April). Infrastructure Needed, Not Fuel, *Modern Power Systems*, http://www.modernpowersystems.com/news/newsinfrastructure-needed-not-fuel.

Patterson, W. (2011). Nuclear Reactions, Paper, Nuclear Power—Past, Present, Future? Oxford Workshop, May 5–6, 2011.

Patton, M. (1990). *Qualitative Evaluation and Research Methods* (2nd Ed.). Newbury Park: Sage.

Patwardhan, A., Azevedo, I., Foran, T., Patankar, M., Rao, A., et al. (2012). Transitions in Energy Systems, in: *Global Energy Assessment—Toward a Sustainable Future* (pp. 1173–1202). Cambridge/Laxenburg: Cambridge University Press/International Institute for Applied Systems Analysis.

Pearce, F. (2012, July 30). Are Fast Breeder Reactors the Answer to Our Nuclear Waste Nightmare? *The Guardian*. https://www.theguardian.com/environment/2012/jul/30/fast-breeder-reactors-nuclear-waste-nightmare.

Pearson, J. (2015). On the Belated Discovery of Fission, *Physics Today*, 68: 6, 40.

Pedersen, J. (2010). Science, Engineering and People with a Mission: Danish Wind Energy in Context 1891–2010, International Schumpeter Society Conference, Aalborg, Denmark, August 2, 2010.

Pedersen, J., and Xinxin, K. (2012). *The Importance of Basic Research for Inventions and Innovations in Wind Energy*, Unpublished paper.

Pedersen, O. (1977). Developments in the Uranium Enrichment Industry, *IAEA Bulletin*, 19: 1, 40–52. https://www.iaea.org/sites/default/files/publications/magazines/bulletin/bull19-1/19104884052.pdf.

People's Daily (PD). (2010, July 22). China Develops 5 Trillion Yuan Alternative Energy Plan. *People's Daily*, http://english.peopledaily.com.cn/90001/90778/90862/7076933.html.

Pereira da Cunha, M. (2010). Sustainability of Sugarcane Bioenergy: Socio-economic Impacts, November 8–9, Presentation, Brasilia, Brazil.

Perez, C. (1983). Structural Change and Assimilation of New Technologies in Economic and Social System, *Futures*, 13, 357–375.

Perez, C. (1985). Microelectronics, Long Waves and World Structural Change: New Perspectives for Developing Countries, *World Development*, 13: 3, 441–463.

Perez, C. (1986). The *New Technologies: An Integrated View*, Working Paper, TOC/TUT No. 19, Technology Governance and Economic Dynamics. Tallinn, Estonia: The Other Canon Foundation, Norway and Tallinn University of Technology.

Perez, C. (1988). Catching Up in Technology: Entry Barriers and Windows of Opportunity (with L. Soete), in: Dosi, G., Freeman, C., Silverberg, G., and Soete, L. (Eds.), *Technical Change and Economic Theory*. London: Pinter.

Perez, C. (1994). Technical Change and the New Context for Development in: Mytelka, L. (Ed.), *South-South Co-operation in a Global Perspective*. Paris: OECD/IEA.

Perez, C. (2001, December). Technological Change and Opportunities for Development as a Moving Target, *CEPAL Review*, 75.

Perez, C. (2004*a*). Technological Revolutions, Paradigm Shifts, and Socio-Institutional Change, in: E. Reinert (Ed.), *Globalization, Economic Development, and Inequality: An Alternative Perspective*. Cheltenham: Edward Elgar.

Perez, C. (2008). A Vision for Latin America: A Resource-based Strategy for Technological Dynamism and Social Inclusion, Globelics Working Paper Series, 08-04. http://www.globelics.org/article/a-vision-for-latin-america-a-resource-based-strategy-for-technological-dynamism-and-social-inclusion/.

Perez, C. (2009*a*). Technological Revolutions and Techno-economic Paradigms, *Cambridge Journal of Economics*, 34: 1, 185–202.

Perez C. (2009*b*). The Double Bubble at the Turn of the Century: Technological Roots and Structural Implications, *Cambridge Journal of Economics*, 33, 779–805.

Perez-Arriaga, I., and Linares, P. (2008). Markets versus Regulation: A Role for Indicative Energy Planning, *Energy Journal*, 149–164.

Pescia, D., Graichen, P., Jacobs, and D. (2015, October). Agora Energiewende and International Energy Transition GmbH, Agora Energiewende: Understanding the Energiewende. www.agora-energiewende.de.

Petersen, E., Troen, I., Frandsen, S., and Hedegaard, K. (1981). *Wind Atlas for Denmark.* Roskilde, Denmark: Riso National Laboratory.

Petrobras. (n.d). Information, http://www.petrobras.com.br/en/. Accessed July 10, 2012.

Petrostrategies. (2012, February). World's Largest Oil and Gas Companies. http://www.petrostrategies.org/.

Petrursson, G. (2008). The Kárahnjúkar 700 MWe Hydro Power Project in NE-Iceland: Challenges during the Construction and Completion Phase, Power and Energy Society General Meeting - Conversion and Delivery of Electrical Energy in the 21st Century, IEEE, http://ieeexplore.ieee.org/document/4596945/?reload=true.

Philipp, S., Gudmundsson, A., and Oelrich, A. (2007). How Structural Geology Can Contribute to Make Geothermal Projects Successful, Proceedings European Geothermal Congress, Unterhaching, Germany, May 30–June 1, 2007.

Pidwirny, M. (2006). Definitions of Systems and Models. Fundamentals of Physical Geography (2nd Edition). http://www.physicalgeography.net/fundamentals/4b.html.

Platts. (2012, January 11). Brazil's Ethanol Imports from US Notch Record High in 2011, *Platts*, http://www.platts.com/RSSFeedDetailedNews/RSSFeed/Petrochemicals/7014446.

Poole, M. (2004). Central Issues in the Study of Change and Innovation, in: Poole, M., and Van den Ven, A. (Eds.), *Handbook of Organizational Change and Innovation.* Oxford: Oxford University Press.

Porter, J., Chirinda, N., Felby, C., and Olsen, J. (2008). Biofuels: Putting Current Practices in Perspective, *Science Magazine*, 320: 5882, 1421–1422.

Porter, M. (1990). *The Competitive Advantage of Nations.* New York: Free Press.

Porter, M. (1996). *On Competition.* Cambridge, MA: Harvard Business Review.

Porter, M. (1998). Clusters and Competition: New Agendas for Companies, Governments, and Institutions on Competition, Working Paper No. 98-080, Boston: Harvard Business School.

Pouget-Abadie, X. (2012). Complementary Safety Assessments of the French Nuclear Facilties, *Comptes Redus Physique*, 13, 359–364, http://adsabs.harvard.edu/abs/2012CRPhy..13..359P.

PR Newswire. (2011). Magma Energy Announces Findings of Special Committee—HS Orka Transaction Fully in Compliance with Icelandic Law, *PR Newswire.* http://www.prnewswire.com/news-releases/magma-energy-announces-findings-of-special-committee---hs-orka-transaction-fully-in-compliance-with-icelandic-law-103139529.html.

PR Newswire. (2010). EPA Reaffirms Sugarcane Biofuel Is Advanced Renewable Fuel with 61% Less Emissions than Gasoline, *PR Newswire*. http://www.prnewswire.com/news-releases/epa-reaffirms-sugarcane-biofuel-is-advanced-renewable-fuel-with-61-less-emissions-than-gasoline-83483922.html.

Pradhan, A. (2012, July 27). India to Establish Nuclear Reactor that Uses Thorium as Fuel: Atomic Energy Commission, *The Times of India*, http://articles.timesofindia.indiatimes.com/2012-06-27/india/32440058_1_thorium-nuclear-reactor-kudankulam.

Pratt, J. (1981). The Ascent of Oil: The Transition from Coal to Oil in Early Twentieth-Century America, in: Perelman, L., Giebelhaus, A., and Yokell, M. (Eds.), *Energy Transitions: Long-Term Perspectives*. Boulder: Westview.

President's Council of Advisors on Science and Technology (PCAST). (1999). *Powerful Partnerships: The Federal Role in International Cooperation on Energy Innovation*. Washington, DC: Office of the President, Executive Office of the United States.

President's Council of Advisors on Science and Technology (PCAST). (2010). *Report to the President on Accelerating the Pace of Change in Energy Technologies through Integrated Federal Energy Policy*. Washington, DC: Office of the President, Executive Office of the United States.

President's Materials Policy Commission. (1952). *Resources for Freedom: Foundations for Growth and Security*. Washington, DC: GPO.

Price, R. (2005). *A Concise History of France* (2nd Ed). Cambridge: Cambridge University Press.

Public Broadcasting Service (PBS). (2009, October 30). Timeline: History of the Electric Car. http://www.pbs.org/now/shows/223/electric-car-timeline.html.

Public Private Infrastructure Advisory Facility (PPIAF), Public Utility Research Center, and World Bank. (2011). Body of Knowledge on Infrastructure Regulation. http://www.regulationbodyofknowledge.org/chapter2/narrative/. Accessed November 20, 2011.

Puerto Rico, J. (2007). Programa de Biocombustíveis no Brasil e na Colômbia: Uma análise da implantação resultados e perspectivas, Thesis, Programa Interunidadesde Pos-Graduacao em Energia, Universidade de São Paulo, Brasil.

Puffert, D. (2010). Path Dependence, Economic History Encyclopedia. http://eh.net/?s=path+dependence.

Puko, T., and Yap, C. (2015, February 26). Falling Chinese Coal Consumption and Output Undermine Global Market, *Wall Street Journal*, http://www.wsj.com/articles/chinas-coal-consumption-and-output-fell-last-year-1424956878.

Quitzow, L, Canzler, W., Grundmann, P., Leibenath, M., Moss, T., et al. (2016). The German Energiewende—What's Happening? *Utilities Policy*, 41, 163–171.

Rabobank. (2012, March 1). The Future of Ethanol—Brazilian and US Perspectives, *PR Newswire*. http://www.prnewswire.com/news-releases/rabobank-report-the-future-of-ethanol---brazilian-and-us-perspectives-141054333.html.

Ragheb, M. (2013). Global Wind Power Status. Paper. http://www.ragheb.co/NPRE%20475%20Wind%20Power%20Systems/Global%20Wind%20Power%20Status.pdf.

Ragnarsson, A. (1996, November). Geothermal Energy in Iceland, *GHC Bulletin*, Klamath Falls: Geo-Heat Center.

Ragnarsson, A. (2003). Utilization of Geothermal Energy in Iceland, International Geothermal Conference, Reykjavik, Iceland, September 2003.

Ragnarsson, A. (2010). Geothermal Development in Iceland 2005–2009, *Proceedings of the 2010 World Geothermal Congress*, Bali, Indonesia, April 25–30, 2010.

Ragnarsson, A. (2015). Geothermal Development in Iceland 2010–2014, *Proceedings of the 2015 World Geothermal Congress*, Melbourne, Australia, April 19–25, 2015.

Rammaaaetlun. (n.d.). The Master Plan for Nature Protection and Energy Utilization, http://www.rammaaaetlun.is/english. Accessed July 20, 2016.

Rascouet, A., and Chmaytelli, M. (2015, October 16). Big Oil Companies Back Agreement to Prevent Climate Change, *Bloomberg*. http://www.bloomberg.com/news/articles/2015-10-16/big-oil-companies-declare-support-for-global-climate-agreement.

Rato, G. (2015, July 7). Motores Flex Precesiam de Mais Eficiencia, *Automotive Business*. http://www.automotivebusiness.com.br/noticia/22288/motores-flex-precisam-de-mais-eficiencia.

Ratti, R., Seol, Y., and Yoon, K. (2011). Relative Energy Price and Investment by European Firms, *Energy Economics*, 33: 5, 721–731.

Raven, R., van den Bosch, S., Weterings, R. (2010). Transitions and Strategic Niche Management, *International Journal of Technology Management*, 51:1, 57–74.

Raven, R. (2012). Analyzing Emerging Sustainable Energy Niches in Europe: A Strategic Niche Management Perspective, in: Verbong, G., and Loorbach, D. (Eds.), *Governing the Energy Transition: Reality, Illusion or Necessity?* New York: Routledge.

Redlinger, R., Dannemand Andersen, P., and Morthorst, P. (2002). *Wind Energy in the 21st Century*. London: Palgrave.

Reed, S. (2015, October 16). Oil and Gas Companies Make Statement in Support of U.N. Climate Goals, *New York Times*. http://www.nytimes.com/2015/10/17/business/energy-environment/oil-companies-climate-change-un.html?_r=0.

Reitun. (2010). Credit Analysis and Rating, Landsvirkjun. http://www2.reitun.is/ . . . / Landsvirkjun/Reitun_Landsvirkjun_CreditRating_j . . .

Renewable Energy Policy Network for the 21st Century (REN21). (2010, 2011, 2012, 2013, 2014, 2015, 2016, 2017). Global Status Report. Paris: REN21. http://www.ren21.net/.

Renewable Energy Research Lab, University of Massachusetts. (1970). Wind Power: Capacity Factor, Intermittency, and What Happens When the Wind Doesn't Blow?, Fact Sheet2a. http://www.ceere.org/rerl/about_wind/RERL_Fact_Sheet_2a_Capacity_Factor.pdf.

Renewable Energy World. (2010, November 22). Measuring Wind Turbine Noise. http://www.renewableenergyworld.com/rea/news/article/2010/11/measuring-wind-turbine-noise.

Renewable Energy World. (2016, June 23). World's Longest Wind Turbine Blade Unveiled in Denmark, http://www.renewableenergyworld.com/articles/2016/06/world-s-longest-wind-turbine-blade-unveiled-in-denmark.html.

Renewable Fuel Association (RFA). (2016). Industry Statistics. http://www.ethanolrfa.org/. Accessed December 15, 2016.

Réseau de Transport d'électricité (RTE). (n.d.). Information. http://www.rte-france.com/en/discover-rte/about-us/organization/by-laws. Accessed July 15, 2012.

Réseau de Transport d'électricité (RTE). (2016). Annual Electricity Report 2015, Cedex: RTE http://www.rte-france.com/en/article/annual-electricity-reports.

Reuters. (2009, January 7). 18 Countries Affected by Russia-Ukraine Row. http://www.reuters.com/article/2009/01/07/uk-russia-ukraine-gas-factbox-idUKTRE5062Q520090107?sp=true).

Reuters. (2012, February 24). Brazil Plans $38 B. Sweetener to Revive Ethanol Sector. http://www.reuters.com/article/2012/02/24/ethanol-brazil-idUSL2E8DO7AH20120224.

Reuters. (2012, July 25). French Nuclear Dismantlement Financing Not Yet Secure—Report. http://www.reuters.com/article/france-nuclear-dismantling-idUSL6E8IO KJI20120725.

Reuters. (2014, March 18). France: Greenpeace Activists Arrested in Break-in, *New York Times*. http://www.nytimes.com/2014/03/19/world/europe/france-greenpeace-activists-arrested-in-break-in.html?_r=0.

Revista Veja. (1979, June 13). O Petroleo de Cana, Veja—Em Profundidade Petróleo. http://veja.abril.com.br/idade/exclusivo/petroleo/130679.html.

Reykjavik Energy. (n.d.). Information, http://www.or.is/. Accessed December 15, 2016.

Rhodes, R. (1993). *Nuclear Renewal: Common Sense About Energy*. New York: Whittle/Viking.

Rhodes, R. (2007). Energy Transitions: A Curious History. Remarks, Center for International Security and Cooperation, Stanford University, September 19, 2007.

Riahi, K., Dentener, F., Gielen, D., Grubler, A., Jewell, J., et al. (2012). Energy Pathways for Sustainable Development, *Global Energy Assessment—Toward a Sustainable Future* (pp. 1203–1306). Cambridge/Laxenburg: Cambridge University Press/International Institute for Applied Systems Analysis.

Richter, A. (2011, January 28). Uncertainty in Iceland's Geothermal Market, *Renewable Energy World*. http://www.renewableenergyworld.com/index.html.

Richter, A. (2016, January 21). Iceland Geosurvey Secures around EUR 4 M in EU Grants for Several Research Projects, Think Geoenergy. http://www.thinkgeoenergy.com/iceland-geosurvey-secures-around-eur-4m-in-eu-grants-for-several-research-projects/.

Richter, D. (1978). National and International Activities in the Field of Underground Disposal of Radioactive Wastes. *IAEA Bulletin*, 20: 4, 30–41. https://www.iaea.org/sites/default/files/publications/magazines/bulletin/bull20-4/20402643041.pdf.

RIDESA. (n.d.). Rede Interuniversitária para o Desenvolvimento do Setor Sucro-energético, Universidade Federal de Goiás, http://www.ridesa.agro.ufg.br/. Accessed July 15, 2016.

Ridley, T., Yee-Cheong, L., and Juma, C. (2006). Infrastructure, Innovation and Development, *International Journal of Technology and Globalization*, 2: 3–4, 268–278.

Righter, R. (1996). *Wind Energy in America*. Norman: University of Oklahoma Press.

Rioux, J. (1989). *The Fourth Republic, 1944–1958*. Cambridge History of Modern France. Cambridge: Cambridge University Press.

Rip, A., Misa, T., and Schot, J. (Eds.). (1995). *Managing Technology in Society: The Approach of Constructive Technology Assessment*. London: Pinter.

Rist, R. (1998). Choosing the Right Policy Instrument at the Right Time, in: Bemelmans-Videc, M., Rist, R., and Vedung, E. (Eds.), *Carrots, Sticks, and Sermons*, New Brunswick: Transaction.

Robinson, M., and Musial, W. (2006, October). *Offshore Wind Technology Overview*. Golden: National Renewable Energy Laboratory.

Rocha, L. (2009). Outlook on Brazil's Sugarcane Industry Competitiveness, *International Sugar Journal*, 111: 1327, 422–427.

Roche, B. (2011). The French Nuclear Program: EDF's Experience. http://apw.ee.pw.edu.pl/tresc/-eng/08-FrenchNucleamProgram.pdf.

Rockwell, T. (2002). *The Rickover Effect*. Lincoln, NE: iUniverse Inc.

Rodrigues, R., and Accarini, J. (2009). Brazil's Biodiesel Program. http://www.dc.itamaraty.gov.br/imagens-e-textos/Biocombustiveis-09ing-programabrasileirobiodiesel.pdf.

Rogers, E. (1995). *Diffusion of Innovations* (4th Ed.) New York: Free Press.

Rogge, K., and Reichardt, K. (2016). Policy Mixes for Sustainability Transitions: An Extended Concept and Framework for Analysis, *Reseach Policy*, 45: 8, 1620–1635.

Rogner, H., Aguilera, R., Archer, C., Bertani, R., Bhattacharya, S., et al. (2012). Energy Resources and Potentials, *Global Energy Assessment—Toward a Sustainable Future* (pp. 423–512). Cambridge/Laxenburg: Cambridge University Press/ International Institute for Applied Systems Analysis.

Rootes, C. (2008). The Environmental Movement, in: Klimke, M., and Scharloth, J. (Eds.), *1968 in Europe*. New York: Palgrave Macmillan.

Rosenberg, N. (1975). *Perspectives on Technology*. Cambridge: Cambridge University Press.

Rosenberg, N. (1994). *Exploring the Black Box: Technology, Economics and History*. Cambridge: Cambridge University Press.

Rosenberg, N., and Frischtak, C. (1983). *Technological Innovation and Long Waves*, Stanford University, Mimeo.

Rosillo-Calle, F., and Cortez, L. (1998). Towards ProÁlcool II: A Review of the Brazilian Bioethanol Programme, *Biomass and Bioenergy*, 14: 2, 115–124.

Rossetto R., Cantarella H., Dias F., Landell M., and Vitti A. (2008). Conservation Management and Nutrient Recycling in Sugarcane with a View to Mechanical Harvesting, *Informações Agronômicas*, 8–13.

Rothkopf, G. (2009). A Blueprint for Green Energy in the Americas, Volume 1. Washington, DC: Inter-American Development Bank. http://www.gartenrothkopf.com/research-and-analysis/custom-research-publications.html.

Runge, C., and Senauer, B. (2007, May–June). How Biofuels Could Starve the Poor, *Foreign Affairs*. https://www.foreignaffairs.com/articles/2007-05-01/how-biofuels-could-starve-poor.

Ruttan, V. (2006). *Is War Necessary for Economic Growth?* Oxford, UK: Oxford University Press.

Sabatier, P., and Jenkins-Smith, H. (1993). *Policy Change and Learning: An Advocacy Coaltion Framework*. Boulder: Westview Press.

Sachs, I., Maimom, D., and Tolmasquim, M. (1987). The Social and Ecological Impact of "Pro-Alcool," *IDS Bulletin*, 18: 39–46. doi: 10.1111/j.1759–5436.1987.mp18001006.x.

Sachs, J., and Warner, A. (1997). *Natural Resource Abundance and Economic Growth*. Cambridge: Center for International Development and Harvard Institute for International Development.

Sachs, J., and Warner, A. (1999). The Big Push and Natural Resource Booms, *Journal of Development Economics*, 59: 1, 43–76.

Sawin, J. (2001). The Role of Government in the Development and Diffusion of Renewable Energy Technologies. PhD Dissertation, Fletcher School of Law and Diplomacy, Tufts University, September 2001.

Shapouri, H., and Salassi, M. (2006). The Economic Feasibility of Ethanol Production from Sugar in the United States. Office of Energy Policy and New Uses (OEPNU), Office of the Chief Economist (OCE), U.S. Department of Agriculture (USDA), and Louisiana State University (LSU).

Samsø Energy Academy. (n.d.). Sustainable Cities. http://sustainablecities.dk/en/city-projects/cases/samsoe-a-role-model-in-self-sufficiency. Accessed July 15, 2016.

Sandalow, D. (2006, May 1). Ethanol: Lessons from Brazil, Brookings, http://www.brookings.edu/research/articles/2006/05/energy-sandalow.

Sandstrom, C. (2010). A Revised Perspective on Disruptive Innovation—Exploring Value, Networks and Business Models, Thesis, Chalmers University of Technology, Goteborg, Sweden.

Sangster, A. (2010). *Energy for a Warming World*. New York: Springer.

Sapp, M. (2016, January 7). Brazilian Biodiesel Production up 85% in 2015, *Biofuels Digest*. http://www.biofuelsdigest.com/bdigest/2016/01/07/brazilian-biodiesel-production-up-85-in-2015/.

Sapp, M. (2013, October 1). Second Phase of Brazilian Ethanol Pipeline Getting Underway, http://www.biofuelsdigest.com/bdigest/2013/10/01/second-phase-of-brazilian-ethanol-pipeline-getting-underway/.

Sarrica, M., Brondi, S., Cottone, P., and Mazzara, B. (2016). One, No One, One Hundred Thousand Energy Transitions in Europe: The Quest for a Cultural Approach, *Energy Research and Social Science*, 13, 1–14.

Sathaye, J., Lucon, O., Rahman, A., Christensen, J., Denton, F., et al. (2011). Renewable Energy in the Context of Sustainable Development, in: Edenhofer, O., Pichs-Madruga, R., Sokona, Y., Seyboth, K., Matschoss, P., et al. (Eds.), *IPCC Special Report on Renewable Energy Sources and Climate Change Mitigation*. Cambridge: Cambridge University Press.

Saumon, D., and Puiseux, L. (1977). Actors and Decisions in French Energy Policy, in: Lindberg. L. (Ed.), *The Energy Syndrome*. Lexington: Lexington Books.

Savic, P. (1989). A Prelude to Fission: France, in: Behrens, J., and Carlson, A. (Eds.), *50 Years with Nuclear Fission*. La Grange Park: American Nuclear Society.

Saving Iceland. (n.d.). Information, http://www.savingiceland.org/. Accessed August 1, 2016.

Sawin, J. (2001). The Role of Government in the Development and Diffusion of Renewable Energy Technologies, PhD Dissertation, Fletcher School of Law and Diplomacy, Tufts University, September 2001.

Sawin, J. (2004*b*). *Mainstreaming Renewable Energy in the 21st Century*, WorldWatch Paper 169. Washington, DC: WorldWatch Institute.

Sawin, J. (2004*b*). National Policy Instruments: Policy Lessons for the Advancement and Diffusion of Renewable Energy Technologies around the World. International Conference for Renewable Energies. http://www.ren21.net/Portals/0/documents/irecs/renew2004/National%20Policy%20Instruments.pdf.

Sawin, J., and Moomaw, W. (2009). Renewable Revolution: Low-Carbon Energy by 2030, WorldWatch Institute and the Renewable Energy and Energy Efficiency Partnership. http://www.worldwatch.org/node/6342.

Scandiffio, M. (2014). Logistics for Ethanol Transport, in: Cortez, L. (Ed.), *Sugarcane Bioethanol – R&D for Productivity and Sustainability*. São Paulo: Blucher.

Shaefer, W. Knowledge and Nature: History as the Teacher of Life Revisited, *Nature and Culture*, 2: 1, 1–9, Spring 2007.

Schaffer, R., Maroun, C., and Rathmann, R. (2011). Brazilian Biofuels Program from the WEL Nexus Perspective, Background Paper for the ERD 2011/2012, November 2011.

Schilling, M., and Chiang, L. (2011). The Effect of Natural Resources on a Sustainable Development Policy, *Energy Policy*, 39, 990–998.

Schneider, M. (2008*a*, June 3). The Reality of France's Aggressive Nuclear Power Push, *Bulletin of Atomic Scientists*, http://thebulletin.org/reality-frances-aggressive-nuclear-power-push-0?qt-most_read_staff_picks=1.

Schneider, M. (2008*b*, December). *Nuclear Power in France*. Paris: Mycle Schneider Consulting.

Schneider, M. (2008*c*). 2008 World Nuclear Industry Status Report: Western Europe, *Bulletin of Atomic Scientists*. http://thebulletin.org/2008-world-nuclear-industry-status-report-western-europe.

Schneider, M. (2009, May). *Nuclear France Abroad*. Paris: Mycle Schneider Consulting.

Schneider, M. (2013). Nuclear Power and the French Energy Transition: It's the Economics, Stupid! *Bulletin of Atomic Scientists*, 69: 1, 18–26.

Schneider, M., and Froggatt, A. (2012). World Nuclear Industry Status Report 2012, http://www.worldnuclearreport.org/.

Schneider, M., and Froggatt, A. (2015). The World Nuclear Industry: Status Report, https://www.worldnuclearreport.org/.

Schneider, M., and Froggatt, A. (2016). The World Nuclear Industry: Status Report, https://www.worldnuclearreport.org/.

Schneider, M., Froggatt, A., and Thomas, S. (2011). World Nuclear Industry Status Report 2010–2011, http://www.worldnuclearreport.org/.

Schneider, M., and Marignac, Y. (2008). Spent Nuclear Fuel Processing in France, Research Report of the International Panel on Fissile Materials, http://www.psr.org/nuclear-bailout/resources/spent-nuclear-fuel.pdf.

Schneider, M., Thomas, S., Froggatt, A., and Koplow, D. (2009). World Nuclear Industry Status Report 2009, http://www.worldnuclearreport.org/.

Schnepf, R. (2007). Agriculture-based Renewable Energy Production, CRS Report RL32712, Washington DC: Congressional Research Service, http://nationalaglaw-center.org/wp-content/uploads/assets/crs/RL32712.pdf.

Schnepf, R. (2010). Agriculture-based Biofuels: Overview and Emerging Issues, CRS Report R41282, Washington DC: Congressional Research Service, http://digitalcom-mons.unl.edu/cgi/viewcontent.cgi?article=1016&context=crsdocs.

Schon, D. (1967). Technology and Change: The New Heraclitus. *Science*, 157: 3795, 1422–1427.

Schon, D. (1971). *Beyond the Stable State*. London: Temple Smith.

Schon, D., and Rein, M. (1995). *Frame Reflection: Toward the Resolution of Intractrable Policy Controversies*. New York: Basic Books.

Schot, J.W., Hoogma, R., and Elzen, B. (1994). Strategies for Shifting Technological Systems, *Futures*, 10: 1060–1076.

Schrek, A. (2011, September 26). Dubai Targets 5% Renewable Energy by 2030, *Bloomsberg Businessweek/Associated Press*. http://www.businessweek.com/ap/finan-cialnews/D9Q0A2N01.htm.

Schumacher, E. (1973). *Small Is Beautiful: Study of Economics As If People Mattered.* London: Blond & Briggs.

Schumpeter, J. (1934). *The Theory of Economic Development.* Cambridge: Harvard University Press.

Schumpeter, J. (1939). *Business Cycles: A Theoretical, Historical and Statistical Analysis of the Capitalist Process*, Volume I. New York: McGraw-Hill.

Schumpeter, J. (1942/1975). *Capitalism, Socialism and Democracy.* New York: Harper.

Scottish Renewables. (2017), Renewables in Numbers, https://www.scottishrenewables.com/sectors/renewables-in-numbers/. Accessed February 1, 2017.

Searchinger, T., Heimlich, R., Houghton, R., Dong, F., Elobeid, A., et al. (2008). Use of US Croplands for Biofuels Increases Greenhouse Gases Through Emissions from Land-Use Change, *Science*, 319: 5867, 1238–1240.

Sedlabanki Islands. (2005). *The Economy of Iceland.* Reykjavik, Iceland: Central Bank of Iceland.

Seelke, C., and Yacobucci, B. (2007). Ethanol and Other Biofuels: Potential for US-Brazil Cooperation, CRS Report RL34191, Washington DC: Congressional Research Service, http://fpc.state.gov/documents/organization/93476.pdf.

Senado Federal-Secretaria de Informação Legislativa. (1975). Diario Official da Uniao, Decreto/Decree 76.593, November 14, 1975, Brasilia, Brasil.

Shafiee, S., and Topal, E. (2009). When Will Fossil Reserves be Diminished? *Energy Policy*, 37: 1, 181–189.

Shah, J., Nagpal, T., Johnson, T., Amann, M., Carmichael, G., et al. (2000). Integrated Analysis for Acid Rain in Asia, *Annual Review of Energy and the Environment*, 25: 339–375.

Shahan, Z. (2012, March 26). Nuclear Power Too Expensive, French Court Finds, *Clean Technica*, https://cleantechnica.com/2012/03/26/nuclear-power-too-expensive-french-court-finds/.

Shankleman, J. (2016, June 2). World's Cheapest Offshore Windfarm Seen Online Early, *Bloomberg.* http://www.bloomberg.com/news/articles/2016-06-02/vattenfall-accelerates-record-offshore-wind-project-in-denmark.

Sheffield, J. (1998). World Population Growth and the Role of Annual Energy Use per Capita, *Technological Forecasting and Social Change* 59, 55–87.

Shell. (2016). Energy Transitions and Portfolio Resilience, Report, http://www.shell.com.

Silva, W. (2007). Biofuels Experience in Brazil, Presentation, Office of International Affairs, EMBRAPA, Georgetown, August 6, 2007.

Silva, W., and Fischetti, D. (2008). *Etanol: A Revolução Verde e Amarela.* Bizz Communição, Brazil.

Simon, C. (2007). *Public Policy: Preferences and Outcomes.* New York: Pearson Longman.

Simon, H. (1957). A Behavioral Model of Rational Choice, in: Simon, H. (Ed.), *Models of Man, Social and Rational: Mathematical Essays on Rational Human Behavior in a Social Setting.* New York: Wiley.

Simon, H. (1990). A Mechanism for Social Selection and Successful Altruism, *Science*, 250: 4988, 1665–1668.

Simon, H. (1991). Bounded Rationality and Organizational Learning, *Organization Science*, 2: 1, 125–134.

Sinclair, P., Isendahl, C., and Barthel, S. (2016). Beyond Rhetoric: Towards a Framework for an Applied Historical Ecology of Urban Planning, in: Isendahl, C., and Stump, D. (Eds.), *The Oxford Handbook of Historical Ecology and Applied Archaeology*. Oxford: Oxford University Press, http://www.oxfordhandbooks.com/view/10.1093/oxfordhb/9780199672691.001.0001/oxfordhb-9780199672691-e-34.

Singleton Jr., R., and Straits, B. (2005). *Approaches to Social Research* (4th Edition). Oxford: Oxford University Press.

Skulason, J., and Hayter, R. (1998). Industrial Location as a Bargain: Iceland and the Aluminium Multinationals 1962–1994, Geografiska. Annaler: Series B, *Human Geography*, 80: 1, 29–48.

Smil, V. (1991). *General Energetics: Energy in the Biosphere and Civilization*. New York: Wiley.

Smil, V. (1994). *Energy in World History*. Boulder: Westview.

Smil, V. (2000). Energy in the Twentieth Century: Resources, Conversions, Costs, Uses, and Consequences, *Annual Review of Energy and the Environment*, 25, 21–51.

Smil, V. (2003). *Energy at the Crossroads: Global Perspectives and Uncertainties*. Cambridge: MIT Press.

Smil, V. (2004). World History and Energy, in: Cleveland, C. (Ed.), *Encyclopedia of Encyclopedia of Energy*. San Diego: Elsevier Science.

Smil, V. (2007). *Energy in Nature and Society: General Energetics of Complex Systems*. Cambridge: MIT Press.

Smil, V. (2010). *Energy Transitions: History, Requirements, Prospects*. Westport: Praegar.

Smil, V. (2012, June 28). A Skeptic Looks at Alternative Energy, *IEEE Spectrum*, 46–52. http://spectrum.ieee.org/energy/renewables/a-skeptic-looks-at-alternative-energy.

Smil, V. (2015). Energy Transitions, Renewables and Rational Energy Use: A Reality Check, *OECD Observer*, 304: 36–37. http://www.oecdobserver.org/news/fullstory.php/aid/5395/Energy_transitions,_renewables_and_rational_energy_use:_A_reality_check.html.

Smil, V. (2016). Examining Energy Transitions: A Dozen Insights Based on Performance, *Energy Research and Social Science*, 22: 194–197.

Smil, V. (2017). *Energy Transitions: Global and National Perspectives* (2nd Edition). Santa Barbara: Praegar.

Smith, A. (2006). Bringing Sustainable Technologies into the Mainstream, in: Monaghan, A., and Steward, F. (Eds.), Catalysing Innovation for Sustainability: Insights form the ESRC Salford: Sustainable Technologies Programme, https://www.seek.salford.ac.uk/user/profile/publications/view.do?publicationNum=3763.

Smith, A., and Raven, R. (2012). What Is Protective Space? Reconsidering Niches in Transitions to Sustainability, *Research Policy*, 41: 6, 1025–1036.

Smith, K. (2000). Innovation as a System Phenomenon: Rethinking the Role of Policy, *Enterprise & Innovation Management Studies*, 1, 73–102.

Smith, K., Balakrishnan, K., Butler, C., Chafe, Z., Fairlie, I., et al. (2012). Energy and Health, *Global Energy Assessment—Toward a Sustainable Future* (pp. 255–324). Cambridge/Laxenburg: Cambridge University Press/International Institute for Applied Systems Analysis.

Soder, L., and Ackermann, T. (2005). Wind Power in Power Systems: An Introduction, in: Ackerman, T. (Ed.), *Wind Power in Power Systems*. Chichester, UK: Wiley.

Söderholm, P., and Sundqvist, T. (2007). Empirical Challenges in the Use of Learning Curves for Assessing the Economic Prospects of Renewable Energy Technologies, *Renewable Energy*, 32, 2559–2578.

Soete, L. (rappateur). (1991). Technology in a Changing World, The Technology/Economy Programme. Paris: OECD.

Solomon, B., and Krishna, K. (2011). The Coming Sustainable Energy Transition: History, Strategies, and Outlook, *Energy Policy*, 39, 7422–7431.

Sonntag-O'Brien, V., and Usher, E. (2004). Mobilising Finance for Renewable Energies, Thematic Background Paper, International Conference for Renewable Energies, Bonn, Germany.

Sorda, G., Banse, M., and Kemfert, C. (2010). An Overview of Biofuel Policies across the World, *Energy Policy*, 38, 6977–6988.

Sorensen, H., Hansen, L., and Larsen, J. (2002). Middelgrunden 40 MW Offshore Wind Farm, Denmark: Lessons Learned, Renewable Realities – Offshore Wind Technologies, Copenhagen: SPOK Consult, October 2002.

Sorenson, B. (1975). Energy and Resources, *Science*, 189: 4199, 255–260.

Sortir du Nucléaire (n.d.). Information. http://www.sortirdunucleaire.org/index.php?menu=english&page=index. Accessed July 15, 2016.

South Centre and Center for International Environmental Law. (2008). The Technology Transfer Debate in the UNFCC: Politics, Patents, and Confusion, *Intellectual Property Quarterly Update*, Q4.

Souza, G., and Sluys, M. (2010). Sugarcane Genomics and Biotechnology, in: Cortez, L. (Ed.), *Sugarcane Bioethanol*. São Paulo: Blucher, 325–332.

Sovacool, B (2008*a*). The Costs of Failure: A Preliminary Assessment of Major Energy Accidents 1907–2007, *Energy Policy*, 36, 1802–1820.

Sovacool, B. (2008*b*). Valuing the Greenhouse Gas Emissions from Nuclear Power: A Critical Survey, *Energy Policy*, 36, 2940–2953.

Sovacool, B. (2010). A Critical Evaluation of Nuclear Power and Renewable Electricity in Asia, *Journal of Contemporary Asia*, 40: 3, 369–400.

Sovacool, B. (2011, March 16). The Dirt on Nuclear Power, *Project Syndicate*. https://www.project-syndicate.org/commentary/the-dirt-on-nuclear-power?barrier=true.

Sovacool, B. (2011*a*). Questioning the Safety and Reliability of Nuclear Power, *GAIA*, 20: 2, 95–103.

Sovacool, B. (2016). How Long Will It Take? Conceptualizing the Temporal Dynamics of Energy Transitions, *Energy Research and Social Science*, 13: 202–215.

Sovacool, B., and Brown, M. (2010). Competing Dimensions of Energy Security: An International Perspective, *Annual Review of Environment and Resources*, 35: 77–108.

Sovacool, B., and Geels, F. (2016). Further Reflections on the Temporality of Energy Transitions: A Response to Critics, *Energy Research and Social Science*, 22: 232–237.

Sovacool, B., Kryman, M., and Laine, E. (2015). Profiling Technological Failure and Disaster in the Energy Sector: A Comparative Analysis of Historical Energy Accidents, *Energy*, 90, 2016–2027.

Sovacool, B., Lindboe, H., and Odgaard, O. (2008, March). Is the Danish Wind Energy Model Replicable for Other Countries? *Electricity Journal*, 21: 2, 27–38. http://dx.doi.org/10.1016/j.tej.2007.12.009.

Sovacool, B., and Sawin, J. (2012). Creating Technological Momentum, *The Whitehead Journal of Diplomacy and International Relations,* 43–57, Summer/Fall 2010.

Sovacool, B., and Valentine, S. (2012). *The National Politics of Nuclear Power*. London: Routledge.

Sperling, D., and Gordon, D. (2008). Advanced Passenger Transport Technologies, *Annual Review of Energy and the Environment*, 33: 63–84.

Spiegel. (2007, March 9). Interview with Danish PM Anders Fogh Rasmussen, Nuclear is 'Not a Renewable Energy,' http://www.spiegel.de/international/interview-with-danish-prime-minister-anders-fogh-rasmussen-nuclear-is-not-a-renewable-energy-a-470841.html.

Sprang, E., Moomaw, W., Gallagher, K., Kirshen, P., and Marks, D. (2014). The Water Consumption of Energy Production: An International Comparison, *Environmental Research Letters,* 9: 10.

Stanford, G., Marsh, G., and Hannum, W. (2009, August 31). Reprocessing Is the Answer, *Bulletin of Atomic Scientists,* http://thebulletin.org/reprocessing-answer.

State of Green. (Ed.). (2015). *Wind Energy Moving Ahead.* Frederiksberg: Danish Wind Industry Association, http://www.stateofgreen.com/wind-energy.

Statistics Iceland. (n.d.). Information, http://www.statice.is/. Accessed January 5, 2017.

Statsministeriet. (n.d.). Governments since 1953, http://www.stm.dk/_p_12693.html. Accessed June 10, 2012.

Stauble, A., and Milius, G. (1970). Geology of Groningen Gas Field, Netherlands, in: Halbouty, M. (Ed.), M 14: Geology of Giant Petroleum Fields, 359–369, Symposium Papers from the Annual Meeting of American Association of Petroleum Geologists/AAPG, April 23–25, 1968, Oklahoma City, OK.

Steering Committee for Formulating a Comprehensive Energy Policy. (2011). Energy Policy for Iceland, Orkustefna fyrir Ísland.

Steingrimsson, B., Bjornsson, S., and Adalsteinsson, H. (2007). Master Plan for Geothermal and Hydropower Development in Iceland. A Presentation at Short Course on Geothermal Development in Central America—Resource Assessment and Environmental Management, Organized by UNU-GTP and LaGeo in San Salvador, El Salvador, November 25—December 1, 2007.

Stevens, P. (2003). National Oil Companies: Good or Bad? Paper presented at National Oil Companies Workshop, World Bank, May 27, 2003.

Stevens, P. (2010). *The 'Shale Gas Revolution': Hype and Reality*, Report. London: Chatham House, https://www.chathamhouse.org/.

Stirling, A. (2014). Transforming Power: Social Science and the Politics of Energy Choices, *Energy Research and Social Science,* 1, 83–95.

Stoffaes, C. (2016). French Plutonium Policy Questioned by Former EDF Official, International Panel on Fissile Materials, http://fissilematerials.org/blog/2016/06/french_plutonium_policy_q.html.

Stokey, E., and Zeckhauser, R. (1978). *A Primer for Policy Analysis.* New York: Norton.

Strapasson, A. (2006, November). Biofuels in Brazil, Presentation, RIO 6, Rio de Janeiro, Brazil. http://www.rio12.com/rio6/programme/second.html.

Sturgeon, T., Memdovic, O., Van Biesebroeck, J., and Gereffi, G. (2009). Globalization of the Automotive Industry, *International Journal of Technological Learning, Innovation and Development,* 2: 1–2.

Suarez, F., and Utterback, J. (1995). Dominant Designs and the Survival of Firms, *Strategic Management*, 16: 6, 415–430.

Subramanya, S., Mustafa, Z., Irwin, D., and Shenoy, P. Energy-Agility: A New Grid-centric Metric for Evaluating System Performance, Limits 2015, Irvine, CA, June 15–16, 2015.

Sulzberger, C. (2004). An Early Road Warrior: Electric Vehicles in Early Years of the Automobile, *IEEE Power and Energy Magazine*, 2: 3, 66–71.

Sundfeld, E. (2009). Bioenergy in Technology Development of Brazil, Presentation, Brasilia, Brasil, October 7, 2009.

Surrey, J., and Huggett, C. (1976). Opposition to Nuclear Power: A Review of International Experience, *Energy Policy*, 4: 4, 286–307.

Sustainable Energy for All. (2011). Information, http://www.sustainableenergyforall.org/, Accessed June 15, 2016.

Suurs, R., and Hekkert, M. (2008). Competition between First and Second Generation Technology, Paper, Druid Conference, June 17–20, 2008, Copenhagen, Denmark.

Sverrisson, F. (2006, November-December). Missing in Action: Iceland's Hydrogen Economy, *World Watch* 19: 6.

Szulanski, G. (1996). Exploring Internal Stickiness: Impediments to the Transfer of Best Practices Within the Firm, *Strategic Management Journal* 17: 10, 27–43.

Tanaka, N. (2009*a*). Energy Security in New Market Realities, IEA Presentation, Petrostocks 2009, Houston, TX, February, 2009.

Tanaka, N. (2009*b*) Energy and Climate Policy, IEA Presentation and Summary Notes, G8 Energy Ministerial Meeting, Rome, Italy, May 24–25, 2009.

Tester, J., Drake, E., Driscoll, M., Golay, M., and Peters, W. (2005). *Sustainable Energy: Choosing Among Options*. Cambridge: MIT Press.

Thais, F. (2014). *France Overview*. Vienna: International Atomic Energy Agency (IAEA).

The Innovation Policy Platform. (n.d.), Absorptive Capacities, The Innovation Policy Platform https://www.innovationpolicyplatform.org/content/absorptive-capacities. Accessed June 15, 2016.

Theodoulou, S. (1995). How Public Policy Is Made, in: Theodoulou, S., and Cahn, M. (Eds.), *Public Policy: The Essential Readings*. Englewood Cliffs: Prentice Hall.

Thordarson, S. (2008). Hundred Years of Space Heating with Geothermal Energy in Iceland. http://samorka.is/doc/1842/Sveinn+%C3%9E%C3%B3r%C3%B0arson,+afm%C3%A6lisdagskr%C3%A1+hitaveitu,+sept+08.pdf.

Thorhallsdottir, T. (2007*a*). Environment and Energy in Iceland: A Comparative Analysis of Values and Impacts (Opnast í nýjum vafraglugga). *Environmental Impact Assessment Review*, 27, 522–544.

Thorhallsdottir, T. (2007*b*). Strategic Planning at the National Level: Evaluating and Ranking Energy Projects by Environmental Impact (Opnast í nýjum vafraglugga). *Environmental Impact Assessment Review*, 27, 545–568.

Thorsteinsson, H., and Tester, J. (2010). Barriers and Enablers of Geothermal District Heating System Development in the US, *Energy Policy*, 38, 803–813.

Tilman, D., Socolow, R., Foley, J., Hill, J., Larson, E., et al. (2009, July 17). Beneficial Biofuels: The Food, Energy, and Environmental Trilemma, *Science*, 325: 5938, 270–271.

Timilsina, G., and Shrestha, A. (2010). Biofuels: Markets, Targets, and Impacts. Policy Research Working Paper 5364, Development Research Group, Environment and Energy Team, Washington DC: World Bank, July 2010, http://www.worldbank.org/.

Tiwari, G., and Ghosal, M. (2007). *Fundamentals of Renewable Energy Sources*. Oxford: Alpha Science Intl.

Todorova, G., and Durisin, B. (2007). Absorptive Capacity: Valuing a Reconceptualization, *Academy of Management Review*, 32: 3, 774–786.

Todtling, F., and Trippl, M. (2011). Regional Innovation Systems, in: Cooke, P., Asheim, B., Boschma, R., Martin, R., Schwartz, D., (Eds.), *Handbook of Regional Innovation and Growth*. London: Edgar Elger.

Tomasson, R. (1980). *Iceland: The First New Society*. Minneapolis: University of Minnesota Press.

Tomei, J., and Helliwell, R. (2016). Food versus Fuel? Going Beyond Biofuels, *Land Use Policy*, 56, 320–326.

Topçu, S. (2006). Nuclear Power: The Mobilization of Scholars to Cons-Associative Expertise, *Nature Science Society*, 14, 249–256.

Topçu, S. (2007). Les Physiciens dans le Mouvement Antinucléaire: Entre Science, *Expertise et Politique*, 102, 89–108.

Topçu, S. (2008). Confronting Nuclear Risks: Counter-expertise as Politics within the French Nuclear Energy Debate, *Nature and Culture*, 3: 2, 91–111.

Tordo, S. Tracy, B., and Arfaa, N. (2008). A Citizen's Guide to National Oil Companies, Technical Report A, Washington, DC: World Bank Group, http://www.world bank.org/.

Tranaes, F. (2000). Danish Wind Energy Cooperatives, Parts 1 and 2. http://ele.aut.ac.ir/~wind/en/articles/coop.htm.

Treganna, F. (2007). Which Sectors Can Be Engines of Growth and Employment in South Africa? Paper, UNU-WIDER Conference on Southern Engines of Global Growth.

Triana, C. (2011). Energetics of Brazilian Ethanol, *Energy Policy*, 39: 8, 4605–4613.

Troen, I., and Petersen, E. (1989). *European Wind Atlas*. Roskilde, Denmark: Risø National Laboratory.

Tsinghua University. (2010). Partitioning of High Level Waste. http://www.tsinghua.edu.cn/publish/ineten/5698/index.html.

Tufte, M., and Thomas, P. (2009). *Participatory Communication: A Practical Guide*. Washington, DC: World Bank.

Turner, E. (2010). Why Has the Number of International Non-Governmental Organizations Exploded since 1960? *Cliodynamics*, 1, 1.

Tvind Internationale Skolcenter. (n.d.). Tvindkraft Vindmill. http://www.tvind.dk/TextPage.asp?MenuItemID=55&SubMenuItemID=160. Accessed June 15, 2012.

Twomey, P., and Gaziulusoy, I. (2014). Review of System Innovation and Transition Theories. Working Paper, Visition and Pathways Project 2040, March 2014, http://www.visionandpathways.com.

Tylecote, A. (1992). *The Long Wave in the World Economy*. London: Routledge.

Tyre, M., and Orlikowski, W. (1996). The Episodic Process of Learning by Using, *International Journal of Technology Management*, 11: 7–8, 790–798.

Uicker, J., Pennock, G., and Shigley, J. (2003). *Theory of Machines and Mechanisms*. New York: Oxford University Press.

UNICA. (2010). 2010 Sustainability Report, São Paulo, Brazil. http://sugarcane.org/resource-library/books/Sustainability.pdf.

UNICA. (2011*a*, November 28). Without Clear Public Policies, Ethanol Cannot Compete in Brazil, Says UNICA, Press Release, http://english.unica.com/br.

UNICA (2011*b*, February 15). Raízen, Fruto da Parceria Shell-Cosan, é Importante Ingrediente para Transformar Etanol em Commodity Global, http://www.unica.com.br/noticia/28739779920341709819/raizen-por-cento2C-fruto-da-parceria-shell-cosan-por-cento2C-e-importante-ingrediente-para-transformar-etanol-em-commodity-global/.

UNICA. (2016). Updated Information on Brazil's Sugarcane Production, http://unica.com.br/?idioma=1. Accessed December 15, 2016.

Union of Concerned Scientists (UCS). (1999). Barriers to the Use of Renewable Energy Technologies, http://www.ucsusa.org/clean_energy/smart-energy-solutions/increase-renewables/barriers-to-renewable-energy.html#.V4UD0bgrLIU.

Union of Concerned Scientists (UCS). (n.d.). Information, http://www.ucsusa.org. Accessed January 15, 2017.

United Kingdom Environment Agency. (n.d.). Radionuclides in the Environment, http://www.environment-agency.gov.uk/cy/ymchwil/llyfrgell/data/34431.aspx. Accessed June 15, 2012.

United Kingdom Renewables, Department of Energy and Climate Change. (2012). http://www.ukrenewables.com/; http://www.decc.gov.uk/. Accessed June 15, 2012.

United Nations. (2007). Press Conference Launching International Biofuels Forum, Department of Public Information, News and Media Division, New York, March 2, 2007. http://www.un.org/News/briefings/docs/2007/070302_Biofuels.doc.htm.

United Nations. (2008). *Handbook on the Least Developed Country Category: Inclusion, Graduation and Special Support Measures*, United Nations publication, Sales No. E.07.II.A.9, http://www.un.org/esa/analysis/devplan/cdppublications/2008cdphandbook.pdf.

United Nations (UN). (2011*b*, May 3). Press Release—2010 Revision of World Population Prospects, http://www.un.org/ . . . /population/ . . . /WPP2010/WPP2010_Volume-I_Comprehensive-Tables.

United Nations. (2014). Country Classification, http://www.un.org/en/development/. Accessed May 10, 2014.

United Nations. (n.d.*a*). Global Compact-Climate Change, https://www.unglobalcompact.org/what-is-gc/our-work/environment/climate. Accessed July 1, 2016.

United Nations (n.d.*b*). Policy Consolidation of the UN and Other Relevant Actors (Low Carbon Development Relevant). http://www.low-carboncity.org/index.php?option=com_flexicontent&view=items&cid=28%3Alow-carbon-policiesintl&id=54%3Aunited-nations&lang=en. Accessed July 1, 2016.

United Nations. (n.d.*c*). UN and Civil Society. http://www.un.org/en/civilsociety/index.shtml. Accessed July 1, 2016.

United Nations Centre for Natural Resources, Energy and Transport (UNCRET). (1980). *State Petroleum Enterprises in Developing Countries*. New York: Pergamon.

United Nations Comtrade (UN Comtrade). (n.d). Data, https://comtrade.un.org/. Accessed December 10, 2016.

United Nations Development Program (UNDP). (2011). *Human Development Report 2011, Explanatory Note on Iceland*. New York: UNDP.

United Nations Development Program (UNDP), UN Department of Economic and Social Affairs, World Energy Council (WEC). (2000). World Energy Assessment. http://www.undp.org/content/undp/en/home/librarypage/environment-energy/sustainable_energy/world_energy_assessmentenergyandthechallengeofsustainability.html.

United Nations Development Program (UNDP), UN Department of Economic and Social Affairs, World Energy Council (WEC). (2004). World Energy Assessment Overview: 2004 Update, http://www.undp.org/content/undp/en/home/librarypage/environment-energy/sustainable_energy/world_energy_assessmentoverview-2004update.html.

United Nations Energy Knowledge Network. (2011, January 8). Ethanol Fuel in Brazil, http://www.un-energy.org/stories/38-ethanol-fuel-in-brazil.

United Nations Environmental Programme (UNEP). (n.d.*a*). Buildings and Climate Change. UNEP Sustainable Buildings and Climate Initiative. Accessed May 10, 2012.

United Nations Environmental Programme (UNEP), Finance Initiative, Climate Wise, The Geneva Association, and MCII. (n.d.*b*). Global Insurance Industry Statement on Adapting to Climate Change in Developing Countries. http://www.unepfi.org/fileadmin/documents/insurance_climatechange_statement.pdf. Accessed July 15, 2016.

United Nations Environmental Programme (UNEP). (2008). Reforming Energy Subsidies: Opportunities to Contribute to the Climate Change Agenda. Paris: UNEP, http://www.unep.org/pdf/pressreleases/reforming_energy_subsidies.pdf.

United Nations Environmental Programme (UNEP). (2011). UNEP Yearbook, http://hqweb.unep.org/yearbook/2011/.

United Nations Environmental Programme (UNEP). (2015). Green Energy Choices: The Benefits, Risks, and Trade-offs of Low-carbon Technologies for Electricity Production. Summary for Policy Makers.

United Nations Environmental Programme (UNEP). (2016). UNEP Inquiry Kicks Off the Year of Green Finance, January 4, 2016, http://www.unep.org/NewsCentre/default.aspx?DocumentID=26862&ArticleID=35818

United Nations Environment Programme (UNEP) and Bloomberg New Energy Finance. (2011). Global Trends in Renewable Energy Investment 2011, http://fs-unep-centre.org/publications/global-trends-renewable-energy-investment-2011.

United Nations Framework Convention on Climate Change (UNFCCC). (2001). Report of the Conference of the Parties on Its Seventh Session, Held at Marrakesh from 29 October to 10 November 2001. Addendum. Part Two: Action Taken by the Conference of the Parties. Volume I., FCCC/CP/2001/13/Add.1, http://unfccc.int/documentation/documents/advanced_search/items/6911.php?priref=600001855.

United Nations News Centre. (2011, October 31). As World Passes 7 Billion Milestone, UN Urges Action to Meet Challenges, http://www.un.org/apps/news/story.asp?NewsID=40257#.V1m_zpErLIU

United Nations Office for Disarmament Affairs. (n.d.). Treaty on the Non-Proliferation of Nuclear Weapons, http://www.un.org/disarmament/WMD/Nuclear/NPT. Accessed June 12, 2012.

United Nations Scientific Committee on the Effects of Atomic Radiation (UNSCEAR). (2008). Sources and Effects of Ionizing Radiation, Volume II, http://www.unscear.org/docs/reports/2008/11-80076_Report_2008_Annex_D.pdf.

United Nations Scientific Committee on the Effects of Atomic Radiation (UNSCEAR). (2010). Report of the United Nations Scientific Committee on the Effects of Atomic Radiation 2010, http://www.unscear.org/docs/reports/2010/UNSCEAR_2010_Report_M.pdf.

United Nations University Geothermal Training Program. (n.d.). Information, http://unugtp.is/, Accessed January 5, 2017.

U.S. Department of Agriculture (USDA). (2010). Foreign Agriculture Service, Brazil Biofuels Annual, Gain Report BR, Washington, DC: USDA, July 30, 2010, https://gain.fas.usda.gov.

U.S. Department of Agriculture (USDA). (2011). Brazil Biofuels Annual, Gain Report BR, July 27, 2011, Washington, DC: USDA, https://gain.fas.usda.gov.

U.S. Department of Agriculture (USDA). (2015). Brazil Biofuels Annual, GAIN Report, Brazil, BR, August 10, 2015, Washington, DC: USDA, https://gain.fas.usda.gov.

U.S. Department of Agriculture (USDA). (2016). GAIN Report BR, Washington, DC: USDA, https://gain.fas.usda.gov.

U.S. Department of Commerce. (2011). Energy Science and Technology. http://www.ntis.gov/products/databases/energy-science-technology/. Accessed May 15, 2012.

U.S. Department of Commerce, Economics and Statistics Administration (US DCESA). (1995, July). Engines of Growth: Manufacturing Industries in the US Economy, Report, Washington DC: US DCESA, http://www.esa.doc.gov/sites/default/files/enginesofgrowth.pdf.

U.S. Department of Energy (DOE). (2015). Energy Transmission, Storage, and Distribution Infrastructure, Quadrennial Energy Review. Washington, DC: US DOE.

U.S. Department of Energy (DOE). (2017). Transforming the Nation's Electricity System, Quadrennial Energy Review. Washington, DC: US DOE.

U.S. Department of Energy (DOE) (n.d.). Information, http://energy.gov/. Accessed January 5, 2017.

U.S. Department of Energy (DOE). (2016, February 6). DOE's Office of Technology Transitions Issues First Call to Launch New Energy Technologies from National Laboratories to Market, http://energy.gov/articles/doe-s-office-technology-transitions-issues-first-call-launch-new-energy-technologies.

U.S. Department of State. (2011). Background Notes: Iceland, November 2011; April 2002; and July 1996, Washington DC. https://www.state.gov/r/pa/ei/bgn/.

U.S. Environmental Protection Agency (EPA). (2007). Greenhouse Gas Impacts of Expanded Renewable and Alternative Fuels Use, http://link.library.in.gov/portal/Greenhouse-gas-impacts-of-expanded-renewable-and/Qk_Uzls1kz4/.

U.S. Environmental Protection Agency (EPA). (2010). Greenhouse Gas Reduction Thresholds, http://www.epa.gov/OMS/renewablefuels/420f10007.htm#7.

U.S. Environmental Protection Agency (EPA). (2011). *Biofuels and the Environment: the First Triennial Report to Congress* (2011 Final Report). US Environmental Protection Agency, Washington, DC: EPA/600/R-10/183F.

U.S. Environmental Protection Agency (EPA). (2012). Commonly Encountered Radionuclides. http://www.epa.gov/rpdweb00/radionuclides/.

U.S. Geological Survey (USGS). (n.d.). Ring of Fire, http://earthquake.usgs.gov/learn/glossary/?term=Ring%20of%20Fire. Accessed July 10, 2016.

U.S. Nuclear Regulatory Commission (NRC). (n.d). Information, http://www.nrc.gov. Accessed July 25, 2016.

Unruh, G. (2000). Understanding Carbon Lock-in, *Energy Policy*, 28: 12, 817–830.

Unruh, G. (2002). Escaping Carbon Lock-in, *Energy Policy*, 30: 4, 317–325.

Upwind. (n.d.). Information. http://www.upwind.eu/. Accessed May 15, 2011.

Urbina, I. (2011*a*, April 16). Chemicals Were Injected into Wells, Report Says, *New York Times*, http://www.nytimes.com/2011/04/17/science/earth/17gas.html.

Urbina, I. (2011*b*, June 25). Insiders Sound an Alarm amid a Natural Gold Rush, *New York Times*, http://www.nytimes.com/2011/06/26/us/26gas.html?hp.

Usher, A. (1929). *A History of Mechanical Inventions*. Cambridge: Harvard University Press.

Utterback, J., and Abernathy, W. (1975). A Dynamic Model of Product and Process Innovation, *Omega*, 3: 6, 639–656.

Utterback, J., and Suarez, F.F. (1993). Innovation, Competition, and Industry Structure, *Research Policy*, 22: 1.

Valdes, C. (2011). Brazil's Ethanol Industry, Economic Research Service Report, BIO-02, June 2011, http://www.ers.usda.gov.

Vaidyanathan, G. (2016, April 4). Fracking Can Contaminate Drinking Water, *Scientific American*, https://www.scientificamerican.com/article/fracking-can-contaminate-drinking-water/.

Valdimarsson, P. (2011). Geothermal Power Plant Cycles and Main Components, Presented at the Short Course on Geothermal Drilling, Resource Development and Power Plants, organized by UNU-GTP and LaGeo, Santa Tecla, El Salvador, January 16–22, 2011.

Valfells, A., Fridleifsson, I., Helgason, T., Ingimarsson, J., Thoroddsson, G. (2004). Sustainable Generation and Utilization of Energy: The Case of Iceland, World Energy Congress, Sydney, Australia, September 5–9, 2004.

Valor International. (2014). Brazil Retains Lowest Ethanol Production Cost, November 4, 2014, http://www.valor.com.br/international/news/3764354/brazil-retains-lowest-ethanol-production-cost.

Van den Wall Bake, J., Junginger, M., Faaij, A., Poot, T., and Walter, A. (2009). Explaining the Experience Curve, *Biomass and Bioenergy*, 33: 4, 644–658.

Van Est, R. (1999). *Winds of Change: A Comparative Study of the Politics of Wind Energy Innovation in California and Denmark*. Utrecht: International Books.

Vanlerberghe, C. (2012). Pr. Pellerin: "L'injustice de Tchernobyl est Réparée, *Le Figaro*, November 21, 2012, http://sante.lefigaro.fr/actualite/2012/11/21/19470-pr-pellerin-linjustice-tchernobyl-est-reparee.

Vasi, I. (2009). *The A to Z of the Petroleum Industry*. Lanham, MD: Scarecrow.

Vasi, I. (2011). *Winds of Change: The Environmental Movement and the Global Development of the Wind Energy Industry*. New York: Oxford University Press.

Vassilou, M. (2009*a*). *Historical Dictionary of the Petroleum Industry*. Lanham, MD: Scarecrow.

Vedung, E. (2005). Policy Instruments: Typologies and Theories, in: Bemelmans-Videc, M., Rist, R., and Vedung, E. (Eds.), *Carrots, Sticks, and Sermons*. New Brunswick: Transaction.

Vendryes, G. (1986). National Report: Observations from France. A Look at Some Reasons Behind the Strong Programme, *IAEA Bulletin*, 28: 3, 52–54, https://www.iaea.org/sites/default/files/28304795254.pdf.

Vendryes, G. (1986, Autumn). Observations from France, *IAEA Bulletin*, https://www.iaea.org/sites/default/files/28304795254.pdf.

Verbong, G., and Geels, F. (2007). The Ongoing Energy Transition: Lessons from a Socio-technical, Multi-level Analysis of the Dutch Electricity System (1960–2004), *Energy Policy*, 35, 1025–1037.

Verbong, G., and Loorbach, D. (Eds.). (2012). Governing the Energy Transition. New York: Routledge.

Verbruggen, A., Moomaw, W., and Nyboer, J. (2011). Annex 1: Glossary, Acronyms, Chemical Symbols and Prefixes, in: Edenhofer, O., Pichs-Madruga, R., Sokona, Y., Seyboth, K., Matschoss, P., et al. (Eds.), *IPCC Special Report on Renewable Energy Sources and Climate Change Mitigation*. Cambridge: Cambridge University Press.

Vinstri Graen. (n.d.). Information, http://www.vg.is/tungumal/english/. Accessed July 20, 2012.

Vitina, A. (2015). Wind Energy Development in Denmark, in: Hand, M. (Ed.), *IEA Wind Task 26—Wind Technology, Cost, and Performance Trends in Denmark, Germany, Ireland, Norway, the European Union, and the United States: 2007–2012*, NREL/TP-6A20-64332 (pp. 16–47). Golden: National Renewable Energy Laboratory.

Von Hippel, E. (1988). *Sources of Innovation*. New York: Oxford University Press.

Von Hippel, E. (2005). *Democratizing Innovation, Creative Commons*, Cambridge: MIT Press.

Von Hippel, E. (2010). The Role of Lead Users in Innovation, in: Teece, D., and Augier, M. (Eds.), *Palgrave Encyclopedia of Strategic Management*. London: Palgrave Macmillan.

Von Hippel, F. (Ed). (2010). The Uncertain Future of Nuclear Energy, Research Report of the International Panel on Fissile Materials. http://fissilematerials.org/library/rr09.pdf.

Von Hippel, E., and Tyre, M. (1995). How Learning by Doing Is Done: Problem Identification in Novel Process Equipment, *Research Policy*, 24: 1–12.

Waggoner, P., and Ausubel, J. (2002). A Framework for Sustainability Science: A Renovated IPAT Identity, *PNAS* 99: 12, 7860–7865, http://www.pnas.org/content/99/12/7860.full.pdf.

Walsh, B. (2009, March 16). Denmark's Winds of Change, *Time Science Magazine*. http://www.time.com/time/magazine/article/0,9171,1883373,00.html.

Walter, A., and Dolzan, P. (2009). Country Report Brazil, *IEA Bioenergy Task 40*, August 2009. http://www.bioenergytrade.org/reports/countryreports/brazil.html

Walter, A., Rosillo-Calle, F., Dolzan, P., Piacente, E., and Borges da Cunha, K. (2007). *Task 40: Sustainable Bio-Energy Trade—Securing Supply and Demand, Deliverable 8, Market Evaluation: Fuel Ethanol, IEA Bioenergy Report*. France: IEA.

Washington Clean Tech Alliance. (2011). President Grimmsson's Talk on Geothermal Energy in Iceland. http://www.ctaw.org/2011/07/president-grimssons-talk-on-geothermal-energy-in-iceland/.

Watkins, J., and Tacchi, J. (Eds.). (2008). *Participatory Content Creation for Development: Principles and Practices*. New Delhi, India: UNESCO.

Webb, C. (2012, July 10). Wind Turbine Blades Push Size Limits, Renewable Energy World. http://www.renewableenergyworld.com/articles/print/volume-15/issue-3/wind-power/wind-turbine-blades-push-size-limits.html.

Weber, M. (1947). *The Theory of Social and Economic Organization* (translated by A. Henderson and T. Parsons). London: Collier Macmillan.

Weber, M. (1958). The Three Types of Legitimate Rule (translated by H. Gerth), *Berkeley Publications in Society and Institutions*, 4: 1, 1–11.

Weinberg, A. (1990). Energy in Retrospect: Is the Past Prologue? *Technological Forecasting and Social Change*, 38, 211–221.

Weiss, C., and Bonvillian, W. (2009*a*). *Structuring an Energy Technology Revolution*. Cambridge: MIT Press.

Weiss, C., and Bonvillian, W. (2009b). Complex, Established "Legacy" Sectors: The Technology Revolutions That Do Not Happen, *Innovations*, 6: 2. 157–187.

Weiss, C., and Bonvillian, W. (2013). Legacy Sectors: Barriers to Global Innovation in Agriculture and Energy, *Technology Analysis & Strategic Management*, 25: 10, 1189–1208.

Welsh, H. (2010). Policy Transfer in the Unified Germany: From Imitation to Feedback Loops, *German Studies Review*, 33: 3, 531–548.

Wene, C. (1999). Experience Curves: Measuring the Performance of the Black Box, in: Wene, C., Voss, A., and Fried, T. (Eds.), *Proceedings IEA Workshop on Experience Curves for Policy Making—The Case of Energy Technologies*. Stuttgart, Germany: Forschungsbericht 67, Institut für Energiewirtschaft und Rationelle Energieanwendung, Universität Stuttgart.

Whitehouse, Office of the Press Secretary. (2007). President Bush and President Lula of Brazil Discuss Biofuel Technology, Washington DC: Whitehouse, March 9, 2007, https://2001-2009.state.gov/p/wha/rls/rm/07/q1/81608.htm.

Wilson, B. (1980). *Systems: Concepts, Methodologies and Applications*. Marblehead: John Wiley.

Wilson, C. (2009). *Meta-analysis of Unit and Industry Level Scaling Dynamics in Energy Technologies and Climate Change Mitigation Scenarios*. IR-09-029. Laxenburg: International Institute for Applied Systems Analysis.

Wilson, C. (2014*a*). Historical Diffusion and Growth of Energy Technologies, in: Grubler, A. and Wilson, C. (Eds.), *Energy Technology Innovation: Learning from Historical Successes and Failures*. Cambridge: Cambridge University Press.

Wilson, C. (2014*b*). Input, Output, and Outcome Metrics for Assessing Energy Innovation, in: Grubler, A. and Wilson, C. (Eds.), *Energy Technology Innovation: Learning from Historical Successes and Failures*, Cambridge: Cambridge University Press.

Wilson, C., and Grubler, A. (2011). Lessons from the History of Technological Change for Clean Energy Scenarios and Policies. *Natural Resources Forum*, 35, 165–184.

Wilson, C., and Grubler, A. (2014*a*). Energy Technology Innovation, in: Grubler, A. and Wilson, C. (Eds.), *Energy Technology Innovation: Learning from Historical Successes and Failures*. Cambridge: Cambridge University Press.

Wilson, C., and Grubler, A. (2014*b*). The Energy Technology System, in: Grubler, A. and Wilson, C. (Eds.), *Energy Technology Innovation: Learning from Historical Successes and Failures*. Cambridge: Cambridge University Press.

Wilson, C., and Grubler, A. (2014*c*). Lessons Learnt from the Energy Technology Innovation System, in: Grubler, A. and Wilson, C. (Eds.), *Energy Technology*

Innovation: Learning from Historical Successes and Failures. Cambridge: Cambridge University Press.

Wind Power Program. (1998). Wind Turbine Power Ouput Variation with Steady Wind Speed. http://www.wind-power-program.com/turbine_characteristics.htm.

Winkler, H. (Ed.). (2006, April). *Energy Policies for Sustainable Development in South Africa*. Cape Town: Energy Research Centre, University of Cape Town, South Africa.

Wiser, R., Yang, Z., Hand, M., Hohmeyer, O., Infield, D., et al. (2011). Wind Energy, in: Edenhofer, O., Pichs-Madruga, R., Sokona, Y., Seyboth, K., Matschoss, P., et al. (Eds.), *IPCC Special Report on Renewable Energy Sources and Climate Change Mitigation*. Cambridge: Cambridge University Press.

Wittgenstein, L. (1969–1975). *On Certainty*. G. Anscombe and G. von Wright (Eds.), translated by D. Paul and G. Anscombe. Oxford: Basil Blackwell, http://web.archive.org/web/20051210213153/http://budni.by.ru/oncertainty.html.

Wood, K. (2012*a*). Wind Turbine Blades: Glass vs. Carbon Fiber, Composites Technology, http://www.compositesworld.com/articles/wind-turbine-blades-glass-vs-carbon-fiber.

Wood, K. (2012*b*). Competition for Carbon Fiber in Longer Blades, Composites Technology, http://www.compositesworld.com/articles/competition-for-carbon-fiber-in-longer-blades.

World Alliance for Citizen Participation. (n.d.). Civil Society Index/CIVICUS, Summary of Conceptual Framework and Research Methodology, http://www.civicus.org/new/media/CSI_Methodology_and_conceptual_framework.pdf.

World Bank. (n.d.*a*). Low Carbon Development, http://web.worldbank.org/WBSITE/EXTERNAL/TOPICS/EXTENERGY2/0,,contentMDK:22844223~pagePK:210058~piPK:210062~theSitePK:4114200,00.html.

World Bank (n.d.*b*). World Development Indicators, http://data.worldbank.org/data-catalog/world-development-indicators. Accessed February 2, 2017.

World Bank. (n.d.*c*). Projects and Operations, http://data.worldbank.org/data-catalog/projects-portfolio. Accessed June 20, 2016.

World Bank. (1980, June 4). Alcohol Production from Biomass Potential and Prospects for Developing Countries, Report No. 3021, Washington, DC: World Bank, http://www.worldbank.org/.

World Bank (1981*a*, April 23). Report and Recommendation of the President of the International Bank for Reconstruction and Development (IBRD) to the Executive Directors on a Proposed Loan to the Federated Republic of Brazil for an Alcohol and Biomass Energy Development Project. Report No. P-3039-BR. Washington, DC: World Bank.

World Bank. (1981*b*, April 22). Staff Appraisal Report: Brazil Alcohol and Biomass Energy Development Project. Report No. 3214-BR. Washington, DC: World Bank.

World Bank. (2008). Rising Food and Fuel Prices: Addressing the Risks to Future Generations, Report, http://documents.worldbank.org/curated/en/2008/10/9905929/rising-food-fuel-prices-addressing-risks-future-generations.

World Bank. (2010a). Placing the 2006/2008 Commodity Price Boom into Perspective. Washington, DC: World Bank.

World Bank. (2010b). World Development Report (chapter 7). Washington, DC: World Bank.

World Bank. (2012, June 21). World Bank to Boost Access to Electricity and Clean Fuels, Renewable Energy and Energy Efficiency, http://www.worldbank.org/en/news/2012/06/21/world-bank-boost-access-electricity-clean-fuels-renewable-energy-efficiency.

World Bank. (2015, April 18). Mobilizing the Billions and Trillions for Climate Finance. http://www.worldbank.org/en/news/feature/2015/04/18/raising-trillions-for-climate-finance.

World Bank/ESMAP. (2009). *Low Carbon Growth Country Studies*. Washington, DC: World Bank.

World Commission on Environment and Development (WCED). (1987). Report of the World Commission on Environment and Development: Our Common Future, http://www.un-documents.net/wced-ocf.htm.

World Economic Forum. (2013). The Green Investment Report, http://reports.weforum.org/green-investing-2013/.

World Economic Forum. (2015). How Energy Companies Are Committing to Tackle Climate Change, reprinted from *Financial Times*, https://www.weforum.org/agenda/2015/10/how-energy-companies-are-committing-to-tackle-climate-change/.

World Energy Council (WEC). (2010*a*). Biofuels: Policies, Standards and Technologies. London: WEC, https://www.worldenergy.org/wp-content/uploads/2012/10/PUB_Biofuels_Policies_Standards_and_Technologies_2010_WEC.pdf.

World Energy Council (WEC). (2010*b*). World Energy Insight. London: WEC, https://www.worldenergy.org/publications/2010/world-energy-insight-2010/.

World Energy Council (WEC). (2010*c*). 2010 Survey of Energy Resources. London: WEC, https://www.worldenergy.org/wp-content/uploads/2012/09/ser_2010_report _1.pdf.

World Energy Council (WEC). (2012). World Energy Perspective: Nuclear Energy One Year After Fukushima. London: WEC, https://www.worldenergy.org/publications/2012/world-energy-perspective-nuclear-energy-one-year-after-fukushima/.

World Energy Council (WEC). (2016). World Energy Resources. London: WEC, https://www.worldenergy.org/publications/2016/world-energy-resources-2016/.

World Health Organization (WHO). (2006). Field Manual for Capacity Assessment of Health Facilities in Responding to Emergencies. WHO Regional Office for Western Pacific, Geneva: WHO, http://www.who.int/en/. Accessed June 10, 2016.

World Nuclear Association (WNA). (n.d.). Information, http://www.world-nuclear.org/info/inf54.html. Accessed January 5, 2017.

World Wildlife Fund (WWF) (n.d.). Cerrado, The Brazilian Savanna. http://wwf.panda.org/what_we_do/where_we_work/cerrado/.

Wright, T. (1936). Factors Affecting the Cost of Airplanes, *Journal of the Aeronautical Sciences*, 3.

Wrigley, E. (1988). *Continuity, Chance and Change: The Character of the Industrial Revolution in England*. Cambridge: Cambridge University Press.

Wustenhagen, R., Wolsink, M., and Burer, M. (2007). Social Acceptance of Renewable Energy Innovation: An Introduction to the Concept, *Energy Policy*, 35: 5, 2683–2691.

Wyman, C. (1999). Biomass Ethanol: Technical Progress, Opportunities and Commercial Challenges, *Annual Review of Energy and the Environment*, 24, 189–226.

Yacobucci, B. (2007). Fuel Ethanol: Background and Public Policy Issues. CRS Report, Paper 31, Washington DC: Congressional Research Service, http://digitalcommons.unl.edu/crsdocs/31.

Yacobucci, B. (2010). *Energy: Ethanol*. Government Series. Alexandria: The Capitol. Net, Inc.

Yacobucci, B., and Bancourt, K. (2010). Calculation of Lifecycle GHG Emissions for the Renewable Portfolio Standard, CRS Report R40460, Washington DC: Congressional Research Service, http://nationalaglawcenter.org/wp-content/uploads/assets/ crs/ R40460.pdf.

Yacobucci, B., and Schnepf, R. (2007). Ethanol and Biofuels: Agriculture, Infrastructure and Market Constraints Related to Expanded Production, CRS Report 33928. Washington, DC: Congressional Research Service, March 16, 2007.

Yep, E. (2010, December 29). India Raises Renewable Energy Target Fourfold, *Wall Street Journal*. http://www.wsj.com/articles/SB10001424052970203513204576048870791325278.

Yep, E. (2011, March 9). India's Widening Energy Deficit, *Wall Street Journal*. http://blogs.wsj.com/indiarealtime/2011/03/09/indias-widening-energy-deficit/.

Yergin, D. (1991). *The Prize: The Epic Quest for Oil, Money & Power*. New York: Free Press.

Yergin, D. (2011). *The Quest: Energy, Security, and the Remaking of the Modern World*. New York: Penguin.

Yin, R. (1994). *Case Study Research* (2nd Ed.). London: Sage.

Yin, R. (2003*a*). *Applications of Case Study Research*. Applied Social Research Methods Series, No. 34. London: Sage.

Yin, R. (2003*b*). *Case Study Research Design and Methods* (3rd Ed.). Thousand Oaks, CA: Sage.

Yost, E., Stanek, J., DeWoskin, R., and Burgoon, L. (2016). Estimating the Potential Toxicity of Chemicals Associated with Hydraulic Fracturing Operations Using Quantitative Structure–Activity Relationship Modeling, *Environmental Science and Technology*, 50: 14, 7732–7742.

Yusuf, N., Kamarudin, S., and Yaakub, Z. (2011). Overview on the Current Trends in Biodiesel Production, *Energy Conversion and Management*, 52: 7, 2741–2751.

Zahariadis, N. (1999). Ambiguity, Time, and Multiple Streams, in: Sabatier, P. (Ed.), *Theories of the Policy Process*. Boulder: Westview.

Zahra, S., and George, G. (2002). Absorptive Capacity: A Review, Reconceptualization, and Extension, *Academy of Management Review*, 27: 2, 185–203.

Zielinski, S. (2016). Iceland Carbon Capture Project Quickly Converts Carbon Dioxide into Stone, *Smithsonian Magazine*, June 9, 2016, http://www.smithsonianmag.com/science-nature/iceland-carbon-capture-project-quickly-converts-carbon-dioxide-stone-180959365/.

INDEX

Note: Page references followed by a "*t*" indicate table; "*f*" indicate figure; "*b*" indicate box; "*n*" indicate note.